Combinatorics

COMBINATORICS

N. Ya. Vilenkin

Translated by A. SHENITZER and S. SHENITZER

YORK UNIVERSITY
TORONTO, ONTARIO
CANADA

1971

ACADEMIC PRESS New York and London

ACADEMIC PRESS, INC.
111 Fifth Avenue, New York, New York 10003

United Kingdom Edition published by
ACADEMIC PRESS, INC. (LONDON) LTD.
Berkeley Square House, London W1X 6BA

LIBRARY OF CONGRESS CATALOG CARD NUMBER: 77-154369

AMS(MOS) 1970 Subject Classification: 05–01

PRINTED IN THE UNITED STATES OF AMERICA

First published in the Russian language under the title

KOMBINATORIKA

by

IZDATEL'STVO "NAUKA"
GLAVNAYA REDAKTSIYA FIZIKO-MATEMATICHESKOĬ LITERATURY
MOSCOW, 1969

Contents

I. General Rules of Combinatorics

II. Samples, Permutations, and Combinations

III. Combinatorial Problems with Restrictions

IV. Distributions and Partitions

V. Combinatorics on a Chessboard

VI. Recurrence Relations

VII. Combinatorics and Series

Foreword

Professor Naum Ya. Vilenkin is not only a distinguished mathematician but also a gifted popularizer of significant mathematics. His "Stories about Sets" (Academic Press, 1968) ranged from a discussion of infinities to a discussion of the dimension of a manifold. His present book on combinatorics is a leisurely tour which takes the reader from very simple combinatorial problems to recurrence relations and generating functions and requires no more than a high school background in mathematics. The book includes a collection of hundreds (439) of problems with solutions.

We feel that Professor Vilenkin's book on combinatorics is another proof of the proposition that it is possible to instruct without boring.

We wish to thank Dr. D. W. T. Bean of York University in Toronto for pointing out a number of errors in the manuscript of the translation.

THE TRANSLATORS

ix

Preface to the Russian Edition

People engaged in almost any area of activity find it necessary to solve problems which require consideration of various arrangements of letters, digits, and other objects. The shop foreman assigns various types of jobs to operators of available machine tools, the agronomist assigns crops to be grown on different fields, the school director plans the curriculum, the research chemist investigates the possible connections between various atoms and molecules, the linguist considers the possible meanings of letters of an unknown alphabet, and so on. The branch of mathematics concerned with the number of different arrangements of given objects subject to various restrictions is called *combinatorics*.

Combinatorics goes back to the 16th century. At that time games of chance played an important role in the lives of members of the privileged classes. People played cards and dice* and won or lost gold and diamonds, palaces and estates, thoroughbreds and precious ornaments. Lotteries were very popular. It is not surprising that in the beginning combinatorial problems were primarily concerned with games of chance; with problems such as the number of ways of obtaining a certain score when throwing two or three dice, or the number of ways of getting two kings in a certain card game. These and similar problems arising from games of chance were the moving force behind the development of combinatorics as well as probability.

The Italian mathematician Tartaglia was among the first to consider the number of outcomes associated with a game of dice. Tartaglia constructed a table of possible scores in a game with r dice without, however, taking into consideration the fact that the same score could be obtained in different ways (for example, $1 + 3 + 4 = 4 + 2 + 2$).

In the 17th century the French scholars Pascal and Fermat pursued theoretical studies of combinatorial problems. The starting point of their studies was likewise games of chance. A problem of great significance connected with games of chance was the problem of division of stakes which the Chevalier de Méré,

---

* In a game of dice the players threw a few cubes whose faces were marked 1 through 6. The winner was the person with the largest score. There existed variants of this game.

xi

an inveterate gambler, put before his friend Pascal: To win a "match" in the game of pitch-and-toss one must win six games. Suppose the game is broken off after one player has won five games and his opponent has won four games; how should the stakes be divided? It was clear that division in the ratio 5–4 was unfair. Pascal used combinatorial methods and solved the problem in the general case when the players must yet win r and s games, respectively, to win the match. Another solution of the problem was given by Fermat.

Further development of combinatorics was due to Jacob Bernoulli, Leibniz, and Euler. They were mostly concerned with applications to various games (lotto, solitaire, and so on). The rapid development of combinatorics in recent years is connected with the growth of interest in problems of finite mathematics. Combinatorial methods are used to solve transportation problems (in particular, problems of composition of timetables) as well as in industrial planning and in scheduling production. Combinatorics is linked to linear programming, statistics, and so on. Combinatorics is also used in coding and decoding and in other areas of information theory.

Combinatorial methods play an important role in certain areas of pure mathematics such as the theory of groups and representation theory, foundations of geometry, nonassociative algebras, and so on.

There are few books on combinatorics in Russian. Outside of elementary books exemplified by public school textbooks, there are only translations of works by M. Hall ("Combinatorial Theory", Ginn, Boston, 1967), J. Riordan ("An Introduction to Combinatorial Analysis," Wiley, New York, 1958), and H. J. Ryser ("Combinatorial Mathematics," Carus Monograph No. 14, Wiley, New York, 1963).

The present book is intended as an interesting and elementary introduction to combinatorics. In spite of the elementary nature of the book, we consider some rather difficult combinatorial problems and introduce the reader to recurrence relations and generating functions.

The first chapter deals with general rules of combinatorics, the rule of sum and the rule of product. The second chapter deals with samples, permutations, and combinations; this traditional high school material is illustrated with some interesting examples. In the third chapter we study combinatorial problems in which the arrangements are subject to various restrictions. In the fourth chapter we consider (ordered and unordered) partitions of numbers and discuss geometric methods used in combinatorics. The fifth chapter is devoted to random walk problems and to a discussion of a number of variants of the arithmetical triangle. In the sixth chapter we discuss recurrence relations. In the seventh chapter we discuss generating functions, and, in particular, the binomial expansion.

The book contains over 400 (solved) problems taken from different sources. Many of the problems come from the books by W. A. Whitworth ("Choice and Chance," Hafner, New York, 1959), J. Riordan ("An Introduction to

Combinatorial Analysis"), A. M. Yaglom and I. M. Yaglom ("Challenging Mathematical Problems with Elementary Solutions," Holden-Day, San Francisco, 1967), from collections of problems given at mathematical olympiads, and so on.

Combinatorics

1

General Rules of Combinatorics

Superstitious Cyclists

The president of a cyclist club looked at the twisted wheel of his bicycle and exclaimed bitterly: "Another 8! And all because the number on my membership card is 008. And now not a month passes without a wheel of my bike turning into a figure 8. The only thing to do is to change that card number and the card numbers of all the other club members. Otherwise they'll say I am superstitious. When numbering the new cards I'll make sure not to use the digit 8."

No sooner said than done. The next day each member of the club got a new card. *How many members were in the club if all 3-digit numbers not containing the digit 8 were used to number the new cards?* (For example, 000 was used, but 836 was not used.)

To solve this problem we first ask for the number of 1-digit numbers other than 8. Clearly, these are the nine numbers 0, 1, 2, 3, 4, 5, 6, 7, 9. Next we find all the 2-digit numbers which do not contain the digit 8. These numbers arise if we write down one of the nine admissible digits and then another. Hence for every admissible 1-digit number we get 9 two-digit numbers. And since there are 9 such 1-digit numbers, there are $9 \cdot 9 = 81$ "eight-free" 2-digit numbers. Specifically, the numbers in question are:

$$
\begin{array}{ccccccccc}
00, & 01, & 02, & 03, & 04, & 05, & 06, & 07, & 09 \\
10, & 11, & 12, & 13, & 14, & 15, & 16, & 17, & 19 \\
20, & 21, & 22, & 23, & 24, & 25, & 26, & 27, & 29 \\
30, & 31, & 32, & 33, & 34, & 35, & 36, & 37, & 39 \\
40, & 41, & 42, & 43, & 44, & 45, & 46, & 47, & 49 \\
50, & 51, & 52, & 53, & 54, & 55, & 56, & 57, & 59 \\
60, & 61, & 62, & 63, & 64, & 65, & 66, & 67, & 69 \\
70, & 71, & 72, & 73, & 74, & 75, & 76, & 77, & 79 \\
90, & 91, & 92, & 93, & 94, & 95, & 96, & 97, & 99
\end{array}
$$

If we follow each of the $9^2 = 81$ "eight-free" 2-digit numbers with one of the nine admissible digits, we obtain $9^2 \cdot 9 = 9^3 = 729$ admissible 3-digit numbers. This means that the club had 729 members. It is not difficult to see that the number of 4-digit "eight-free" numbers is $9^4 = 6561$.

The members of another cyclist club were even more superstitious. Since the the digit 0 resembles an elongated wheel, they decided to use only the eight digits 1, 2, 3, 4, 5, 6, 7, 9. *How many members were in the club if the membership cards were numbered with all the admissible 3-digit numbers?*

This problem is similar to the previous problem except that instead of nine we have eight admissible digits. This means that the number of members was $8^3 = 512$.

Samples with Repetitions

The problem about the cyclists is of the following type: We are given objects of n types. We make up all possible arrangements of k such objects. In making up an arrangement, we are free to use objects of the same type. Two arrangements are regarded as different if they contain different numbers of elements of a certain type or if their elements are differently ordered. *We are required to compute the number of such arrangements.*

Arrangements of this type are called *k-samples with repetitions of elements of n types*. The number of such arrangements is denoted by the symbol $\bar{A}_n^k$. In the first problem about cyclists, we dealt with elements of nine types (all the digits other than 8), and each sample (that is, each admissible card number) consisted of three elements. We saw that in that case the number of samples was $\bar{A}_9^3 = 9^3$. It is natural to expect that if there are n types of elements and each sample involves k elements, then the number of samples with repetitions is n^k. A formal proof follows.

We show that the number of k-samples with repetitions of elements of n types is equal to

$$\bar{A}_n^k = n^k. \tag{1}$$

We keep n fixed and use induction on k, the number of elements in the sample. For $k = 1$, every sample consists of 1 element, and to obtain different samples we must take elements of different types. Since there are n types of elements, the number of the samples in question is n. Thus, in accordance with Eq. (1), we have $\bar{A}_n^1 = n$.

Now we suppose that the equality $\bar{A}_n^{k-1} = n^{k-1}$ holds, and we consider k-samples with repetitions. All such samples can be obtained as follows: Starting with any $(k-1)$-sample $(a_1, ..., a_{k-1})$ we add on to it an element a_k of one of the n types. The result is the k-sample $(a_1, ..., a_{k-1}, a_k)$. It is clear that each $(k-1)$-sample gives rise to as many k-samples as there are types of elements, namely, n. Also, this procedure for generating k-samples does not leave out any k-sample nor does it yield the same k-sample twice (if $(a_1, ..., a_{k-1}) \neq (b_1, ..., b_{k-1})$ or $a_k \neq b_k$, then $(a_1, ..., a_k) \neq (b_1, ..., b_k)$). It follows that there are n times as many k-samples with repetitions of elements of n types as there are $(k-1)$-samples with repetitions of elements of n types, that is, $\bar{A}_n^k = n\bar{A}_n^{k-1}$. Therefore

$$\bar{A}_n^k = n \cdot n^{k-1} = n^k.$$

This proves Eq. (1) for all values of k.

Equation (1) occurs in many situations. A few of these situations are discussed below.

Systems of Numeration

In recording numbers, we usually make use of the base ten. We can also make use of other bases such as the base two, the base three, and so on. If we use n as a base, then in recording numbers we use n digits.*With n as a base, let us compute the number of natural numbers (at this point it is convenient to regard zero as a natural number) which can be recorded using exactly k digits. If we allow a zero in the first position, then every number consisting of k digits can be viewed as a k-sample with repetitions of elements of n types. By Formula (1) there are n^k such numbers.

However, in recording a natural number we never start with a zero. Therefore we must reduce n^k by the total number of numbers whose records (to the base n) start with a zero. If we remove in each of these numbers the leading zero, then we obtain a number consisting of $k - 1$ digits (conceivably starting with zero).

* The term "digit" refers to each of the numbers $0, 1, ..., n - 1$.

By Formula (1), there are n^{k-1} such numbers. It follows that if we use the base n, then there are

$$n^k - n^{k-1} = n^{k-1}(n - 1)$$

numbers consisting of k digits.

For example, in the decimal system, there are $10^3 \cdot 9 = 9000$ four-digit numbers; in fact, 9000 is the difference between the 10,000 numbers from 0 to 9999 and the 1000 numbers from 0 to 999.

The above result can be derived in another way. If we use the base n, then, in recording a number made up of k digits, we can take as the first entry one of $n - 1$ digits, and for each of the $k - 1$ succeeding entries one of the n digits. It follows readily that there are $(n - 1)n^{k-1}$ required numbers.

Combination Lock

Safes and vaults are locked by means of combination locks which open only after one has dialed a "secret word." Such a word is dialed by means of one or more disks on which are stamped letters (or numbers). Suppose that there are 12 letters on a disk and the secret word consists of 5 letters. *How many times can a man who does not know the secret word fail to dial it?*

In view of Formula (1) the total number of combinations is

$$12^5 = 248,832.$$

This means that the maximal number of failures is 248,831. Incidentally, safes are usually constructed so that after the first failure to open them an alarm is sounded.

The Morse Code

The Morse code is used to transmit telegrams. In this code letters, numbers, and punctuation marks are denoted by means of dots and dashes. The number of dots and dashes used to denote a particular letter ranges from one to five.

Why go up to five? Is it not possible to transmit all messages using blocks of, say, no more than four symbols? It turns out that this is impossible. This conclusion is implied by the formula for the number of samples with repetitions. In fact, Formula (1) shows that $\bar{A}_2^1 = 2$, that is, that using 1 dot or dash we can transmit only 2 letters. Using dots and dashes in groups of 2, 3, and 4, we can transmit $2^2 = 4$, $2^3 = 8$, and $2^4 = 16$ letters, respectively. While it is true that

$$2 + 4 + 8 + 16 = 30$$

and the number of letters in the Latin alphabet is only 26, there remain numbers and punctuation marks to be transmitted. This shows that we cannot stop at blocks of 4 dots and dashes. On the other hand, the use of blocks of 5 dots and dashes yields an additional 32 symbols for a total of 62. This number is more than sufficient for the purposes of telegraphy.

Signaling at Sea

At sea one sometimes uses flags for signaling. To each letter of the alphabet there corresponds a definite position of each of the two flags used by the flagman. As a rule the flags are on either side of the flagman's body. However, this is not always the case. Why the exception ? The answer to this question is found by consulting the formula for the number of samples with repetitions. Each of the 2 flags can occupy one of 5 positions (up, half way up, out, halfway down and down). With the flags on either side of the flagman's body the number of distinct signals is $\bar{A}_5^2 = 5^2 = 25$. On the other hand, we need 26 signals to represent the 26 letters in the Latin alphabet and an extra signal to separate words. That is why some letters are signaled with both flags to one side of the flagman's body.

Electronic Computers

Electronic computers can solve a large variety of problems. The same machine can be used to decipher an inscription in an unknown language, to design a dam, or to process data on the flight of a rocket. What makes a computer so versatile ? Mainly the fact that all of these problems can be reduced to operations on numbers. Still, how can a computer solve so many problems with so many different numerical data ? How many numbers can one store in a computer ?

To answer this question we consider, by way of an example, the computer "Arrow." Its memory consists of 2048 cells each containing 43 binary bits. Each bit represents 0 or 1. This gives $43 \cdot 2048 > 87,000$ different locations, each of which can be filled in two ways (0 or 1). It follows from Formula (1) that the number of states of our computer exceeds $2^{87,000}$. The size of this number boggles the imagination. Suffice it to say that the number of neutrons which fit into a sphere of radius equal to the distance from the earth to the most distant of all known nebulae is under 2^{500}.

It would take an army of one hundred thousand typists nine years to print out all the numbers that can turn up in a single memory cell. (These figures are based on the assumption that a typist works 7 hours a day and takes 10 seconds to type a 43-digit number.)

Genetic Code

A remarkable discovery of 20th century biology was the solution of the mystery of the genetic code. Scientists managed to explain how hereditary information is perpetuated. It turned out that such information is carried by the giant molecules of deoxyribonucleic acid, or DNA. Molecules of DNA differ in the sequence of four nitrogen bases: adenine, thymine, guanine, and cytosine. These bases determine the structure of the proteins of the organism. The proteins are built out of 20 amino acids. The code for each amino acid involves three nitrogen bases.

The role of the number 3 is clear. The number of 2-samples (with repetitions) of the 4 nitrogen bases is $4^2 = 16$. On the other hand, the number of amino acids to be coded is 20. Hence the need for 3-samples. At the same time it would be interesting to know how nature uses the excess of information represented by the difference between the available $4^3 = 64$ three-samples and the 20 three-samples needed for the coding of 20 amino acids.

Each chromosome contains a few tens of millions of nitrogen bases. The number of arrangements of these bases is fantastically large; specifically, if N is the number of bases, then the number of arrangements is 4^N. A tiny fraction of these arrangements would account for the variety of all living forms during the period of existence of life on earth. It must, of course, be borne in mind that only a tiny fraction of the arrangements which are possible in theory lead to viable living organisms.

General Rules of Combinatorics

We shall see in the sequel that there are many types of combinatorial problems. However, most problems can be solved using two basic rules, the rule of sum and the rule of product.

It is often possible to separate the arrangements under consideration into several classes so that every arrangement belongs to just one class. Then, obviously, the *total number of arrangements is equal to the sum of the arrangements in the different classes*. This assertion is known as *the rule of sum*. Another formulation of this rule states:

If an object A can be selected in m ways and another object B can be selected in n ways, then the choice of "either A or B" can be effected in m + n ways.

When applying the rule of addition in the latter formulation, we must see to it that none of the ways of choosing A coincides with some way of choosing B (or, to use the earlier terminology, that no arrangement is common to two classes). Otherwise the rule of sum is not applicable. In fact, if the number of coincidences is k, then the number of ways of choosing A or B is $m + n - k$.

The second rule, *the rule of product*, is somewhat more involved. When making up an arrangement of two elements we may know that the first element can be selected in m ways and that for each choice of the first element the second element can invariably be selected in n ways. Then the pair of elements can be selected in mn ways. In other words:

If an object A can be selected in m ways and if, following the selection of A, an object B can be selected in n ways, then the pair (A, B), A first, B second, can be selected in mn ways.

To prove the rule of product we observe that each of the m ways of choosing A can be combined with one of the n ways of choosing B. This gives mn ways for choosing the pair (A, B).

The rule of product is made particularly clear by Table 1.

TABLE 1

$$(A_1, B_{11}), \quad ..., \quad (A_1, B_{1n})$$
$$(A_2, B_{21}), \quad ..., \quad (A_2, B_{2n})$$
$$. . . . .$$
$$(A_i, B_{i1}), \quad ..., \quad (A_i, B_{in})$$
$$. . . . .$$
$$(A_m, B_{m1}), \quad ..., \quad (A_m, B_{mn})$$

Here $A_1, ..., A_m$ denote m ways of selecting A, and $B_{i1}, ..., B_{in}$ denote n ways of selecting B after A has been selected in the ith way. It is clear that our table gives all ways of selecting the pair (A, B) and that it consists of mn elements.

If the manner of selection of B is independent of the manner of selection of A, then in place of Table 1, we obtain the simpler Table 2.

TABLE 2

$$(A_1, B_1), \quad (A_1, B_2), \quad ..., \quad (A_1, B_n)$$
$$(A_2, B_1), \quad (A_2, B_2), \quad ..., \quad (A_2, B_n)$$
$$. . . . .$$
$$(A_m, B_1), \quad (A_m, B_2), \quad ..., \quad (A_m, B_n)$$

Obviously, we may wish to make arrangements of more than 2 elements. This leads to the following problem:

What is the number of k-arrangements if the first element can be of one of n_1 different types, the second of one of n_2 different types,..., the kth of one of n_k different types. (Here two arrangements are regarded as different if they have elements of different types in one or more corresponding positions.)

This problem is solved in the manner of the problem about the cyclists.

The first element can be selected in n_1 ways. Each first element can be combined with an element of one of n_2 types to give $n_1 n_2$ pairs. In turn, each pair can be combined with an element of one of n_3 types to give $n_1 n_2 n_3$ triples. Continuing this process we obtain $n_1 n_2 \cdots n_k$ arrangements of the required type.

In the problem about the cyclists we were required to select 3 elements (the number of hundreds, the number of tens, and the number of ones). At each step we could choose one of 9 admissible digits. This gave us $9 \cdot 9 \cdot 9 = 729$ numbers.

The following is a somewhat more difficult problem of this type:

What is the number of ordered triples consisting of 1 of 4 geometric figures, a letter, and a digit? (See Fig. 1.)

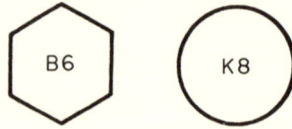

Fig. 1

There are 4 choices for the first element, 26 for the second, and 10 for the third. This gives a total of $4 \cdot 26 \cdot 10 = 1040$ arrangements.

A Domino Problem

Combinatorial problems in which the number of choices at each step depends on the elements already selected represent a more difficult class of problems. One such problem is the following:

In how many ways is it possible to select, in stated order, a pair of matching dominos out of a total of 28? (Two dominos are matched if they have a common half-face.)

The first domino can be selected in 28 ways. Of these, 7 are "doubles" (that is, 00, 11, 22, 33, 44, 55, 66) and 21 are "nondoubles" (for example, 05, 13, and so on). If the first choice is a double, then it can be matched in 6 ways (for example, if the first choice is 11, then the possible second choices are 01, 12, 13, 14, 15, 16). If the first choice is a nondouble, then it can be matched in 12 ways (for example, if the first choice is 35, then the possible second choices are 03, 13, 23, 33, 34, 36, 05, 15, 25, 45, 55, 56). By the rule of product there are $7 \cdot 6 = 42$ choices in the first case and $21 \cdot 12 = 252$ choices in the second. By the rule of sum the total number of choices is $42 + 252 = 294$.

Note that, with order taken into account, each pair of dominos is counted twice (for example, there is the pair 01, 16 and the pair 16, 01). If we ignore the order in which the dominos in a pair are picked, then the number of matched pairs is half of the total just computed, that is, 147.

Crew of a Spaceship

If, as in the previous example, the number of choices at any stage depends on earlier choices, then it is convenient to represent the process of formation of an arrangement by means of a "tree." We start with a point and lead from it as many segments as there are initial choices (thus each segment corresponds to a single element). From the endpoint of each segment we lead as many segments as there are second-stage choices corresponding to the first-stage choices represented by that segment, and so on. The result of this construction is a tree from which one can easily read off the number of solutions of our problem. An illustration follows.

When selecting the crew of a spaceship it is necessary to consider the psychological compatibility of the prospective astronauts under conditions of space travel. It may well happen that individually acceptable candidates may not be suited for the long trip together. Now let us suppose that we are to select a 3-man crew—a commander, an engineer, and a doctor—for a spaceship. Let a_1, a_2, a_3, a_4 denote 4 candidates for the post of commander; b_1, b_2, b_3, 3 candidates for the post of engineer; and c_1, c_2, c_3, 3 candidates for the post of doctor. Tests show that commander a_1 is compatible with engineers b_1, b_3 and doctors c_2, c_3; commander a_2 with engineers b_1, b_2 and all the doctors; commander a_3 with engineers b_1, b_2 and doctors c_1, c_3; and commander a_4 with all the engineers and doctor c_2. Also, engineer b_1 is incompatible with doctor c_3, engineer b_2 with doctor c_1, and engineer b_3 with doctor c_2. *What is the number of compatible crews for the spaceship?*

The relevant tree is shown in Fig. 2. This tree shows that there are only ten

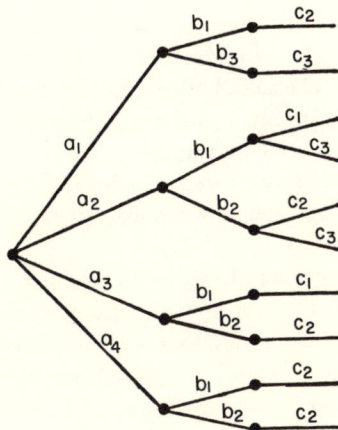

Fig. 2

compatible crews (as against $4 \cdot 3 \cdot 3 = 36$ crews which could be chosen if we ignored compatibility considerations).

A Checkers Problem

We propose to solve the following problem:

In how many ways can one place a black checker and a white checker on a checkerboard so that the white checker can capture the black checker?

According to the rules of the game, the checkers are placed on the black squares. To capture a checker, one must jump over it to an empty square as illustrated in Fig. 3.

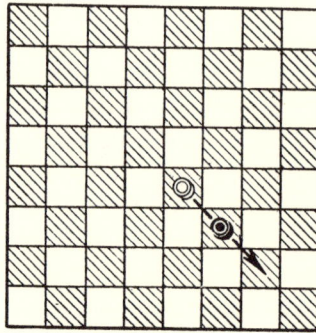

Fig. 3

When a checker reaches any of the squares on the opposite extreme side of the board it becomes a king. A king can jump one or more checkers which share a diagonal with it and do not occupy terminal squares.

The difficulty of our problem consists in the fact that the number of positions in which the black checker can be captured by the white checker changes with the position of the white checker. For example, if the white checker occupies the square $a1$, then there is only 1 position in which the black checker can be captured. If the white checker is in the square $c3$, then there are 4 positions in which the black checker can be captured. Finally, if the white checker is crowned and occupies the square $h8$, then the number of positions in which the black checker can be captured is 6.

It is clear that the simplest way to solve our problem is to mark each position of the white checker with the corresponding number of positions in which the black checker can be captured and to add these numbers. The numbers in question are indicated on the board in Fig. 4a. Their sum is 87. This means that there are 87 ways of placing the two checkers.

Interchanging the roles of the checkers in the statement of our problem has no effect on the answer. On the other hand, the number of ways in which one can place a white checker and a black checker on a checker board so that each checker can capture the other is less than 87. For example, if the white checker is at the edge of the board, then it cannot be captured regardless of the position of the black checker. Therefore, all black squares at the edge of the board must be marked with zeros. By marking the remaining squares with appropriate numbers, as in Fig. 4b, and adding these numbers, we find that the checkers can be placed in 50 ways.

Finally we compute the number of locations of a black and white pair of checkers in which neither checker can capture the other. One way of solving this problem is to proceed as before by placing the white checker on each black square, computing in each case the number of ways of placing the black checker so that neither checker can capture the other, and adding the resulting numbers. However, it is simpler to use the "teakettle principle"* and to reduce the problem to one solved earlier. To do this we first compute the number of ways of placing a pair of checkers on the checker board. The white checker can be placed on any one of the 32 black squares. Subsequently, the black checker can be placed on any one of the unoccupied 31 squares. By the rule of product, the two checkers can be placed on the board in $32 \cdot 31 = 992$ ways. In 87 of the 992 cases the white checker can capture the black checker. In another 87 cases

---

* A mathematician asked a physicist: "Suppose you were given an empty teakettle and an unlit gas plate. How would you bring water to a boil?" "I'd fill the teakettle with water, light the gas, and set the teakettle on the plate." "Right," said the mathematician, "and now please solve another problem. Suppose the gas were lit and the teakettle were full. How would you bring the water to a boil?" "That's even simpler. I'd set the teakettle on the plate." "Wrong," exclaimed the mathematician. "The thing to do is to put out the flame and empty the teakettle. This would reduce our problem to the previous problem!"

That is why, whenever one reduces a new problem to problems already solved, one says in jest that one is applying the "teakettle principle."

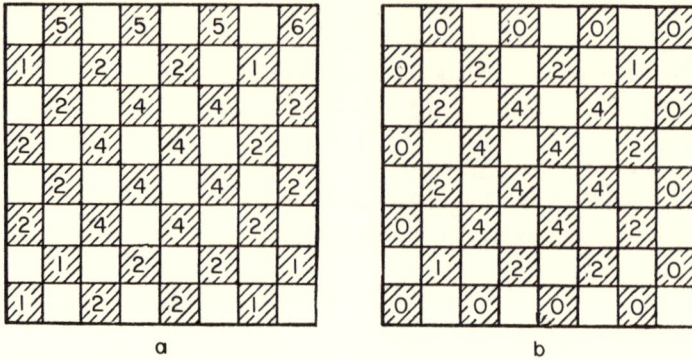

a   b

FIG. 4

the black checker can capture the white. It is therefore necessary to reject $2 \cdot 87 = 174$ cases. At the same time we must note that in so doing we would be rejecting some cases twice (namely, the cases in which each piece can capture the other). We know that there are 50 such cases. It follows that the number of cases in which neither checker can capture the other is

$$992 - 174 + 50 = 868.$$

How Many People Do Not Know a Foreign Language?

The method we used in the preceding problem is frequently used to solve combinatorial problems. Here is another illustration of the use of this method.

A certain research institute employs 67 people. Of these 47 speak French, 35 speak German, and 23 speak both French and German. How many members of the institute speak neither French nor German?

In order to solve this problem we must partition the set of workers at the institute into classes without common elements. One such class consists of those who (in addition to English) speak only French, another consists of those who speak only German, a third consists of those who speak French and German, and a fourth consists of those who speak neither French nor German (see Fig. 5). We know that there are 23 people in the third class. Since there are 47 people who speak French, it follows that the number of people who speak only French is $47 - 23 = 24$. Similarly, the number of people who speak only German is $35 - 23 = 12$. It follows that the number of people who speak at least one of these languages is $23 + 24 + 12 = 59$. Since there are 67 workers at the institute, the number of people who speak neither Franch nor German is $67 - 59 = 8$.

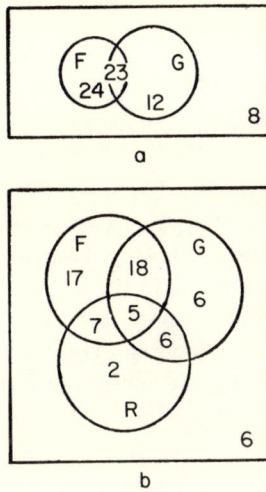

FIG. 5

This result can be written as follows:

$$8 = 67 - (23 + 24 + 12).$$

Note that 24 is the difference between 47 and 23, and 12 is the difference between 35 and 23. Hence

$$8 = 67 - 23 - (47 - 23) - (35 - 23) = 67 - 47 - 35 + 23.$$

Now the law of formation is clear. We subtract from the total number of workers at the institute the number of workers who speak French and the number of workers who speak German. But this means that some of the workers, namely, the 23 polyglots who speak both French and German, are "subtracted" twice. We remedy this by adding the number 23 and so obtain the number of people who speak neither French nor German.

We consider a more complicated version of this problem by including one more language. Suppose that 20 people speak Russian, 12 speak French and Russian, 11 speak German and Russian, and 5 people speak all three languages. It is clear that the number of people who speak just French and Russian (and so no German) is $12 - 5 = 7$. The number of people who speak just German and Russian is $11 - 5 = 6$. Hence, the number of people who speak just Russian is $20 - 7 - 6 - 5 = 2$. These 2 people belong to the class of 8 who do not speak French or German. Hence there are $8 - 2 = 6$ people who do not speak French, German, or Russian.

This result can be written as follows:

$$6 = 8 - 2 = 67 - 47 - 35 + 23 - (20 - 7 - 6 - 5)$$
$$= 67 - 47 - 35 + 23 - 20 + (12 - 5) + (11 - 5) + 5$$
$$= 67 - 47 - 35 - 20 + 23 + 12 + 11 - 5.$$

Now the law of formation is clear. To begin with, we subtract from the total number of workers the number of workers who speak at least 1 of the 3 languages. But then some of the workers, namely, those who speak 2 languages, are "subtracted" twice. That is why we add the numbers 23, 12, and 11 of workers who speak different pairs of (and possibly all 3) languages. This, in turn, implies that the workers who speak all 3 languages are "subtracted" and then "added" 3 times. Since their number must be subtracted from the total, it is necessary to subtract the number 5.

The Principle of Inclusion and Exclusion

Our example enables us to formulate a general rule. Suppose that we have a set of N objects and a set of n properties α_1, α_2, ..., α_n. Some of our N objects may have none of the n properties and some may have one or more of these properties. We use the symbol $N(\alpha_i \alpha_j \cdots \alpha_k)$ to denote the number of objects which have the properties α_i, α_j, ..., α_k (and possibly additional properties). If we wish to stress the fact that we are concerned with elements which lack a certain property, we prime the corresponding α. For example, $N(\alpha_1 \alpha_2 \alpha_4')$ denotes the set of elements which have the properties α_1 and α_2 and do not have the property α_4 (the question of the remaining properties is left open). In line with this convention, $N(\alpha_1' \alpha_2' \cdots \alpha_n')$ denotes the set of elements with none of the properties α_1, α_2, ..., α_n.

The general rule mentioned in the beginning of this paragraph states that

$$N(\alpha_1' \alpha_2' \cdots \alpha_n') = N - N(\alpha_1) - N(\alpha_2) - \cdots - N(\alpha_n)$$
$$+ N(\alpha_1 \alpha_2) + N(\alpha_1 \alpha_3) + \cdots + N(\alpha_1 \alpha_n) + \cdots + N(\alpha_{n-1} \alpha_n)$$
$$- N(\alpha_1 \alpha_2 \alpha_3) - \cdots - N(\alpha_{n-2} \alpha_{n-1} \alpha_n) + \cdots$$
$$+ (-1)^n N(\alpha_1 \alpha_2 \cdots \alpha_n). \qquad (2)$$

The above sum is taken over all combinations of the properties α_1, α_2, ..., α_n (without regarding their order). A summand involving an even number of properties enters with a plus sign, and a summand involving an odd number of properties enters with a minus sign. Relation (2) is referred to as the *principle of inclusion and exclusion*. This name reflects the fact that we exclude all elements which have at least one of the properties α_1, α_2 ,..., α_n , include all elements which have at least two of these properties, exclude all elements which have at least three of these properties, and so on.

To prove Formula (2) we use induction on the number of properties. In the case of a single property α the formula is obviously true. Namely, an object either has this property or does not have it. Therefore,

$$N(\alpha') = N - N(\alpha).$$

Suppose Formula (2) is true for $n - 1$ properties, that is, suppose that

$$
\begin{aligned}
N(\alpha'_1\alpha'_2 \cdots \alpha'_{n-1}) = {} & N - N(\alpha_1) - \cdots - N(\alpha_{n-1}) \\
& + N(\alpha_1\alpha_2) + \cdots + N(\alpha_{n-2}\alpha_{n-1}) \\
& - N(\alpha_1\alpha_2\alpha_3) - \cdots - N(\alpha_{n-3}\alpha_{n-2}\alpha_{n-1}) + \cdots \\
& + (-1)^{n-1} N(\alpha_1\alpha_2 \cdots \alpha_{n-1}).
\end{aligned}
\tag{3}
$$

We may use Formula (3) with any number of objects. In particular, this formula holds for the set of $N(\alpha_n)$ of objects which have the property α_n . If we replace N by $N(\alpha_n)$ in Formula (3), we obtain

$$
\begin{aligned}
N(\alpha'_1\alpha'_2 \cdots \alpha'_{n-1}\alpha_n) = {} & N(\alpha_n) - N(\alpha_1\alpha_n) - \cdots - N(\alpha_{n-1}\alpha_n) \\
& + N(\alpha_1\alpha_2\alpha_n) + \cdots + N(\alpha_{n-2}\alpha_{n-1}\alpha_n) \\
& - N(\alpha_1\alpha_2\alpha_3\alpha_n) - \cdots \\
& + (-1)^{n-1} N(\alpha_1\alpha_2 \cdots \alpha_{n-1}\alpha_n)
\end{aligned}
\tag{4}
$$

(to get (4) from (3) one takes in each set corresponding to a summand in (3) only those elements which have the property α_n).

Now we subtract Eq. (4) from Eq. (3). The difference between the right-hand side of Eq. (4) and the right-hand side of Eq. (3) is just the right-hand side of Eq. (2). The difference between the left-hand side of Eq. (4) and the left-hand side of Eq. (3) is equal to

$$N(\alpha'_1\alpha'_2 \cdots \alpha'_{n-1}) - N(\alpha_1\alpha'_2 \cdots \alpha'_{n-1}\alpha_n). \tag{5}$$

Now, $N(\alpha'_1\alpha'_2 \cdots \alpha'_{n-1})$ represents the number of objects which do not have the properties α_1 , α_2 ,..., α_{n-1} and possibly have the property α_n . $N(\alpha'_1\alpha'_2 \cdots \alpha'_{n-1}\alpha_n)$ represents the number of objects which do not have the properties α_1 , α_2 ,..., α_{n-1} , but definitely have the property α_n . It follows that the difference in (5) is just the number of objects which have none of the properties α_1 , α_2 ,..., α_{n-1} , α_n . In other words,

$$N(\alpha'_1\alpha'_2 \cdots \alpha'_{n-1}) - N(\alpha'_1\alpha'_2 \cdots \alpha'_{n-1}\alpha_n) = N(\alpha'_1\alpha_2 \cdots \alpha'_{n-1}\alpha'_n).$$

This proves (2) for the case when the number of properties is n.

Having proved the validity of Eq. (2) for $n = 1$ and the fact that the validity of (2) for $n - 1$ implies its validity for n, we conclude that (2) is valid for any number of properties.

The following is a convenient symbolic representation of Eq. (2):

$$N(\alpha'\beta' \cdots \omega') = N(1 - \alpha)(1 - \beta) \cdots (1 - \omega). \tag{6}$$

Here a typical term $N\alpha\beta \cdots \lambda$ in the product on the right is to be written as $N(\alpha\beta \cdots \lambda)$. For example, we write $N(\alpha\beta\delta\omega)$ for $N\alpha\beta\delta\omega$.

Find the Error!

A certain class monitor stated the following facts about the students: "There are 45 students in the class. Twenty-five of the students are boys. Thirty students are A or B students. Of these 16 are boys. Twenty-eight students take up sports. Of these 18 are boys and 17 are A or B students. Fifteen boys are A or B students and take up sports."

A few days later the class teacher, who happened to teach mathematics, told the monitor that his data were incorrect. Let us see how the teacher arrived at his conclusion. To this end we compute the number of girls who do not take up sports and occasionally get a C (or less). Let α_1 be the property of being male, α_2 the property of being a good (A or B) student, and α_3 the property of being interested in sports. We wish to compute $N(\alpha_1'\alpha_2'\alpha_3')$. According to the data,

$$N(\alpha_1) = 25, \qquad N(\alpha_2) = 30, \qquad N(\alpha_3) = 28, \qquad N(\alpha_1\alpha_2) = 16,$$
$$N(\alpha_1\alpha_3) = 18, \qquad N(\alpha_2\alpha_3) = 17, \qquad N(\alpha_1\alpha_2\alpha_3) = 15.$$

By the principle of inclusion and exclusion,

$$N(\alpha_1'\alpha_2'\alpha_3') = 45 - 25 - 30 - 28 + 16 + 18 + 17 - 15 = -2.$$

This inadmissible result shows that the data are indeed incorrect.

The Sieve of Erastothenes

One of the great mysteries of mathematics is the distribution of primes in the realm of natural numbers. Sometimes two primes are separated by just one composite number (for example, 17 and 19, 29 and 31) and sometimes by a million. By now mathematicians have a fairly good idea as to the number of primes in the first N natural numbers. Their computations utilize, among others, a method due to the Greek mathematician Erastothenes who lived in Alexandria in the third century B.C.

Erastothenes had very wide scientific interests. These interests included mathematics, astronomy, and other disciplines. This wide range of interests invited a measure of superficiality. His contemporaries referred to him, rather unkindly, as "the universal runner-up" (a runner-up to Euclid in mathematics, to Hipparchus in astronomy, and so on).

The mathematical problem that concerned Erastothenes was the problem of finding all the primes in the sequence of natural numbers from 1 to N. (Erastothenes regarded 1 as a prime. Modern mathematicians find it convenient to regard 1 as a special number, neither prime nor composite.) In this connection he invented the following procedure: Remove all multiples of 2 other than 2. Keep the first remaining natural number exceeding 2, namely, the prime 3. Remove all multiples of 3 other than 3. Keep the first remaining natural number exceeding 3, namely, the prime 5. Remove all multiples of 5 other than 5, and so on. The retained numbers are the primes. Since in Erastothenes' time people wrote on wax tablets and punched out rather than erased numbers, the result of applying Erastothenes' procedure to a wax tablet was rather like a sieve. Hence the name "sieve of Erastothenes" given to Erastothenes' method for finding primes.

We now compute how many numbers from 1 to 100 are not divisible by 2, 3, and 5, that is, how many of these numbers remain after the first three steps of Erastothenes' procedure. The problem is solved using the principle of inclusion and exclusion.

Let α_1 be the property of being divisible by 2, α_2 the property of being divisible by 3, and α_3 the property of being divisible by 5. Then $\alpha_1\alpha_2$ denotes divisibility by 6, $\alpha_1\alpha_3$ divisibility by 10, and $\alpha_2\alpha_3$ divisibility by 15. Also, $\alpha_1\alpha_2\alpha_3$ denotes divisibility by 30. We wish to find out how many numbers from 1 to 100 are not divisible by 2, 3, or 5, that is, have none of the properties α_1, α_2, α_3. By Eq. (2)

$$N(\alpha_1'\alpha_2'\alpha_3') = 100 - N(\alpha_1) - N(\alpha_2) - N(\alpha_3)$$
$$+ N(\alpha_1\alpha_2) + N(\alpha_1\alpha_3) + N(\alpha_2\alpha_3) - N(\alpha_1\alpha_2\alpha_3).$$

To find how many numbers from 1 to N are divisible by n we must divide N by n and take the integer part of the quotient. Hence

$$N(\alpha_1) = 50, \qquad N(\alpha_2) = 33, \qquad N(\alpha_3) = 20,$$
$$N(\alpha_1\alpha_2) = 16, \qquad N(\alpha_1\alpha_3) = 10, \qquad N(\alpha_2\alpha_3) = 6, \qquad N(\alpha_1\alpha_2\alpha_3) = 3,$$

Therefore

$$N(\alpha_1'\alpha_2'\alpha_3') = 32$$

Thus 32 numbers between 1 and 100 are not divisible by 2, 3, or 5. These numbers and the numbers 2, 3, 5 are the 35 numbers which remain after the first three steps of Erastothenes' procedure.

Of the first 1000 natural numbers, 335 remain after the first three steps of Erastothenes' procedure. This follows from the fact that in this case

$$N(\alpha_1) = 500, \qquad N(\alpha_2) = 333, \qquad N(\alpha_3) = 200,$$
$$N(\alpha_1\alpha_2) = 166, \qquad N(\alpha_1\alpha_3) = 100, \qquad N(\alpha_2\alpha_3) = 66, \qquad N(\alpha_1\alpha_2\alpha_3) = 33.$$

II

Samples, Permutations, and Combinations

In Chap. I we considered certain general rules for the solution of combinatorial problems. These rules enable us to solve a large variety of problems. Nevertheless, certain problems occur so frequently that it is more convenient to solve them using ready-made formulas instead of general rules. (In geometry, too, it is frequently more convenient to solve problems by making use of appropriate theorems instead of relying on the axioms.) One such formula is the formula developed in the beginning of Chap. I which asserts that the number of k-samples with repetitions of elements of n types is n^k. We are about to compute the number of such samples without repetitions. We begin with the following problem.

Soccer Tournament

Seventeen teams participate in a soccer tournament. The first team is awarded a gold medal, the second a silver medal, and the third a bronze medal. In how many ways can the medals be distributed?

This problem can be solved using the rule of product: Any one of the 17 teams can win the gold medal. Once the gold medal has been awarded to a certain team, only one of 16 teams can be awarded the silver medal. Similarly, the bronze medal can go to one of the remaining 15 teams. Application of the rule of product tells us that the medals can be awarded in $17 \cdot 16 \cdot 15 = 4080$ ways.

Samples without Repetitions

The problem just solved belongs to the class of combinatorial problems involving samples without repetitions. In such problems we are given n distinct

18

objects and we are required to compute the number of arrangements of these objects taken k at a time; here two arrangements are regarded as different if they differ in composition or in the order of their elements.

Arrangements of this type are called *k-samples without repetitions*. Their number is denoted by the symbol A_n^k. We note that we can choose the first element of a k-sample without repetitions in n ways, the second element in $n-1$ ways, the third element in $n-2$ ways,..., the kth element in $n-k+1$ ways. By the rule of product, the number of k-samples without repetitions of n elements is equal to

$$A_n^k = n(n-1)\cdots(n-k+1). \tag{1}$$

Learned Society

We will apply the formula just derived to solve the following problem: *A certain learned society has 25 members. The members of the society are to elect a president, a vice president, a secretary, and a treasurer. In how many ways is it possible to select the 4 officers if no member of the society can hold more than one office at a time?*

Clearly, in this case, a particular person is chosen for a particular office (Smith for president, Brown for vice president, Clark for secretary, and Wood for treasurer). In mathematical terms, we are to find the number of samples (without repetitions) of 25 elements taken 4 at a time. This number is equal to

$$A_{25}^4 = 25 \cdot 24 \cdot 23 \cdot 22 = 303,600.$$

Permutations

Two k-samples without repetitions of n objects can differ in composition and in the order of their elements. However, if $k = n$, n-samples can differ only in the order of their elements. n-samples of n objects are called *permutations of n elements* or, briefly, *n-permutations*. (When there is no danger of misunderstanding, one simply speaks of *permutations*.)

In other words, n-permutations are samples without repetitions of n elements which contain all the n elements. One could also say that the permutations of n elements are all the possible n-arrangements each of which contains every element once, with two such arrangements differing only in the order of their elements. The number of n-permutations is denoted by P_n. The value of P_n is obtained from the formula for the number of samples without repetitions. Specifically,

$$P_n = A_n^n = n(n-1)\cdots 2 \cdot 1. \tag{2}$$

In other words, the number of permutations of n objects is equal to the product of all the natural numbers from 1 to n. This product is denoted by the symbol $n!$ (read: n-factorial). In particular, $1! = 1$.

In the sequel, we shall make use of the symbol $0!$. It is natural to define $0! = 1$ (and not $0! = 0$).

To see that the definition $0! = 1$ is reasonable, note that for $n > 1$ we have the equality

$$n! = n(n-1)!$$

For $n = 1$, the left-hand side of this equality reduces to 1 and the right-hand side to $1 \cdot 0!$. If our equality is to hold for $n = 1$, then we must put $0! = 1$.

One more remark: It is not difficult to see that

$$A_n^k = \frac{n!}{(n-k)!}.$$ (3)

In fact, the numerator and denominator of the fraction in Eq. (3) contain the factors $1, 2, 3, ..., n - k$. Reduction gives $A_n^k = n(n-1) \cdots (n-k+1)$ as in Eq. (1) above.

A Rook Problem

In how many ways can 8 rooks be placed on a (conventional) chessboard so that no rook can attack another?

Suppose the rooks are placed on the chessboard in the required manner. Then there is exactly 1 rook in each row and in each column. Let a_1 be the number of the occupied square in the first row, a_2 the number of the occupied square in the second row,..., a_8 the number of the occupied square in the eighth row. Then $(a_1, a_2, ..., a_8)$ is a permutation of the numbers $1, 2, ..., 8$. (Clearly, the numbers $a_1, a_2, ..., a_8$ are all different; otherwise, 2 rooks would end up in the same column.) Conversely, to a given permutation $a_1, a_2, ..., a_8$ of the numbers $1, 2, ..., 8$, there corresponds a placement of the rooks such that no rook can attack another. (For example, Fig. 6 illustrates a placement of the rooks corresponding to the permutation 7, 5, 4, 6, 1, 3, 2, 8.) Hence, the required number of placements of the rooks is equal to the number of permutations of the numbers $1, 2, ..., 8$, that is, P_8. Now,

$$P_8 = 8! = 1 \cdot 2 \cdot 3 \cdot 4 \cdot 5 \cdot 6 \cdot 7 \cdot 8 = 40{,}320.$$

This means that there are 40,320 ways of placing 8 rooks on a chessboard so that no rook can attack another.

Using the same reasoning, we can show that *there are $n!$ ways of placing n rooks on an n by n chessboard so that no rook can attack another.*

Our problem has an altogether different answer if the n rooks are numbered.

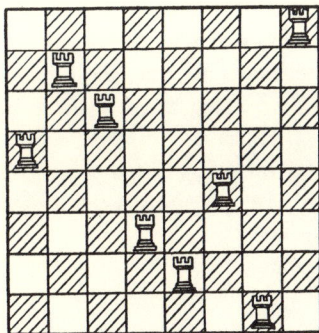

To find the answer in this case we observe that for each way of placing the unnumbered rooks there are $n!$ ways of placing the numbered rooks (the result of permuting the latter). This means that there are $(n!)^2$ ways of placing n numbered rooks on an $n \times n$ chessboard so that no rook can attack another.

The same answer can be obtained using the rule of product. The first rook can be placed in one of n^2 squares. If we ignore the row and column determined by the occupied square, we are left with an $(n-1)$ by $(n-1)$ chessboard with $(n-1)^2$ squares. This means that the second rook can be placed in one of $(n-1)^2$ squares. Similarly, the third rook can be placed in one of the $(n-2)^2$ squares, and so on. In all there are

$$n^2(n-1)^2 \cdots 1^2 = (n!)^2$$

ways of placing the rooks.

Linguistic Problems

Linguists study living and dead languages. They are frequently faced with the task of deciphering inscriptions in unknown languages. Suppose a linguist found a text employing 26 unknown symbols. The 26 symbols stand for 26 letters representing 26 sounds. *In how many ways can one pair off the sounds and the letters?*

To answer this question, consider a particular arrangement of the symbols in the text. Every correspondence between the symbols and the sounds determines a definite permutation of the sounds. Since there are $P_{26} = 26!$ permutations of the 26 sounds, the sounds and the letters can be paired off in 26! ways.

The number 26! is approximately equal to $4 \cdot 10^{26}$. No human being or even electronic computer can check out which pairing makes sense. Hence the need to reduce the number of possibilities. Frequently it is possible to separate symbols denoting vowels froms symbols denoting consonants (combinations of vowels and consonants occur more frequently then combinations of vowels or combinations of consonants). Suppose it has been possible to identify 7 symbols for vowels and 19 symbols for consonants. *By what factor does this reduce the number of possibilities?* There are 7! ways of arranging the symbols for the vowels, and 19! ways of arranging the symbols for the consonants. This means that the number of possibilities is now equal to 7! · 19!. In turn, this means that the number of possibilities has been reduced by a factor of 26!/7! · 19! ≈ 650,000. True, this is significant improvement, but 7! · 19! is still a gigantic number.

To reduce the number of possibilities further, one sometimes computes the frequencies with which individual symbols occur. By comparing these frequencies with the frequencies of occurrence of letters in languages related to the language of the text it may be possible to guess the meaning of some symbols. Other symbols may be identified by comparing the text with a translation (ancient rulers liked to advertise their feats in a number of languages).

Suppose that as a result of such efforts it was possible to identify 4 vowels and 13 consonants. What is the remaining number of possibilities? Clearly, it is 3! · 6! = 4320. This number of possibilities can be systematically checked out using electronic computers.

Similar difficulties are encountered by cryptographers (experts in deciphering codes).

Dancing in a Circle

Seven girls dance in a circle. In how many different ways can they form a circle?
If the girls were standing still, the answer would be 7! = 5040, the number of permutations of 7 objects. But the girls are moving and so what counts is their mutual positions and not their positions relative to the surrounding objects. This means that we must identify permutations which differ from one another by a rotation. Since each permutation gives rise to 6 new ones by rotation, it follows that the number 5040 must be divided by 7. This gives 5040/7 = 720 different permutations of 7 girls dancing in a circle.

More generally, if n objects are arranged in a circle and we agree to identify arrangements which differ by a rotation, then we obtain $(n - 1)!$ different arrangements.

Now let us compute the number of different necklaces which can be made out of 7 different beads. By analogy with the problem just solved, we might think that the number of different necklaces is 720. However, a necklace can not only be rotated but also turned over (Fig. 7). Hence the answer to our problem is 720/2 = 360.

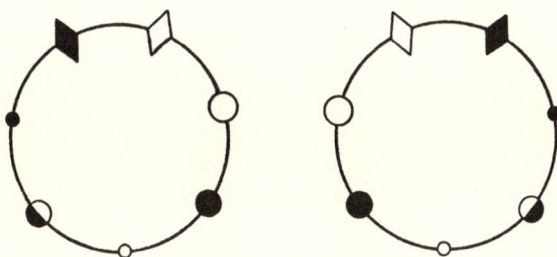

Fig. 7

Permutations with Repetitions

So far we have considered permutations of distinct objects. If some of the permuted objects are of the same type, then the number of permutations decreases since some of the permutations coincide.

To illustrate this situation consider the word "mart." This word gives rise to 24 different permutations:

mart	ramt	mtra	rtma
matr	mtar	ratm	rtam
mrat	rmat	rmta	mrta
tram	tmar	amtr	artm
tarm	atrm	atmr	tamr
amrt	armt	tram	tmra

Now consider the word "mama." To obtain the permutations of this word we could change "r" to "m" and "t" to "a" in the 24 permutations of the word "mart." But then some of the latter permutations would coincide. Specifically, the permutations "mart," "ramt," "mtra," "rtma" would all reduce to the word "mama." Similarly, the permutations in the second row would all reduce to the word "maam," and so on. We see that the 24 permutations can be divided into groups of four such that changing "r" to "m" and "t" to "a" reduces all permutations in the same group to the same permutation. In the table above, the groups in question coincide with the rows. Hence the number of different permutations of the word "mama" is equal to 24/4 = 6. The 6 different permutations of "mama" are

mama,   maam,   mmaa,   amam,   aamm,   amma.

The general problem can be formulated as follows:

There are k different types of objects. What is the number of n-permutations of n_1 elements of the first type, n_2 elements of the second type,..., n_k elements of the kth type?

Each permutation consists of $n = n_1 + n_2 + \cdots + n_k$ elements. If all of these elements were different, then we would have $n!$ permutations. Since some elements are of the same type, there are actually fewer than $n!$ permutations. To obtain the number of different permutations consider the permutation

$$\underbrace{aa \cdots a}_{n_1} \underbrace{bb \cdots b}_{n_2} \cdots \underbrace{xx \cdots x}_{n_k}, \tag{4}$$

where the elements of the first kind come first, the elements of the second kind come next,..., the elements of the kth kind come last. The elements of the first kind can be permuted in $n_1!$ ways. Since these elements are of the same type, this has no effect on the permutation in (4). Similarly, the permutation in (4) is not affected by the $n_2!$ permutations of the elements of the second type,..., the $n_k!$ permutations of the elements of the kth type. (For example, permuting the first two or the last two letters in "mmaa" has no effect on the arrangement of all the four letters.)

The permutations of the blocks of objects of the same type are independent of one another. Therefore (by the rule of product), the elements of the permutation in (4) can be interchanged in $n_1! \, n_2! \cdots n_k!$ ways without affecting it. The same is true of any other permutation of our n objects. Hence the $n!$ permutations can be arranged in groups of $n_1! \, n_2! \cdots n_k!$ equal permutations. But then the number of different permutations with repetitions of $n = n_1 + n_2 + \cdots + n_k$ objects is

$$P(n_1, n_2, ..., n_k) = \frac{n!}{n_1! \, n_2! \cdots n_k!}. \tag{5}$$

Using Formula (5) we can readily answer the following question: *What is the number of permutations of the letters in the word "Mississippi?"* Since "m" appears once, "i" appears four times, "s" appears four times, "p" appears twice, and there are 11 letters in all, we have, by Formula (5),

$$P(4, 4, 2, 1) = \frac{11!}{4! \cdot 4! \cdot 2! \cdot 1!} = 34{,}650.$$

Anagrams

Until the 17th century there were no scientific journals. Scholars learned of the work of their colleagues from books or from private communications. This caused great difficulties when it came to publication of new results. Years might pass before a book was printed. On the other hand, if one relied on a private communication, then there was the danger of the recipient claiming the result as his own. If this happened it was not easy to prove that such was indeed the case. Then there was the possibility that the recipient of the letter found nothing new in it; that he had thought about the problem, solved it, and was about to write his colleague a similar letter.

All this gave rise to many priority arguments. Even at the end of the 17th century there were still protracted priority arguments involving Newton and Leibniz (as to which of the two was the first to discover the differential and integral calculus), Newton and Hooke (as to which of the two was the first to formulate the law of universal gravitation), and so on.

Archimedes once found it necessary to resort to a ruse to settle a priority argument. When he wrote about his results to some scholars in Alexandria and the latter claimed his results as their own, Archimedes sent them an additional letter containing truly remarkable formulas for areas and volumes of certain figures. The scholars again said that they had known these formulas long ago and that Archimedes was not telling them anything new. But then it turned out that Archimedes trapped them; the remarkable formulas were false!

To safeguard their claims to priority and to prevent premature disclosure of their results, scholars frequently put the essence of a discovery in a brief phrase, rearranged the letters in the phrase and sent the rearranged phrase, an anagram, to their colleagues. When the full result appeared in print it included the key to the anagram. Anagrams were also used in political arguments. For example, when the French king Henry III was murdered, the name of the murderer, Brother Jacques Clément, appeared in the form of the anagram: C'est l'enfer qui m'a créé (I am a creature of hell). The king's opponents would not be outdone and made the name Henri de Valois into the anagram Vilain Herodés (Herod-like villainy). When Christian Huygens (1629–1695) discovered the ring of Saturn, he made up the anagram

aaaaaaa, ccccc, d, eeeee, g, h, iiiiiii, lll,

mm, nnnnnnnnn, oooo, pp, q, rr, s, ttttt, uuuuu.

An appropriate rearrangement of these letters yields the sentence
"Annulo cingitur tenui, plano, nusquam cohaerente, ad eclipticam inclinato."
(Surrounded by a thin ring, flat, suspended nowhere, inclined to the ecliptic.)

Anagrams were not always an effective means of guarding secrets. When Huygens discovered the first satellite of Saturn (Titan) and found that it rotated about that planet in 15 days, he put the discovery in the form of an anagram and sent it to his colleagues. One of them, Wallis, a master cryptographer, deciphered the anagram, made it into a new anagram and sent it to Huygens. After the scholars exchanged solutions of the anagrams, it looked as if Wallis made the discovery before Huygens. Later Wallis admitted the joke; he had made it to show that anagrams were of little use in protecting secrets. Huygens was not amused.

We compute the number of permutations required to guarantee the solution of the first anagram of Huygens. The anagram consists of 7 a's, 5 c's, 1 d, 5 e's, 1 g, 1 h, 7 i's, 3 l's, 2 m's, 9 n's, 4 o's, 2 p's, 1 q,2 r's, 1 s, 5 t's, and 5 u's; a total of 61 letters. In view of Formula (5), the number of permutations is

$$\frac{61!}{7!\ 5!\ 1!\ 5!\ 1!\ 1!\ 7!\ 3!\ 2!\ 9!\ 4!\ 2!\ 1!\ 2!\ 1!\ 5!\ 5!}\ .$$

This huge number is approximately equal to 10^{60}.

The job of going through these permutations is so big that an electronic computer which does a million operations in a second could not cope with it in all the time of the existence of the solar system.

In a sense, a human being could solve this problem more easily than a machine. After all, a human being would consider only those permutations which contain meaningful words, would take into consideration laws of morphology, and so on, and this would greatly reduce the number of required trials. Most importantly, he has a rough idea of the questions considered by his correspondent. With all that, the job is cumbersome, to put it mildly.

Combinations

In some arrangements the order of the elements is immaterial. For example, if out of 20 semifinalists in a chess tournament three players enter the finals, then what counts is being one of the three finalists. Frequently, the winner turns out to be a player who did not rank highest in the semifinals. Similarly, if 17 teams participate in a soccer tournament and 13 reach the finals, then there is little comfort in ranking 14th rather than 17th.

The term *combination* applies to arrangements in which the order of the elements is immaterial. We speak of *k-combinations of n elements* and regard two such combinations as different only if they differ in composition. The number of k-combinations of n elements is denoted by the symbol C_n^k.

There is a simple relation between the number C_n^k of k-combinations and the number A_n^k of k-samples. In fact, permuting the elements of each k-combination, we obtain $k!$ distinct k-samples. This means that

$$k!\ C_n^k = A_n^k\ .$$

Hence

$$C_n^k = \frac{A_n^k}{k!} = \frac{n!}{(n-k)!\,k!}.$$ (6)

permutations of just derived coincides with the formula for the numberfo
The formula k elements of one type and $n - k$ elements of another type:

$$P(k, n - k) = \frac{n!}{k!\,(n-k)!}.$$

In other words,

$$C_n^k = P(k, n - k).$$ (7)

Equation (7) can be deduced directly without involving the formula for the number of samples. To this end we order the n objects out of which we are to make up the various k-combinations and assign to each combination a code consisting of n ones and zeros. In making up the code for a particular combination, we write a 1 if a certain element enters the combination and a 0 if it does not enter the combination.
(For example, the 5-combination {a, c, f, h, i} of the 10 letters a, b, c, d, e, f, g, h, i, j is coded as 1010010110, and the code 0111001001 stands for the 5-combination {b, c, d, g, j}.) It is clear that every k-combination determines an arrangement of k ones and $n - k$ zeros, and every arrangement of k ones and $n - k$ zeros determines a k-combination. Also, different arrangements of ones and zeros determine different combinations. This means that the number of k-combinations of n elements coincides with the number of permutations of k elements of one type (ones) and $n - k$ elements of another type (zeros).
By means of Formula (6) we can easily solve the problems discussed in the beginning of this section. The number of different outcomes of the semifinals in the chess tournament is

$$C_{20}^3 = \frac{20!}{3!\,17!} = 1140.$$

The number of "sad" outcomes of the football tournament is

$$C_{17}^4 = \frac{17!}{4!\,13!} = 2380.$$

The following is another problem involving combinations:
In how many ways can 8 rooks be set out on a chessboard? In distinction to the rook problem considered on p. 20, we no longer impose the condition that no rook can attack another. It follows that all we need do is count the number of

ways in which it is possible to choose 8 of the 64 squares on a chessboard. The number in question is

$$C_{64}^3 = \frac{64!}{8!\,56!} = 4{,}328{,}284{,}968.$$

Similarly, the number of ways in which it is possible to set out k rooks on a "chessboard" with m rows and n columns is

$$C_{mn}^k = \frac{(mn)!}{k!\,(mn-k)!} \,.$$

If we replaced the k identical rooks with k different pieces, then the position of each piece would matter. In that case, instead of counting combinations, we would be counting samples, and the required answer would be

$$A_{mn}^k = \frac{(mn)!}{(mn-k)!} \,.$$

The Genoa Lottery

At one time there flourished the so-called Genoa lottery. In some places the Genoa lottery has survived to this day. The essentials of the Genoa lottery were as follows: Participants bought tickets on which there appeared numbers from 1 to 90. One could also buy tickets with two, three, four, or five numbers. On a designated day, five tokens were selected in a chance drawing from a bag of tokens bearing numbers from 1 to 90. The winners were owners of tickets bearing exclusively numbers appearing on the selected tokens. (The popular game of lotto is a variant of the Genoa lottery.) For example, a ticket with the numbers 8, 21, 49 was a winning ticket if the selected tokens were numbered 3, 8, 21, 37, 49, and a losing ticket if the selected tokens were numbered, say, 3, 7, 21, 49, 63 (the number 8 appeared on the ticket but not on the selected tokens).

A winning ticket could have on it one, two, three, four, or five numbers, and its owner received, accordingly, 15, 270, 5500, 75,000, 100,000 times the price of the ticket. Many tried to get rich quickly by buying tickets with two or three numbers, but there were practically no winners. The real winners were the owners of the lottery.

To see what was involved we shall try to compute the ratio of "lucky" outcomes to all outcomes for different types of tickets. The total number of outcomes can be found by means of Formula (6). Indeed, a drawing was a selection of 5 out of 90 tokens in the bag without regard to order, and, so, a combination of 90 things taken 5 at a time. The total of such combinations is

$$C_{90}^5 = \frac{90!}{5!\,85!} = \frac{90 \cdot 89 \cdot 88 \cdot 87 \cdot 86}{1 \cdot 2 \cdot 3 \cdot 4 \cdot 5}.$$

Now suppose a player bought a lottery ticket with 1 number on it. In how many cases did he win? If he is to win, then the number on his ticket must coincide with 1 of the 5 numbers on the 5 tokens taken from the bag. The numbers on the remaining 4 tokens can be arbitrary. These 4 numbers are chosen from among 89 numbers. It follows that the number of lucky combinations is

$$C_{89}^4 = \frac{89 \cdot 88 \cdot 87 \cdot 86}{1 \cdot 2 \cdot 3 \cdot 4}.$$

But then the ratio of lucky combinations to all combinations is

$$\frac{C_{89}^4}{C_{90}^5} = \frac{5}{90} = \frac{1}{18}.$$

This means that the player's chances of winning are 1 in 18. Put differently, he pays for 18 tickets and wins 15 times the price of a ticket; the cost of 3 tickets is pocketed by the owners of the lottery.

Clearly, this does not mean that a player wins precisely once in 18 drawings. Sometimes he may lose 20 or 30 times in a row and sometimes he may win twice in a row. What counts is the average number of wins computed over a long period of time for a large number of players. Failure to understand this point may result in a faux pas of the kind attributed to a doctor who said to his patient: "Your illness is fatal in 9 out of 10 cases. The last 9 of my patients who suffered from your illness died. This means that you are bound to get well!"

Now let us compute the chances of winning when the ticket has 2 numbers. In this case 3 of the 5 numbers on the tokens selected from the bag are arbitrary.

Since these 3 numbers are chosen from among 88, the total of lucky outcomes is in this case equal to

$$C_{88}^3 = \frac{88 \cdot 87 \cdot 86}{1 \cdot 2 \cdot 3},$$

and the ratio of lucky outcomes to all outcomes is

$$\frac{C_{88}^3}{C_{90}^5} = \frac{4 \cdot 5}{90 \cdot 89} = \frac{2}{801}.$$

Since the winner receives 270 times the price of the ticket, it follows that, for every 801 two-number tickets sold, the lottery owners pocket the cost of 261 tickets. Clearly, a 2-number ticket is even less of a bargain for the player than a 1-number ticket.

In the case of 3-, 4-, and 5-number tickets, the winning chances of the players are even smaller. In the case of 3-number tickets, the ratio of lucky outcomes to all outcomes is

$$\frac{C_{87}^2}{C_{90}^5} = \frac{3 \cdot 4 \cdot 5}{90 \cdot 89 \cdot 88} = \frac{1}{11,748},$$

in the case of 4-number tickets it is

$$\frac{C_{86}^1}{C_{90}^5} = \frac{2 \cdot 3 \cdot 4 \cdot 5}{90 \cdot 89 \cdot 88 \cdot 87} = \frac{1}{511,038},$$

and in the case of 5-number tickets it is

$$\frac{1}{C_{90}^5} = \frac{1 \cdot 2 \cdot 3 \cdot 4 \cdot 5}{90 \cdot 89 \cdot 88 \cdot 87 \cdot 86} = \frac{1}{43,949,268}.$$

The winner receives, respectively, 5500, 7500, and 1,000,000 times the price of a ticket. The reader can readily compute the losses of the players in all of these cases.

Buying Pastry

A pastry shop sells 4 kinds of pastries: napoleons, éclairs, shortcakes, and cream puffs. How many different sets of 7 pastries can one buy?

This problem is different from the problems we have solved so far. Since the order in which the pastries are placed in a box is immaterial, the problem we are dealing with is closer to combinations than to permutations. On the other hand, the problem differs from problems involving combinations by the fact that repetitions are allowed (for example, one may buy 7 éclairs). In such cases it is natural to speak of combinations with repetitions.

To solve this problem we code each purchase using zeros and ones. Specifically, we write as many ones as there are napoleons in the purchase. Then we write a zero to separate napoleons from éclairs and follow it with as many ones as there are éclairs. Then we write another zero (this means that if no éclairs were purchased, then the code contains two successive zeros). After that we write as many ones as there are shortcakes, a zero, and finally, as many ones as there are cream puffs. If, for example, the purchase consisted of 3 napoleons, 1 éclair, 2 shortcakes, and 1 cream puff, then the corresponding code is 1110101101. The code for 2 napoleons and 5 shortcakes is 1100111110. Clearly, different purchases determine different arrangements of 7 ones and 3 zeros. Conversely, every arrangement of 7 ones and 3 zeros describes a purchase. For example, the arrangement 0111011110 describes the purchase of 3 éclairs and 4 shortcakes.

We see that the number of different purchases is equal to the number of permutations with repetitions of 7 ones and 3 zeros. In view of Formula (5) on p. 24, this number is

$$P(7, 3) = \frac{10!}{7! \, 3!} = \frac{10 \cdot 9 \cdot 8}{1 \cdot 2 \cdot 3} = 120.$$

There is another way of obtaining the same result. We arrange the pastries in each purchase in the following order: napoleons, éclairs, shortcakes, and cream puffs. Then we assign to each éclair its "position number" increased by 1, to each shortcake its position number increased by 2, and to each cream puff its position number increased by 3 (the position numbers of the napoleons are unchanged). For example, suppose a purchase consisted of 2 napoleons, 3 éclairs, 1 shortcake, and 1 cream puff. Then the pastries are numbered 1, 2, 4, 5, 6, 8, 10. It is clear that the largest number in any sequence is 10 (the last puff is numbered $7 + 3 = 10$) and the smallest number is 1 (this is the number assigned to the

first napoleon). Also, the numbers in a sequence are distinct. Conversely, every
increasing sequence of 7 of the 10 numbers from 1 to 10 describes a purchase.
For example, the sequence 2, 3, 4, 5, 7, 8, 9 describes the purchase of 4 éclairs
and 3 shortcakes. To see that this is the case, we subtract from the given numbers
the numbers 1, 2, 3, 4, 5, 6, 7. The result is 1, 1, 1, 1, 2, 2, 2, that is, 4 ones and
3 twos. Now, 1 was added to the position numbers of the éclairs, and 2 was
added to the position numbers of the shortcakes. But then, as asserted, the given
sequence describes the purchase consisting of 4 éclairs and 3 shortcakes.

Note that all of our sequences are increasing. This means that each sequence is
uniquely determined by its elements. It follows that the number of our
7-sequences is equal to the number of 7-combinations of 10 numbers (from 1 to
10). The number of such combinations is

$$C_{10}^7 = \frac{10!}{7!\,3!} = 120,$$

in agreement with our earlier result.

Combinations with Repetitions

As mentioned earlier, the preceding problem belongs to the class of problems
involving *combinations with repetitions*. These problems can be formulated as
follows: There are n different types of objects. We are required to compute the
number of k-arrangements of these objects without regard to order (in other
words, two arrangements are different only if they differ in the number of
elements of at least one type).

The general problem is solved very much like the problem about pastries.
Namely, each arrangement is coded by means of one and zeros, with each type
represented by as many ones as there are elements of this type in the arrangement
and with different types separated by zeros (if objects of certain types are absent
then we write two or more zeros in a row). It follows that the number of ones
in each code is equal to the number of objects in an arrangement, and the number
of zeros in each code is one less than the number of types of objects. Thus each
k-arrangement is represented by a permutation with repetitions of k ones and
$n - 1$ zeros. Also, different arrangements are represented by different permutations with repetitions, and each permutation with repetitions represents a
definite arrangement. This means that the number $\bar{C}_n^k$ of k-combinations with
repetitions of elements of n types is equal to the number $P(k, n - 1)$ of permutations with repetitions of $n - 1$ zeros and k ones. Since

$$P(k, n - 1) = \frac{(k + n - 1)!}{k!\,(n - 1)!}$$

we have

$$\bar{C}_n^k = \frac{(k+n-1)!}{k!\,(n-1)!} = C_{n+k-1}^k.$$

The same formula can be derived in a different way. We group the elements in each combination by type. Then we number the elements in accordance with their positions in the combination except that we increase these numbers by 1 for elements of the second type, by 2 for elements of the third type, and so on. In this way each of our combinations with repetitions determines a k-combination without repetitions of the numbers $1, 2,..., n + k - 1$. This again shows that

$$\bar{C}_n^k = C_{n+k-1}^k = \frac{(n+k-1)!}{k!\,(n-1)!}. \tag{8}$$

There are problems involving combinations with repetitions where each combination must include elements belonging to r fixed types, $r \leqslant n$. Such problems can be easily reduced to problems already solved. To guarantee inclusion of elements of the required r types, we fill the first r positions of each k-combination with elements of the r types in question, and fill the remaining $k - r$ positions in any manner whatever with elements of n types. This means that the number of required combinations is equal to the number of $(k - r)$-combinations with repetitions of elements of n types, that is,

$$\bar{C}_n^{k-r} = C_{n+k-r-1}^{k-r}.$$

In particular, if $n \leqslant k$ and each k-combination must contain at least one element of each of the n types, then the number of combinations is C_{k-1}^{k-n}.

The Soccer Tournament Revisited

We studied problems involving samples, permutations, and combinations. Frequently different types of arrangements occur in the same problem.

Consider, for example, a soccer tournament with 17 participating teams. By the rules of the tournament the first 3 teams are awarded gold, silver, and bronze medals and the last 4 teams drop to a lower league. We shall say of two outcomes of the tournament that they are *essentially the same* if the teams which recieve the gold, silver, and bronze medals, respectively, and the teams which drop to a lower league are the same in both cases. *We are to compute the number of essentially different outcomes of the tournament.*

We know that the medals can be awarded in $A_{17} = 17 \cdot 16 \cdot 15$ ways (see p. 18). When it comes to relegating the last 4 of the remaining 14 teams to a lower league, the order of the last 4 teams is irrelevant and so the number of possibilities is $C_{14}^4 = 14!/(4!\ 10!)$. By the rule of product, the number of essentially different outcomes of the tournament is

$$A_{17}^3 \cdot C_{14}^4 = 17 \cdot 16 \cdot 15 \cdot \frac{14!}{4!\ 10!} = \frac{17!}{4!\ 10!} = 4{,}084{,}080.$$

The same result can be obtained by means of a different argument. If we rule out ties, then there are $P_{17} = 17!$ outcomes of the tournament. Permutations of the teams rated 4 through 13 and permutations of the teams rated 14 through 17 lead to essentially the same outcomes of the tournament. The number of such permutations is $10! \cdot 4!$. It follows that the number of different outcomes of the tournament is $17!/(10!\ 4!)$.

Suppose the outcome of the tournament is to be communicated by means of a telegram consisting of k dots and dashes. *What is the smallest value of k?* We know that the number of arrangements of k dots and dashes is 2^k. It follows that the least value of k must be large enough to satisfy the inequality

$$2^k \geqslant 4{,}084{,}080.$$

This means that $k \geqslant 22$. In other words, we must use at least 22 symbols to communicate the outcome of the tournament.

Clearly, in communicating the results of a competition one does not make use of computations of this kind. Nevertheless, it is easy to think of situations (such as sending photographs from a space ship) in which transmission of information involves great technical difficulties and each symbol is "worth its weight in gold." Then one must consider the various possibilities and select the one that is most economical. These problems are studied in a branch of mathematics called *information theory*.

Properties of Combinations*

The numbers C_n^k have many remarkable properties. These properties can be proved in many ways. In some cases it is easiest to make use of the relation

$$C_n^k = \frac{n!}{k!\ (n-k)!}.\tag{9}$$

In other cases we resort to a combinatorial argument. Specifically, we compute

---

* The rest of this chapter may be omitted in a first reading. In the sequel, we make use of the relations $C_n^k = C_n^{n-k}$ and $C_n^k = C_{n-1}^{k-1} + C_{n-1}^k$ proved here.

the number of arrangements of a certain kind and break them up into disjoint classes. Then we find the number of arrangements in each class. By adding the number of arrangements in each class we obtain once more the number of all arrangements. This leads to the required relation.

We begin with the very simple relation

$$C_n^k = C_n^{n-k}. \tag{10}$$

This relation follows directly from Formula (9). In fact, if in (9) we replace k with $n - k$, then $n - k$ becomes $n - (n - k) = k$, and all that happens is that the factors in the denominator are interchanged. It is also possible to give an easy proof of (10) without making use of the expression for the number of combinations. We pair off each k-combination of n different elements with the $(n - k)$-combination of the $n - k$ remaining elements. In this pairing, different k-combinations determine different $(n - k)$-combinations, and conversely. It follows that $C_n^k = C_n^{n-k}$.

It is not much more difficult to prove the relation

$$C_n^k = C_{n-1}^{k-1} + C_{n-1}^k. \tag{11}$$

We break up the set of k-combinations of n elements $a_1, a_2, ..., a_{n-1}, a_n$ into two classes. The first class consists of all combinations which contain a_n and the second class consists of all combinations which do not contain this element. If we remove from an arbitrary combination in the first class the element a_n, then we are left with a $(k - 1)$-combination of the elements $a_1, a_2, ..., a_{n-1}$. The number of such combinations is C_{n-1}^{k-1}, so that the number of combinations in the first class is C_{n-1}^{k-1}. The combinations in the second class are k-combinations of the $n - 1$ elements $a_1, ..., a_{n-1}$. The number of such combinations is C_{n-1}^k. Since each of the C_n^k k-combinations of the elements $a_1, ..., a_n$ belongs to exactly one of these classes, we are led to the relation (11).

A similar argument yields the relation

$$C_n^0 + C_n^1 + C_n^2 + \cdots + C_n^n = 2^n. \tag{12}$$

We recall that 2^n is the number of n-samples with repetitions of elements of 2 types. We separate these samples into classes with the samples in the kth class consisting of k elements of the first type and $n - k$ elements of the second type. The samples in the kth class are precisely the permutations of k elements of the first type and $n - k$ elements of the second type. We know that the number of such permutations is $P(k, n - k)$, and $P(k, n - k) = C_n^k$ (see pp. 24 and 27). This means that the number of samples in all the classes is $C_n^0 + C_n^1 + \cdots + C_n^n$. On the other hand, this number is 2^n. This proves (12).

In just this way it is possible to prove that

$$\sum_{n_1+n_2+n_3=n} P(n_1, n_2, n_3) = 3^n,$$ (13)

where the sum is taken over all partitions of the number n into three ordered summands (this means that if the sum includes $P(n_1, n_2, n_3)$, then it also includes $P(n_2, n_3, n_1)$, and so on). For proof we consider all n-samples of elements of 3 types and separate them into classes of samples composed of the same elements (that is, each sample belonging to the same class contains the same number of elements of the first, second, and third type).

More generally we have the relation

$$\sum_{n_1+\cdots+n_k=n} P(n_1, ..., n_k) = k^n$$ (14)

where the sum is taken over all ordered partitions of the number n into k summands.

Next we consider m-combinations with repetitions made up of elements of $n+1$ types, say, $n+1$ letters $a, b, c, ..., x$. The number of such combinations is $\bar{C}_{n+1}^m = C_{n+m}^m$. We separate these combinations into classes. A combination is put into the kth class if and only if it contains exactly k copies of the letter a. The remaining $m-k$ places in the combination can be filled with letters taken from among the n letters $b, c, ..., x$. Thus the number of combinations in the kth class is equal to the number of $(m-k)$-combinations with repetitions of elements of n types, that is, $C_{n+m-k-1}^{m-k}$. It follows that the number of all the combinations is

$$C_{n+m-1}^m + C_{n+m-2}^{m-1} + \cdots + C_n^1 + C_{n-1}^0.$$

On the other hand, we know that this number is C_{n+m}^m. This proves that

$$C_{n-1}^0 + C_n^1 + C_{n+1}^2 + \cdots + C_{n+m-1}^m = C_{n+m}^m.$$ (15)

If in Eq. (15) we change n to $n+1$ and m to $m-1$ and make use of Eq. (10), then we obtain the relation

$$C_n^n + C_{n+1}^n + C_{n+2}^n + \cdots + C_{n+m-1}^n = C_{n+m}^{n+1}.$$ (16)

For $n = 1, 2, 3$, relation (16) yields the relations

$$1 + 2 + \cdots + m = \frac{m(m+1)}{2},$$ (17)

$$1 \cdot 2 + 2 \cdot 3 + \cdots + m(m+1) = \frac{m(m+1)(m+2)}{3},$$ (18)

$$1 \cdot 2 \cdot 3 + 2 \cdot 3 \cdot 4 + \cdots + m(m+1)(m+2) = \frac{m(m+1)(m+2)(m+3)}{4}.$$ (19)

Using Formulas (17)–(19), we can easily find the sum of the squares and the sum of the cubes of the natural numbers from 1 to m. Formula (18) can be rewritten as follows:

$$1^2 + 2^2 + \cdots + m^2 + 1 + 2 + \cdots + m = \frac{m(m+1)(m+2)}{3}.$$

By Formula (17), $1 + 2 + \cdots + m = m(m+1)/2$. Therefore,

$$1^2 + 2^2 + \cdots + m^2 = \frac{m(m+1)(m+2)}{3} - \frac{m(m+1)}{2}$$

$$= \frac{m(m+1)(2m+1)}{6}. \tag{20}$$

Similar manipulation of Formula (19) yields the relation

$$1^3 + 2^3 + \cdots + m^3 = \frac{m^2(m+1)^2}{4}. \tag{21}$$

We leave it to the reader to obtain in this way formulas for sums of higher powers of the natural numbers.

m-combinations with repetitions of elements of n types can be separated into classes according to the number of types of elements in a given combination. In other words, the first class will contain combinations of elements of exactly 1 type, the second class will contain combinations of elements of exactly 2 types,..., the nth class will contain combinations of elements of exactly n types (clearly, if $m < n$, then there are only m classes).

Let us calculate the number of combinations in each class. Consider a particular selection of k types. The number of m-combinations with repetitions of elements actually representing each of the selected k types is $C_{m-1}^{m-k} = C_{m-1}^{k-1}$ (see p. 33). The number of ways of selecting k out of n types is C_n^k. By the rule of product the number of combinations in the kth class is $C_n^k C_{m-1}^{k-1}$. By adding the number of combinations in each class, we obtain the number of m-combinations with repetitions of elements of n types. This number is C_{m+n-1}^m. It follows that

$$C_n^1 C_{m-1}^0 + C_n^2 C_{m-1}^1 + \cdots + C_n^n C_{m-1}^{n-1} = C_{m+n-1}^m. \tag{22}$$

If $m < n$, then the last term in the sum in (22) will be $C_n^m C_{m-1}^{m-1}$. It is convenient to replace the C_n^k in each summand in (22) with C_n^{n-k}. This yields the relation

$$C_n^{n-1} C_{m-1}^0 + C_n^{n-2} C_{m-1}^1 + \cdots + C_n^0 C_{m-1}^{n-1} = C_{m+n-1}^{n-1}. \tag{23}$$

In each of the summands in the left-hand side of (23), the sum of the upper indices is $n - 1$ and the sum of the lower indices is $n + m - 1$. Also, the upper indices vary and the lower indices stay fixed. A somewhat different form of (23) is

$$C_p^0 C_{n-p}^m + C_p^1 C_{n-p}^{m-1} + \cdots + C_p^m C_{n-p}^0 = C_n^m. \tag{23'}*$$

We now deduce an analogous formula in which the upper as well as the lower indices vary. To this end we consider all the m-combinations with repetitions of p different vowels and $n - p$ different consonants. We separate these combinations into classes and put into the kth class all combinations consisting of k vowels and $m - k$ consonants. Now we compute the number of combinations in the kth class. To do this we think of each combination in this class as composed of a k-combination (with repetitions) of p vowels and an $(m - k)$-combination (with repetions) of $n - p$ consonants. It follows that the number of combinations in the kth class is $C_{k+p-1}^k C_{m+n-p-k-1}^{m-k}$. But then the number of all the combinations under consideration is

$$C_{p-1}^0 C_{m+n-p-1}^m + C_p^1 C_{m+n-p-2}^{m-1} + \cdots + C_{m+p-1}^m C_{n-p-1}^0 .$$

On the other hand, these combinations are all the m-combinations with repetitions of elements of n types and the number of such combinations is C_{m+n-1}^m. By equating the two expressions, we obtain the relation

$$C_{p-1}^0 C_{m+n-p-1}^m + C_p^1 C_{m+n-p-2}^{m-1} + \cdots + C_{m+p-1}^m C_{n-p-1}^0 = C_{m+n-1}^m . \tag{24}$$

If we apply the identity $C_r^q = C_r^{r-q}$ to all the terms in Eq. (24), then we obtain the formula

$$C_{p-1}^{p-1} C_{m+n-p-1}^{n-p-1} + C_p^{p-1} C_{m+n-p-2}^{n-p-1} + \cdots + C_{m+p-1}^{p-1} C_{n-p-1}^{n-p-1} = C_{m+n-1}^{n-1} .$$

in which only the lower indices in the sum on the left-hand side vary. If in this formula we change p to $p + 1$, n to $n + 2$, and m to $m - n$, then we obtain the formula

$$C_p^p C_{m-p}^{n-p} + C_{p+1}^p C_{m-p-1}^{n-p} + \cdots + C_{m-n+p}^p C_{n-p}^{n-p} = C_{m+1}^{n+1} . \tag{24'}$$

We see that in the sum on the left-hand side of (24') the upper indices are

---

* One way of going from (23) to (23') is to replace $n - 1$ by q, $m - 1$ by p and $n + m - 1$ by r in (23). Then we get the relation

$$C_p^0 C_{r-p}^q + C_p^1 C_{r-p}^{q-1} + \cdots + C_p^q C_{r-p}^0 = C_r^q .$$

This relation reduces to (23') if we replace r by n and q by m. (Translators) fixed, the lower indices vary, the sum of the upper indices is n, and the sum of the lower indices is m.

We note a special case of the relation (23). If in (23′) we put $n - p = m$, then we obtain the relation

$$C_p^0 C_m^0 + C_p^1 C_m^1 + \cdots + C_p^m C_m^m = C_{p+m}^m . \tag{25}$$

For $p = m$, (25) reduces to

$$(C_p^0)^2 + (C_p^1)^2 + \cdots + (C_p^p)^2 = C_{2p}^p . \tag{26}$$

It is possible to generalize the relations just obtained. To this end we consider a set of elements of q types. There are n_1 elements of the first type, n_2 elements of the second type ,..., n_q elements of the qth type. We assume that the elements of any particular type are all different (for example, the type of an element is determined by its color and all the elements which have the same color differ in shape).

We consider all m-combinations of elements of our set and separate them into classes by composition, that is, by the number of elements of the first type, of the second type,..., of the qth type in the combination. Hence each class is characterized by q nonnegative numbers $(m_1, m_2 ,..., m_q)$ satisfying the inequalities $0 \leqslant m_i \leqslant n_i$ and the equality $m_1 + m_2 + \cdots + m_q = m$.

By the rule of product, the number of combinations in the class characterized by $(m_1, m_2 ,..., m_q)$ is $C_{n_1}^{m_1} C_{n_2}^{m_2} \cdots C_{n_q}^{m_q}$. Summing over all classes, we obtain the relation

$$\sum C_{n_1}^{m_1} C_{n_2}^{m_2} \cdots C_{n_q}^{m_q} = C_n^m , \tag{27}$$

where $n = n_1 + n_2 + \cdots + n_q$, and the summation extends over all ordered arrangements $(m_1, m_2,..., m_q)$ of the nonnegative numbers $m_1, m_2,..., m_q$ such that $m_1 + m_2 + \cdots + m_q = m$.

If we consider combinations with repetitions, then we get the analogous relation

$$\sum C_{n_1+m_1-1}^{m_1} C_{n_2+m_2-1}^{m_2} \cdots C_{n_q+m_q-1}^{m_q} = C_{n+m-1}^m , \tag{28}$$

where, as before, $n = n_1 + n_2 + \cdots + n_q$ and the summation extends over the ordered arrangements $(m_1, m_2 ,..., m_q)$.

One more property of combinations is based on the relation

$$C_n^k C_{n-k}^{m-k} = C_m^k C_n^m . \tag{29}$$

We give a combinatorial proof of (29). To this end we take n different objects, select k of the n objects, and then select $m - k$ of the remaining $n - k$ objects. This yields an m-combination of n elements. For a fixed k, this process can be carried out in $C_n^k C_{n-k}^{m-k}$ ways. It is easy to see that each of the C_n^m combinations is obtained in C_m^k ways. This proves (29).

We write down (29) for $k = 0,..., m$ and add the results. Since, by Formula (12),

$$C_m^0 + C_m^1 + \cdots + C_m^m = 2^m,$$

we conclude that

$$C_n^0 C_n^m + C_n^1 C_{n-1}^{m-1} + \cdots + C_n^m C_{n-m}^0 = 2^m C_n^m,$$

or

$$C_n^0 C_n^{n-m} + C_n^1 C_{n-1}^{n-m} + \cdots + C_n^m C_{n-m}^{n-m} = 2^m C_n^m. \qquad (30)$$

Special Case of the Inclusion and Exclusion Formula

Many properties of combinations can be deduced from the inclusion and exclusion formula (see p. 14). Consider the special case when the number $N(\alpha_1 ,..., \alpha_k)$ of elements with properties $\alpha_1 ,..., \alpha_k$ depends not on the properties involved but on their number, that is, suppose that

$$N(\alpha_1) = \cdots = N(\alpha_n),$$
$$N(\alpha_1\alpha_2) = N(\alpha_1\alpha_3) = \cdots = N(\alpha_{n-1}\alpha_n),$$
$$N(\alpha_1\alpha_2\alpha_3) = N(\alpha_1\alpha_2\alpha_4) = \cdots = N(\alpha_{n-2}\alpha_{n-1}\alpha_n),$$

and so on. Let $N^{(1)}$ denote the common value of the numbers $N(\alpha_1),..., N(\alpha_n)$. Then $N(\alpha_1) + \cdots + N(\alpha_n) = nN^{(1)} = C_n^1 N^{(1)}$. Similarly, if we define $N^{(2)} = N(\alpha_1\alpha_2)$, then it follows that

$$N(\alpha_1\alpha_2) + N(\alpha_1\alpha_3) + \cdots + N(\alpha_{n-1}\alpha_n) = C_n^2 N^{(2)}.$$

Quite generally, we obtain the relation

$$N(\alpha_1\alpha_2 \cdots \alpha_k) + \cdots + N(\alpha_{n-k+1} \cdots \alpha_n) = C_n^k N^{(k)}, \qquad (31)$$

where the sum on the left-hand side extends over all combinations of n properties taken k at a time.

It follows that in this case the inclusion and exclusion formula takes the form

$$N^{(0)} = N - C_n^1 N^{(1)} + C_n^2 N^{(2)} - \cdots + (-1)^n C_n^n N^{(n)}. \qquad (32)$$

Alternating Sums of Combinations

Now we deduce further properties of combinations. The new relations are similar to the relations deduced earlier except for the fact that in the new relations the summands enter with alternating signs (a plus sign is followed by a minus sign which, in turn, is followed by a plus sign, and so on).

The simplest relation of this type is

$$C_n^0 - C_n^1 + C_n^2 - \cdots + (-1)^n C_n^n = 0. \tag{33}$$

This relation is a consequence of (11). For proof, note that $C_n^0 = C_{n-1}^0 = 1$. This allows us to replace the first summand in (33) with C_{n-1}^0. Now by (11) we have

$$C_{n-1}^0 - C_n^1 = -C_{n-1}^1, \qquad -C_{n-1}^1 + C_n^2 = C_{n-1}^2,$$

and so on. This makes it clear that the sum on the left-hand side of (33) actually reduces to zero, as asserted.

It is possible to give a combinatorial proof of (33). We write down all k-combinations, $k = 1,\ldots, n$, of n letters $a_1,\ldots, a_n$ and subject them to the following transformation: we put the letter a_1 in those combinations which do not contain it and remove it from those combinations which contain it. It is easy to see that in this way we again obtain all the combinations of n letters and that each combination is obtained exactly once. Obviously, our transformation changes a combination with an even number of elements into a combination with an odd number of elements, and conversely. This means that there are as many combinations with an even number of elements (including the empty combination without any elements) as there are combinations with an odd number of elements. But this is precisely the statement of (33).

Next we prove the more complex relation

$$C_n^0 C_n^m - C_n^1 C_{n-1}^{m-1} + C_n^2 C_{n-2}^{m-2} - \cdots + (-1)^m C_n^m C_{n-m}^0 = 0. \tag{34}$$

For proof, we consider m-combinations of n elements $a_1,\ldots, a_n$. We use the symbol $(a_1,\ldots, a_k)$, $1 \leqslant k \leqslant n$, to indicate that a combination contains the elements $a_1,\ldots, a_k$. The number $N^{(k)} = N(a_1,\ldots, a_k)$ of such combinations is C_{n-k}^{m-k} (in these combinations k places are taken by the elements $a_1,\ldots, a_k$, and there are $n - k$ candidates for the remaining $m - k$ places). The number N of m-combinations is C_n^m. The number $N^{(0)}$ of combinations without any of the properties $(a_1),\ldots, (a_n)$ is zero (every combination has some elements). Hence $N = C_n^m$, $N^{(0)} = 0$, $N^{(k)} = C_{n-k}^{m-k}$. Substituting these values in Formula (32) we obtain the relation (34).

In much the same way we show that

$$C_n^0 C_{n+m-1}^m - C_n^1 C_{n+m-2}^{m-1} + C_n^2 C_{n+m-3}^{m-2} - \cdots + (-1)^n C_n^n C_{m-1}^{m-n} = 0,$$
$$\text{for} \quad m \geqslant n,$$

$$C_n^0 C_{n+m-1}^m - C_n^1 C_{n+m-2}^{m-1} + \cdots + (-1)^m C_n^m C_{n-1}^0 = 0,$$
$$\text{for} \quad m < n. \tag{35}$$

For proof we consider all m-combinations with repetitions of elements of n types $a_1, a_2, ..., a_n$. We use the symbol (a_k), $1 \leqslant k \leqslant n$, to indicate that a combination contains elements of type a_k (and perhaps other elements). Then $N(a_1, ..., a_k)$ stands for the number of combinations (with repetitions) containing elements of types $a_1, ..., a_k$. Such a combination can be thought of as a combination with repetitions in which k places are taken up by elements representing each of the k types $a_1, ..., a_k$, and the remaining $m - k$ places are taken up by elements of n types. This means that the number $N(a_1, ..., a_k)$ coincides with the number of $(m - k)$-combinations with repetitions of elements of n types, that is, $N(a_1, ..., a_k) = C_{n+m-k-1}^{m-k}$. The number of m-combinations with repetitions of n types is C_{n+m-1}^m. There are no combinations with none of the properties (a_k), $1 \leqslant k \leqslant n$. In other words, $N^{(0)} = 0$, $N = C_{n+m-1}^m$, $N^{(k)} = C_{n+m-k-1}^{m-k}$. Substitution of these values in (32) yields (35).

Finally we show that for $m < n$

$$n^m - C_n^1(n-1)^m + C_n^2(n-2)^m - \cdots + (-1)^{n-1}C_n^{n-1} \cdot 1^m = 0. \quad (36)$$

For proof we consider all m-samples with repetitions of elements of n types and use the symbol (a_k), $1 \leqslant k \leqslant n$, to indicate that a sample does not contain elements of type a_k. Then $N(a_1, ..., a_k)$ denotes the number of m-samples with repetitions which do not contain elements of types $a_1, ..., a_k$, that is, are made up of elements of $n - k$ types $a_{k+1}, ..., a_n$. Since the number of such samples is $(n - k)^m$, it follows that $N^{(k)} = N(a_1, ..., a_k) = (n - k)^m$. The number of all m-samples with repetitions of elements of n types is n^m. Finally, observe that an m-sample without any of the properties $(a_1), ..., (a_n)$ would contain elements of all n types. In view of the assumption that $m < n$, this is impossible. Hence $N^{(0)} = 0$. Substitution of these values for $N^{(0)}$, N, and $N^{(k)}$ in (32) yields (36).

We have established many properties of the numbers C_n^k. These properties can be obtained in other ways. In Chap. V we will present geometric ways of proving these results and in Chap. VII we will present the most powerful method of proof, known as the method of generating functions. Using the method of generating functions it is possible to prove not only the results of this chapter but also many other interesting results.

III

Combinatorial Problems with Restrictions

So far we have considered problems in which no restrictions were imposed on the order of the elements in an arrangement. In the case of samples and permutations, all ways of ordering the elements were permitted. In the case of combinations, order was irrelevant. Now we propose to discuss problems in which the order of the elements is subject to restrictions.

Lions and Tigers

An animal tamer marches 5 lions and 4 tigers into the arena. In how many ways can he line up the animals if a tiger must not be followed by another tiger?

The 5 lions can be lined up in 5! = 120 ways. A tiger can be placed ahead of the lions, between two lions (the lions are supposed to be reasonably far apart), or behind the lions. This gives a total of 6 places for the 4 tigers. Since the order of the tigers is relevant, the number of ways in which they can be placed is equal to the number of samples of 6 things taken 4 at a time, that is, $A_6^4 = 360$. But this means that there are 120 · 360 = 43,200 ways of lining up the animals.

With n lions and k tigers, the problem would admit

$$P_n A_{n+1}^k = \frac{n! \, (n+1)!}{(n-k+1)!}$$

solutions. Of course, we must have $k \leqslant n+1$, or else there will inevitably be two tigers in a row.

Construction of a Staircase

Points A and B are to be joined by a staircase (Fig. 8). The distance from A to C is 4.5 meters. The distance from C to B is 1.5 meters. The height of each step is 30 centimeters. The width of each step is an integer multiple of 50 centimeters. In how many ways can the staircase be constructed?

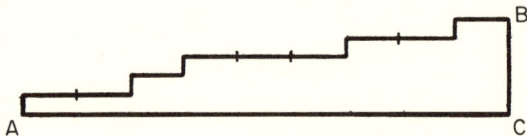

Fig. 8

Clearly, the staircase is to have 5 steps. Since 4.5/0.5 = 9, there are 10 spots where a step can be constructed. In other words, we must choose 5 out of 10 spots. This can be done in

$$C_{10}^5 = \frac{10!}{5! \, 5!} = 252$$

ways.

In general, if there are k steps and if there are $n+1$ spots where a step can be constructed, then the staircase can be constructed in C_{n+1}^k ways.

This problem is similar to the problem about the animals and the animal tamer. The tamer must not place two tigers in a row, and the carpenter must not double the height of a step. On the other hand, there is a basic difference between the two problems: To the animal tamer it *is* important in what order the animals appear (for example, it is one thing to put the tiger Shah ahead of the others and another thing to put the tiger Akbar ahead of the others), whereas to the carpenter it makes no difference at which of the admissible spots a step is constructed. That is why the carpenter has fewer choices than the animal tamer. In fact, if the staircase were 1.2 meters high and 2.5 meters long, there would be 4 steps and 6 spots where the steps could be constructed. Then the carpenter would have $C_6^4 = 15$ choices. In a corresponding situation the animal tamer has 43,200 choices. This, of course, is due to the fact that the tamer can permute the 5 lions in $5! = 120$ ways and the 4 tigers in $4! = 24$ ways which means that he can arrange his animals in $15 \cdot 120 \cdot 24 = 43,200$ ways.

In its general form the staircase problem is equivalent to the following problem:

In how many ways is it possible to arrange n zeros and k ones so that no 2 ones occur together?

To see this we first take another look at Fig. 8. The broken line in Fig. 8 can be thought of as a representation of the 5-step staircase of "horizontal length" 9 discussed in the problem. It consists of 5 vertical segments any two of which are separated by one or more of 9 horizontal segments. If we label each vertical segment by a one and each horizontal segment by a zero, then we obtain a sequence (10010100010010) of 5 ones any two of which are separated by one or more of 9 zeros. In the more general case of a k-step staircase of "horizontal length" n, the corresponding broken line consists of k vertical segments any two of which are separated by one or more of n horizontal segments, and the corresponding sequence consists of k ones any two of which are separated by one or more of n zeros.

In view of the equivalence of our two problems, it follows that the number of sequences of k ones and n zeros with no two consecutives ones is C_{n+1}^k .

A Bookshelf

There are 12 books on a bookshelf. In how many ways can 5 of these books be selected if a selection must not include two neighboring books?

This problem reduces to the problem just solved. We can describe a selection by assigning ones to the selected books and zeros to the remaining books. This means that a selection is described by a sequence of 5 ones and 7 zeros. Since we must not select two neighboring books, our sequences do not include two consecutive ones. But the number of sequences of 5 ones and 7 zeros which do not contain two consecutive ones is $C_8^5 = 56$.

In general, suppose that there are n books on a shelf and that we are to choose k of them without choosing neighboring books. Then this can be done in C_{n-k+1}^k ways. It follows that the problem is solvable only if $n - k + 1 \geqslant k$, that is, only if $n \geqslant 2k - 1$.

Knights of the Round Table

Twelve knights are seated at King Arthur's Round Table. Each of the 12 knights regards his immediate neighbors as foes. Five knights must be chosen to free an enchanted princess. In how many ways can one select a compatible group of knights?

This problem is similar to the bookshelf problem except that the knights are seated in a circle and not in a row. However, it is not difficult to reduce the problem to one in which the knights are seated in a row. We choose one knight, say, Sir Lancelot. Then we put all admissible selections of 5 knights which include Sir Lancelot in one class and the remaining selections in another and compute the number of selections in each of these two classes.

If Sir Lancelot is to set out to free the enchanted princess, then he must not be accompanied by his immediate neighbors at the table. This means that his 4 companions must be selected from among the remaining 9 knights. Since Lancelot's immediate neighbors are not among the 9 knights, any selection of 4 of the 9 knights will do provided that it does not include foes, that is, immediate neighbors. Observe that if we disregard Sir Lancelot and his two immediate neighbors, then we may think of the remaining 9 knights as being seated in a row. But then 4 of these 9 knights can be selected in $C_6^4 = 15$ ways.

To compute the number of selections in the second class, we observe that this time Sir Lancelot does not take part in the expedition, so that we can ignore his presence at the Round Table. But then we are dealing with 11 knights who, for all intents and purposes, are seated in a row. From these 11 knights we must select 5 members of the expedition subject to the restriction that the selection must not include immediate neighbors. This can be done in $C_7^5 = 21$ ways. It follows that the required number of selections is $15 + 21 = 36$.

In general, *if n knights are seated at a round table and we are to select k of them without including immediate neighbors, then this can be done in* $C_{n-k-1}^{k-1} + C_{n-k}^{k}$ *ways.*

The proof of this assertion is essentially a repetition of the earlier argument. Sir Lancelot is included in C_{n-k-1}^{k-1} selections and is not included in C_{n-k}^{k} selections.

It is easy to show that

$$C_{n-k-1}^{k-1} + C_{n-k}^{k} = \frac{n}{n-k} C_{n-k}^{k}.$$

For example, if $n = 12$, $k = 5$, then

$$\frac{12}{7} C_7^2 = \frac{12}{7} \cdot 21 = 36.$$

A Girl Has a Date

A film comedy with this title dealt with the misadventures of two vacationers who forgot their passparts at home. It was decided to send them the passports by mail. The girl at the post office had a date and was in such a rush that she managed to put each of the two passports in the wrong envelope. Of course, it could have been worse. If the girl had handled five passports instead of two, then five rather than two wretches would have had to spend the night on the hard benches of a park in the spa.

Come to think of it, this is not necessarily so. After all, the girl could have, by chance, put some of the passports in the right envelopes. It is of interest to see *in how many ways she could commit the perfect blunder, that is, send each passport to the wrong address.*

This problem can be formulated as follows: How many permutations of the 5 numbers, 1, 2, 3, 4, 5, leave none of the numbers fixed ? To solve this problem we make use of the principle of inclusion and exclusion (p. 14). We write (α) to indicate that a permutation leaves the number α fixed and denote by N_α the number of permutations with this property. Similarly, we denote by $N_{\alpha\beta}$ the number of permutations with properties (α) and (β), that is, the number of permutations which leave α and β fixed. The sense of $N_{\alpha\beta\gamma}$, and so on, is clear. Finally, we denote by $N^{(0)}$ the number of permutations with none of the properties (1), (2), (3), (4), (5), that is, the number of permutations which leave none of the numbers 1, 2, 3, 4, 5 fixed. By the principle of inclusion and exclusion

$$N^{(0)} = N - N_1 - N_2 - N_3 - N_4 - N_5$$
$$+ N_{12} + \cdots + N_{45} - N_{123} - \cdots - N_{345}$$
$$+ N_{1234} + \cdots + N_{2345} - N_{12345}, \tag{1}$$

where $N = P_5$ is the number of permutations of 5 elements (see p. 14).

Our problem is simplified by the fact that the properties (1), (2), (3), (4), and (5) are on a par, so that $N_1 = N_2 = N_3 = N_4 = N_5$. Similarly, $N_{12} = N_{23} = \cdots = N_{45}$ (it makes no difference whether the fixed numbers are 1 and 2 or 3 and 4), and so on.

The number of unordered pairs which can be selected from among the numbers 1, 2, 3, 4, 5 is C_5^2 [we count unordered pairs because the property (1, 2), say, is the same as the property (2, 1)]. Similarly, there are C_5^3 unordered triples, C_5^4 unordered quadruples, and C_5^5 unordered quintuples. If we denote by $N^{(k)}$, $k = 1, 2, 3, 4, 5$, the number of permutations leaving k prescribed numbers fixed, then Formula (1) takes the form

$$N^{(0)} = N - C_5^1 N^{(1)} + C_5^2 N^{(2)} - C_5^3 N^{(3)} + C_5^4 N^{(4)} - C_5^5 N^{(5)}. \tag{2}$$

To complete the solution of the problem, we must compute the values of $N^{(k)}$, $k = 1, 2, 3, 4, 5$. $N^{(1)}$ denotes the number of permutations which leave, say, 1 fixed. If 1 stays fixed, then the remaining numbers can be permuted in $P_4 = 24$ ways. Hence $N^{(1)} = P_4$. Again, if, say, 1 and 2 remain fixed, then the remaining numbers can be permuted in $P_3 = 6$ ways. Hence $N^{(2)} = P_3$. Similarly,

$$N^{(3)} = P_2 = 2, \qquad N^{(4)} = P_1 = 1, \qquad \text{and} \qquad N^{(5)} = P_0 = 1.$$

Substituting these values in (2), we find that

$$N^{(0)} = P_5 - C_5^1 P_4 + C_5^2 P_3 - C_5^3 P_2 + C_5^4 P_1 - C_5^5 P_0$$
$$= 120 - 5 \cdot 24 + 10 \cdot 6 - 10 \cdot 2 + 5 \cdot 1 - 1 \cdot 1 = 44.$$

This means that in 44 out of 120 cases none of the people will receive his passport.

In much the same way we can compute the number of cases in which just 1 person receives his passport. For every choice of a lucky person the remaining 4 unlucky ones can have their passports interchanged in

$$P_4 - C_4^1 P_3 + C_4^2 P_2 - C_4^3 P_1 + C_4^4 P_0 = 9$$

cases. Since each of the 5 people may be the lucky one, there are $5 \cdot 9 = 45$ cases in which just 1 person receives his passport.

We leave it to the reader to verify that in 20 cases just 2 people receive their passports, in 10 cases just 3, in 0 cases just 4 (if 4 out of 5 people receive their passports, then so does the fifth person), and in 1 case all 5 people receive their passports.

Another way of summarizing our results is to say that of the 120 permutations of 5 elements, 44 have no fixed elements, 45 have just 1 fixed element, 20 have just 2 fixed elements, 10 have just 3 fixed elements, and 1 has 5 fixed elements.

A Telepathic Séance

Some people claim to be able to read minds at a distance. Such claims were tested by picking so-called Zener figures (Fig. 9) in a definite order and asking the telepathist to guess in what order the figures were picked.

FIG. 9

Suppose the 5 figures are selected without repetitions. Then the number of permutations of these figures is 5! = 120. At a séance one of the permutations of the figures is chosen and the telepathist is asked to name the figure appearing in each position. The success of the telepathist is measured by the number of correctly guessed figures. The calculations which we carried out on pp. 48–49 imply that the results of random guessing would be approximately as follows:

no figure guessed correctly in 44 out of 120 cases, 1 figure guessed correctly in 45 cases, 2 figures guessed correctly in 20 cases, 3 figures guessed correctly in 10 cases, and all 5 figures guessed correctly in 1 case. It follows that in random guessing the number of correctly identified figures is equal, on the average, to

$$\frac{45 + 20 \cdot 2 + 10 \cdot 3 + 5}{120} = 1$$

per permutation. Also, this result does not depend on the number of different figures involved in a permutation. In other words, in the case of a permutation of *n* distinct figures, the expected number of correctly identified figures is equal, on the average, to 1 per permutation. If a person systematically manages to guess correctly a larger number of figures, then this merits careful scrutiny in order to determine whether one is faced with a hoax (a rather frequent occurrence) or a special endowment.

We will now try to determine how the average number of correctly identified figures changes if we allow repetitions. To allow repetitions is to go from permutations to samples with repetitions. The number of samples of *n* elements in which no element is in its "lawful" position is $(n-1)^n$. In fact, the first position can be filled with any element other than the first, the second position can be filled with any element other than the second, and so on. In other words, each position can be filled with $n-1$ candidates. By the rule of product we conclude that the number of possible arrangements is indeed equal to $(n-1)^n$.

Next we compute in how many cases just 1 element is in its lawful position. Let this be, say, the first element. Then for each of the remaining $n-1$ positions there are $n-1$ candidates (all the elements other than the "lawful occupant" of the particular position). But then the number of samples in which just the first element is in lawful position is $(n-1)^{n-1}$. Since the same count applies to each of the *n* positions, it follows that the number of samples with just 1 element in lawful position is $n(n-1)^{n-1}$. A similar argument shows that the number of samples with just *k* elements in lawful position is $C_n^k(n-1)^{n-k}$.

For example, if there are 5 distinct elements, then we obtain the following result: there are $4^5 = 1024$ samples with repetitions in which all the elements are displaced, $5 \cdot 4^4 = 1280$ samples with just 1 element in lawful position, $10 \cdot 4^3 = 640$ samples with just 2 elements in lawful position, $10 \cdot 4^2 = 160$ samples with just 3 elements in lawful position, $5 \cdot 4 = 20$ samples with just 4 elements in lawful position, and $1 \cdot 4^0 = 1$ sample with all elements in lawful position. All of these classes of samples add up to

$$1024 + 1280 + 640 + 160 + 20 + 1 = 3125$$

samples. On the other hand,

$$\bar{A}_5^5 = 5^5 = 3125.$$

In random guessing the number of correctly identified figures is again equal, on the average, to

$$\frac{1280 + 640 \cdot 2 + 160 \cdot 3 + 20 \cdot 4 + 1 \cdot 5}{3125} = 1$$

per permutation. In other words, in random guessing one can expect to identify correctly 1 out of 5 figures, and this outcome does not depend on whether or not repetitions of figures are allowed. On the other hand, the following table shows that the distribution of the number of correctly identified figures is different in the two cases:

Number of correctly identified figures	Without repetitions	With repetitions
0	0.366	0.328
1	0.375	0.410
2	0.167	0.205
3	0.083	0.051
4	0,000	0.006
5	0.009	0.000

Derangements*

The techniques of the previous section enable us to solve the following problem: *Find the number D_n of permutations of n elements in which no element stays in its original position.* The solution is given by the formula

$$D_n = P_n - C_n^1 P_{n-1} + C_n^2 P_{n-2} - \cdots + (-1)^n C_n^n$$
$$= n! \left[1 - \frac{1}{1!} + \frac{1}{2!} - \cdots + \frac{(-1)^n}{n!} \right]. \tag{3}$$

A reader familiar with the theory of series will note that the expression in brackets is a partial sum of the series for e^{-1}.

It is convenient to extend Formula (3) to the case $n = 0$. The natural definition is $D_0 = 1$.

The number $D_{n,r}$ of permutations in which just r elements remain in their original positions and the remaining $n - r$ elements change their positions is given by

$$D_{n,r} = C_n^r D_{n-r} . \tag{4}$$

* This section can be omitted in a first reading of the book.

Indeed, the r elements which stay in their original positions can be selected in C_n^r ways. For every choice of the r fixed elements, the remaining $n - r$ elements may be permuted in all possible ways but must not end up in their original positions. This can be done in D_{n-r} ways. Thus, by the rule of product, the number of required permutations is equal to $C_n^r D_{n-r}$.

We separate all permutations into classes of permutations with the same number of fixed elements. Since there are in all $n!$ permutations, we obtain the relation

$$n! = \sum_{r=0}^{n} D_{n,r} = \sum_{r=0}^{n} C_n^r D_{n-r}. \qquad (5)$$

Another relation connecting the numbers $n!$ and $D_{n,r}$ is obtained by counting the number of fixed elements in the $n!$ permutations of n elements $a_1, ..., a_n$. This count can be effected in two ways. One way is to note that the element a_1 stays fixed in $P_{n-1} = (n-1)!$ permutations and the same is true for each of the remaining elements $a_2, ..., a_n$. This means that the number of fixed elements is $n(n-1)! = n!$ Another way is to note that the number of permutations in the rth class, that is, permutations with just r fixed elements, is $D_{n,r}$. But then the number of fixed elements in the rth class is $rD_{n,r}$. Summing over all the classes we obtain $\sum_{r=0}^{n} rD_{n,r}$ fixed elements. This proves the relation

$$n! = \sum_{r=0}^{n} rD_{n,r} = \sum_{r=0}^{n} rC_n^r D_{n-r}. \qquad (5')$$

The principle of inclusion and exclusion enables us to solve the following problem: *Find the number of permutations of n elements in which r prescribed elements are displaced* (and the remaining are either displaced or remain fixed). The required number is given by the formula

$$n! - C_r^1(n-1)! + C_r^2(n-2)! - \cdots + (-1)^r(n-r)!. \qquad (6)$$

Subfactorials*

The numbers D_n are sometimes referred to as *subfactorials*. These numbers share many properties with ordinary factorials. For example, for ordinary factorials we have the relation

$$n! = (n-1)[(n-1)! + (n-2)!]. \qquad (7)$$

(Proof: $(n-1)[(n-1)! + (n-2)!] = (n-1)(n-2)! \, n = n!$.) We show that for subfactorials we have the analogous relation

$$D_n = (n-1)[D_{n-1} + D_{n-2}]. \qquad (8)$$

---

* This section may be omitted in a first reading of the text.

To prove (8) we replace D_{n-1} and D_{n-2} with their expansions based on Formula (3). Then it is easy to see that

$$(n-1)[D_{n-1} + D_{n-2}] = (n-1)[(n-1)! + (n-2)!]$$

$$\times \left[1 - \frac{1}{1!} + \frac{1}{2!} - \frac{1}{3!} + \cdots + \frac{(-1)^{n-2}}{(n-2)!}\right]$$

$$+ (-1)^{n-1}(n-1).$$

By Formula (7),

$$(n-1)[(n-1)! + (n-2)!] = n!.$$

Also,

$$(-1)^{n-1}(n-1) = n! \left[\frac{(-1)^{n-1}}{(n-1)!} + \frac{(-1)^n}{n!}\right].$$

Therefore,

$$(n-1)[D_{n-1} + D_{n-2}] = n! \left[1 - \frac{1}{1!} + \frac{1}{2!} - \frac{1}{3!} + \cdots + \frac{(-1)^{n-2}}{(n-2)!}\right.$$

$$\left. + \frac{(-1)^{n-1}}{(n-1)!} + \frac{(-1)^n}{n!}\right] = D_n.$$

The following, purely combinatorial, proof of relation (8) is due to Euler. We consider the permutations all of whose elements are displaced. In these permutations the first position is filled by any element different from the first. Since the number of such elements is $n-1$, our D_n permutations can be separated into $n-1$ classes according to the elements in the first position. Clearly, each of these classes contains the same number of permutations.

We compute the number of permutations in one of these classes, say, the class in which the first position is filled by the second element. This class can be separated into a subclass consisting of permutations in which the first element occupies the second position, and a subclass consisting of the remaining permutations. A permutation in the first subclass is characterized by the fact that its first two elements are interchanged and the remaining $n-2$ elements are displaced. This means that the first subclass consists of D_{n-2} permutations.

Now we show that the second subclass consists of D_{n-1} permutations. The second subclass consists of all permutations in which the first element is not in the second position and the remaining elements are displaced. If, for the moment, we regard the second position as the "lawful" position of the first element, then it is clear that the first, third, fourth,..., nth element are displaced. Since the number of these elements is $n-1$, it follows that the number of permutations in the second subclass is D_{n-1}. But then our class consists of $D_{n-2} + D_{n-1}$ permutations. Multiplying this number by $n-1$ (the number of classes into which we split our D_n permutations) we obtain relation (8).

Formula (8) implies that

$$D_n - nD_{n-1} = -[D_{n-1} - (n-1)D_{n-2}].$$

Applying this relation an appropriate number of times, we obtain the relation

$$D_n - nD_{n-1} = (-1)^{n-2}[D_2 - 2D_1].$$

Since $D_2 = 1$ and $D_1 = 0$, this relation reduces to

$$D_n = nD_{n-1} + (-1)^n. \qquad (9)$$

Formula (9) resembles the relation $n! = n(n-1)!$ for factorials.

The following table gives the values of subfactorials for the first 12 natural numbers:

n	D_n	n	D_n	n	D_n	n	D_n
1	0	4	9	7	1854	10	1 334 961
2	1	5	44	8	14 833	11	14 684 570
3	2	6	265	9	133 496	12	176 214 841

Caravan in the Desert

A caravan of 9 camels travels across the desert. The journey lasts many days and each of the travelers finds it boring to see the same camel in front of him. In how many ways is it possible to permute the camels so that each camel is preceded by a camel different from the previous one?

Such permutations exist. One way of realizing such a permutation is to reverse the order of the camels so that the last camel comes first, and so on. (This scheme is suggested by the Arab proverb: "On the return journey the lame camel is the leader.")

To solve the problem, we consider the camels in their original order and then, starting with the last camel, assign to them the numbers 1, 2, 3, 4, 5, 6, 7, 8, 9. Our problem now is to find the number of permutations of the numbers 1, 2, 3, 4, 5, 6, 7, 8, 9 which do not contain a single one of the pairs (1, 2), (2, 3), (3, 4), (4, 5), (5, 6), (6, 7), (7, 8), (8, 9). To solve this problem we again apply the principle of inclusion and exclusion.

First we compute the number of permutations containing the pair (1, 2). It is convenient to think of this pair as a single new element. Then we are dealing with 8 rather than 9 elements, and the number of permutations containing (1, 2) is P_8. Clearly, the same count is valid for each of the 8 pairs.

Now we consider permutations containing 2 prescribed pairs. If the 2 pairs in question share an element (for example, (1, 2) and (2, 3)), then we view the 3 distinct elements in the 2 pairs as a single new element. If the 2 pairs are disjoint (for example, (1, 2) and (5, 6)), then we regard them as 2 new elements. In either case we are dealing with 7 rather than 9 elements, and the 7 elements can be permuted in P_7 ways. Also, 2 out of 8 pairs can be selected in C_8^2 ways. More generally, there are P_{9-k} permutations containing k prescribed pairs, and k out of 8 pairs can be selected in C_8^k ways. By the rule of product, there are $C_8^k P_{9-k}$ permutations containing just k pairs. By the principle of inclusion and exclusion, the number of permutations which do not contain any of the 8 pairs is

$$P_9 - C_8^1 P_8 + C_8^2 P_7 - C_8^3 P_6 + C_8^4 P_5 - C_8^5 P_4 + C_8^6 P_3 - C_8^7 P_2 + C_8^8 P_1$$

$$= 8! \left[9 - \frac{8}{1!} + \frac{7}{2!} - \frac{6}{3!} + \frac{5}{4!} - \frac{4}{5!} + \frac{3}{6!} - \frac{2}{7!} + \frac{1}{8!} \right] = 148{,}329.$$

An analogous argument shows that the number of permutations of the n numbers 1, 2, 3,..., n which do not contain any of the $n - 1$ pairs (1, 2), (2, 3),..., $(n - 1, n)$ is given by the formula

$$E_n = P_n - C_{n-1}^1 P_{n-1} + C_{n-1}^2 P_{n-2} - C_{n-1}^3 P_{n-3} + \cdots + (-1)^{n-1} C_{n-1}^{n-1} P_1$$

$$= (n - 1)! \left[n - \frac{n - 1}{1!} + \frac{n - 2}{2!} - \frac{n - 3}{3!} + \cdots + \frac{(-1)^{n-1}}{(n - 1)!} \right]. \qquad (10)$$

Using factorials, we can give (10) a striking form. We write each term in the sum on the right-hand side of (10) as a sum of two terms:

$$\frac{(-1)^k (n - k)}{k!} = \frac{(-1)^k n}{k!} + \frac{(-1)^{k-1}}{(k - 1)!}.$$

Then

$$E_n = n! \left[1 - \frac{1}{1!} + \frac{1}{2!} - \cdots + \frac{(-1)^{n-1}}{(n - 1)!} + \frac{(-1)^n}{n!} \right]$$

$$+ (n - 1)! \left[1 - \frac{1}{1!} + \frac{1}{2!} - \cdots + \frac{(-1)^{n-2}}{(n - 2)!} + \frac{(-1)^{n-1}}{(n - 1)!} \right].$$

(The extra term in each of the two sums on the right-hand side is innocuous: after multiplication by $n!$ and $(n - 1)!$ these terms reduce to $(-1)^n$ and $(-1)^{n-1}$, respectively, and $(-1)^n + (-1)^{n-1} = 0$.) But then

$$E_n = D_n + D_{n-1}. \qquad (11)$$

Thus the number of permutations of 1, 2, 3,..., n which do not contain any of the pairs $(1, 2), (2, 3),..., (n - 1, n)$ is $D_n + D_{n-1}$.

The same argument shows that the number of permutations of n elements which do not contain $r \leqslant n - 1$ preassigned pairs is

$$P_n - C_r^1 P_{n-1} + C_r^2 P_{n-2} - \cdots + (-1)^r C_r^r P_{n-r} . \tag{12}$$

The answer is different if the number of disallowed pairs is greater than $n - 1$. For example, suppose that, in addition to the pairs $(1, 2), (2, 3),..., (n - 1, n)$, we disallow the pair $(n, 1)$. Reasoning as before, we find that the answer is given by the formula

$$F_n = P_n - C_n^1 P_{n-1} + C_n^2 P_{n-2} - \cdots + (-1)^k C_n^k P_{n-k} + \cdots + (-1)^{n-1} C_n^{n-1} P_1$$
$$= n! \left[1 - \frac{1}{1!} + \frac{1}{2!} - \cdots + \frac{(-1)^{n-1}}{(n - 1)!} \right] = n D_{n-1} . \tag{13}$$

To see this, note that in this case there are n disallowed pairs and that no permutation contains all the n pairs (if a permutation contains the pairs $(1, 2)$, $(2, 3),..., (n - 1, n)$, say, then its first element is 1 and its last element is n, and so it cannot contain the pair $(n, 1)$). But then the last term in (13) is $(-1)^{n-1} C_n^{n-1} P_1$ and not $(-1)^n C_n^n P_0 = (-1)^n$.

It would be of interest to derive the answer $F_n = n D_{n-1}$ by purely combinatorial means.

On the Merry-Go-Round

n children go on a merry-go-round. They decide to switch seats so that each child has a new companion in front of him. In how many ways can this be done?

This problem is similar to the caravan problem solved earlier. However, now there are n disallowed pairs $((1, 2), (2, 3),..., (n - 1, n)$, and $(n, 1))$ and some permutations, for instance the original one, contain all of them. Also, in the present case we do not distinguish between permutations which differ by a rotation, so that now k elements yield only $P_{k-1} = (k - 1)!$ essentially different permutations. Making these allowances and using the principle of inclusion and exclusion, we see that the required number of permutations is

$$Q_n = P_{n-1} - C_n^1 P_{n-2} + C_n^2 P_{n-3} - \cdots + (-1)^{n-1} C_n^{n-1} P_0 + (-1)^n C_n^n . \tag{14}$$

It is easy to see that this relation can be written as

$$Q_n = D_{n-1} - D_{n-2} + D_{n-3} - \cdots + (-1)^{n-3} D_2 . \tag{15}$$

Indeed, since $C_n^k - C_{n-1}^{k-1} = C_{n-1}^k$, Formula (14) implies that for $n > 1$

$$Q_n + Q_{n-1} = P_{n-1} - C_{n-1}^1 P_{n-2} + C_{n-1}^2 P_{n-3} - \cdots + (-1)^{n-1} ,$$

and this, in turn, is equal to D_{n-1} (see p. 53). Thus, $Q_n + Q_{n-1} = D_{n-1}$. By Formula (14), $Q_2 = 0$. Adding the chain of equalities

$$Q_n + Q_{n-1} = D_{n-1},$$

$$-Q_{n-1} - Q_{n-2} = -D_{n-2},$$

$$Q_{n-2} + Q_{n-3} = D_{n-3},$$

$$\cdot \cdot \cdot \cdot \cdot \cdot \cdot \cdot \cdot \cdot \cdot \cdot \cdot \cdot$$

$$(-1)^{n-3}Q_3 = (-1)^{n-3}D_2,$$

we obtain relation (15).

Queue at the Box Office

m + k moviegovers line up for tickets at the box office. m of the people have dollar bills and k have half-dollar coins. A ticket is 50 cents. To begin with there is no money in the box office. This means that some queues are "bad" in the sense that at some point the cashier is unable to make change and other queues are "good" in the sense that no such snag develops. We are to compute the number of good queues.

If, for example, $m = k = 2$, then there are only 2 good queues: *hdhd* and *hhdd* (*h* denotes a half-dollar and *d* a dollar). The remaining 4 queues are bad: in the queues *ddhh*, *dhdh*, and *dhhd* already the first moviegoer cannot get change and in the queue *hddh* the third moviegoer cannot get change.

For small values of *m* and *k*, it is easy to sort out the good and bad queues, but for large values of *m* and *k* this is not possible. After all, the number of permutations of *m* dollars and *k* half-dollars is

$$P(m, k) = \frac{(m + k)!}{m! \, k!}.$$

If, say, $m = k = 20$, then

$$P(20, 20) = \frac{40!}{20!\,20!},$$

which is more than one hundred billion.

In mathematical terms, our problem is to find the number of permutations of m d's and k h's such that for every r, $1 \leqslant r \leqslant m + k$, the number of h's in the first r terms of the permutation is not less than the number of d's (in other words, if one is to avoid a snag, then the number of half-dollar coins must not be smaller than the number of dollar bills).

It is clear for permutations of the required type to exist we must have $m \leqslant k$; otherwise there will not be enough half-dollar coins to give to the people with dollar bills. We, therefore, assume that $0 \leqslant m \leqslant k$.

It is convenient to compute first the number of bad permutations (in our case, permutations corresponding to queues in which the cashier is unable to make change). Then, to get the solution of our problem, we need only subtract this number from the number $P(m, k) = C_{m+k}^m$ of all permutations of m d's and k h's.

We will prove that the number of bad permutations of m d's and k h's is $P(m - 1, k + 1) = C_{m+k}^{m-1}$, the number of permutations of $m - 1$ d's and $k + 1$ h's.

For proof, consider a bad permutation of m d's and k h's. Suppose a snag develops at a certain position. Then this position must be occupied by a d, and the preceding positions must be occuped by equal numbers of d's and h's; in other words, the number of the "snag position" has the form $2s + 1$, and the preceding positions are occupied by s d's and s h's.

Let us put an h in front of our permutation (protesters in the queue must be told that this will help make change). The result is a permutation of m d's and $k + 1$ h's which starts with an h and has the property that half of its first $2s + 2$ letters are d's and half are h's (there were s h's and $s + 1$ d's and we added an h).

Our next move is bound to displease the owners of dollar bills and to please the owners of half-dollar coins. We restrict ourselves to the first $2s + 2$ positions, give each owner of a half-dollar coin a dollar bill in exchange for his coin, and each owner of a dollar bill a half-dollar coin in exchange for his dollar bill (For example, in the line up

$$hhdhdhddhd\mathbf{d}hhdhhd$$

we hit a snag at the position marked by $\mathbf{d}$. If we put an h in front of this lineup and carry out our exchange procedure, we end up with the lineup

$$dddhdhdhhdh\mathbf{h}hhdhhd.)$$

Since in the first $2s + 2$ positions the number of dollar bills was equal to the number of half-dollar coins, our exchange procedure left the number of bills and the number of coins unchanged. This means that we end up with a permutation of m d's and $k + 1$ h's starting with a d. In other words, we have associated with every bad sequence of m d's and k h's a definite sequence of m d's and $k + 1$ h's starting with a d.

Now we show that in this way we obtain all sequences of m d's and $k + 1$ h's which start with a d. Consider a sequence of this type. Since $m \leqslant k$, some segment of our sequence (beginning with the first position) must contain equal numbers of d's and h's. If in this segment of the sequence we change d's to h's and h's to d's and discard the initial h, then we end up with a bad sequence of bills and coins; in fact, the snag occurs at the point at which in the original sequence the number of d's and the number of h's became equal for the first time.

We have shown that the number of bad sequences of dollar bills and half-dollar coins is equal to the number of permutations of m d's and $k + 1$ h's starting with an h. If we reject the first letter in each sequence, then we obtain all permutations of $m - 1$ d's and $k + 1$ h's. The number of such permutations is

$$P(m - 1, k + 1) = C_{m+k}^{m-1} .$$

In other words, the number of bad permutations is C_{m+k}^{m-1}. Since the number of permutations of m d's and k h's is C_{m+k}^{m}, it follows that the number of good permutations is

$$C_{m+k}^{m} - C_{m+k}^{m-1} = \frac{k - m + 1}{k + 1} C_{m+k}^{m} . \tag{16}$$

In particular, if $k = m$, that is, in case of queues in which there are as many dollar bills as there are half-dollar coins, things will go smoothly in $[1/(k + 1)]C_{2k}^{k}$ cases, and snags will develop in $[k/(k + 1)]C_{2k}^{k}$ cases. This means that the larger the value of k the smaller the percentage of good queues.

Having solved our problem, we consider a problem which is closely related to it. Suppose the cashier thought ahead and *when the sale of tickets started he had on hand q half-dollar coins. Consider all possible queues of moviegoers, m of whom have dollar bills and k of whom have half-dollar coins. In how many cases will the cashier avoid snags over making change?*

It is clear that if $m \leqslant q$, then no snag can develop; the cashier has enough half-dollar coins on hand to make change even if all the people in the queue present dollar bills. It follows that we can restrict ourselves to the case when

$$q < m \leqslant k + q.$$

Furthermore, we can assume that the cashier acquired q half-dollar coins because someone put q people with half-dollar coins at the head of the queue. This allows us to formulate our problem as follows:

A queue consists of $k + q$ people with half-dollar coins and m people with dollar bills. The first q people have half-dollar coins. In how many cases will no one have to wait for change?

The solution of this problem is very similar to the solution of the special case $q = 0$ discussed above. First we count the bad queues. In each bad queue a snag occurs at a position occupied by a man with a dollar bill who is preceded by $2s$ people half of whom have dollar bills and half of whom have half-dollar coins. We put in front of the queue a man with a half-dollar coin. We restrict our attention to the first $2s + 2$ people in the queue and change their dollar bills to half-dollar coins and conversely. The result is a permutation of m dollar bills and $k + q + 1$ half-dollar coins with the first $q + 1$ positions occupied by dollar bills. Each such permutation is derived from just one bad permutation of dollar bills and half-dollar coins. It follows that there are as many bad permutations as there are permutations of m dollar bills and $k + q + 1$ half-dollar coins where the first $q + 1$ positions are taken by dollar bills. In counting permutations of the latter type, we can set aside the first $q + 1$ dollar bills. But then we get all permutations of $m - q - 1$ dollar bills and $k + q + 1$ half-dollar coins. The number of such permutations is $P(m - q - 1, \ k + q + 1) = C_{m+k}^{m-q-1}$. This means that there are C_{m+k}^{m-q-1} bad permutations. Since there are C_{m+k}^{k} permutations all told, it follows that the number of good permutations is

$$C_{m+k}^{m} - C_{m+k}^{m-q-1}. \tag{17}$$

The following results are closely related to the results discussed above:

If $m < k$, then the number of permutations of m d's and k h's with more h's than d's in front of each letter (other than the first) is

$$C_{m+k-1}^{m} - C_{m+k-1}^{m-1} = \frac{k - m}{m} C_{m+k-1}^{m}. \tag{18}$$

If $m = k$, then the number of permutations of m d's and k h's with more h's than d's in front of each letter (other than the first) is $(1/k)C_{2k-2}^{k-1}$.

Suppose $k > 1$. The permutations in each of these two problems start with two h's. Setting the first of these h's aside reduces the first of our problems to the first problem in this section and the second of our problems to the special case of the first problem in this section treated immediately after (16). This being so, the answers to the present problems are obtained from the answers to the earlier problems by replacing k by $k - 1$.

Problem of Two Rows

It often happens in combinatorics that two apparently unrelated problems are really two versions of the same problem. The following problem illustrates such a situation.

In how many ways is it possible to line up 2n people of different height in 2 rows of n people each so that the people in each row are lined up according to height and the person in the first row is invariably taller than his counterpart in the second row?

We show that this problem reduces to (a special case of) the problem of the queue in front of the box office. After lining up the people in 2 rows in accordance with the requirements, we give to each person in the first row a half-dollar coin and to each person in the second row a dollar bill. Then we arrange the 2 rows of people into 1 row by height. The result is a queue in which n people have dollar bills and n people have half-dollar coins. We claim that this queue is good in the sense that it will file past the cashier without a snag. In fact, consider a person who occupies the kth position in the second row. Then of the people with dollar bills only $k - 1$ are taller than he, and of the people with half-dollar coins at least k are taller than he (namely the first k people in what was originally the first row). But this means that by the time he gets to the box office, the cashier is bound to have at least one half-dollar coin and so will be able to make change. Conversely, start with a good queue of n people with dollar bills and n people with half-dollar coins of which we assume, without loss of generality, that its members are arranged by height. The reader will verify that if the owners of the half-dollar coins are lined up by height in one row and placed in the back of the first row, then the resulting 2-row lineup conforms to the requirements of our problem. This means that the number of arrangements conforming to the requirements of our problem is equal to the number of good permutations of n d's and n h's, that is, to $[1/(n + 1)]C_{2n}^n$.

Additional Properties of Combinations*

We are now in a position to deduce additional properties of the numbers C_m^k (see p. 34). To this end we separate the "bad" permutations of m d's and k h's into classes. We saw that the number of the first "snag position" is of the form $2s + 1$, that this position is occupied by a letter d, and that this d is preceded by s d's and s h's which form a "good" permutation. We put in the sth class all bad permutations in which the snag occurs at the position $2s + 1$. Clearly, s can take on the values $0, 1, 2,..., m - 1$.

We compute the number of permutations in the sth class. The first $2s$ elements of a permutation in this class form a good permutation of s d's and s h's, and the number of such permutations is $[1/(s + 1)]C_{2s}^s$. Furthermore, there is a letter d in the $(2s + 1)$th position followed by an arbitrary permutation of $m - s - 1$ d's and $k - s$ h's. The number of such permutations is $P(m - s - 1$,

---

* This section can be omitted in a first reading of the text.

$k - s) = C_{m+k-2s-1}^{m-s-1}$. By the rule of product, the number of permutations in the sth class is

$$\frac{1}{s+1} C_{2s}^s C_{m+k-2s-1}^{m-s-1} .$$

Since there are C_{m+k}^{m-1} bad permutations and m classes, we obtain, for $m \leqslant k$, the relation

$$C_0^0 C_{m+k-1}^{m-1} + \frac{1}{2} C_2^1 C_{m+k-3}^{m-2} + \frac{1}{3} C_4^2 C_{m+k-5}^{m-3} + \cdots + \frac{1}{m} C_{2m-2}^{m-1} C_{k-m+1}^0 = C_{m+k}^{m-1} . \quad (19)$$

This relation is a special case of the relation

$$\sum_{s=p}^{m-1} [C_{2s-p}^s - C_{2s-p}^{s-p-1}] C_{m+k+p-2s-1}^{m-s-1} = C_{m+k}^{m-p-1}, \quad (20)$$

where $p < m \leqslant p + k$ (we assign to C_s^{-1} the value zero). The proof of (20) is similar to the proof of (19) except that we are separating into classes bad permutations of m d's and $k + p$ h's which start with p d's (see p. 60).

Next we consider relations obtained by separating good permutations into classes. The number of good permutations of k d's and k h's is $[1/(k + 1)]C_{2k}^k$. We know that after the $2k$ people in a good queue have all bought tickets, the cashier is left without half-dollar coins. In the case of some good permutations, it may happen that the cashier is left with no half-dollar coins before the end of the queue, and what saves the day is that the next moviegoer had a half-dollar coin. We break up the good permutations into classes and put into the sth class all the permutations in which the cashier is first left without half-dollar coins after serving the moviegoer occupying the $2s$th position, $s = 1, 2,..., k$.

We compute the number of permutations in the sth class. Such a permutation is made up of two parts. The first $2s$ letters form a permutation of s d's and s h's with each letter preceded by more h's than d's (otherwise equality of d's and h's would set in before position $2s$). We saw that the number of such permutations is $(1/s)C_{2s-2}^{s-1}$ (see p. 60). After the sale of the first $2s$ tickets, the cashier is left without half-dollar coins. It follows that in order for the rest of the queue to pass without a snag, the remaining $k - s$ d's and $k - s$ h's must form a good permutation. The number of such permutations is $[1/(k - s + 1)]C_{2k-2s}^{k-s}$ (see p. 59). By the rule of product the number of elements in the sth class is

$$\frac{1}{s(k - s + 1)} C_{2s-2}^{s-1} C_{2k-2s}^{k-s} .$$

Since the k classes contain a total of $[1/(k + 1)]C_{2k}$ good permutations, it follows that

$$\sum_{s=1}^k \frac{k+1}{s(k+s-1)} C_{2s-2}^{s-1} C_{2k-2s}^{k-s} = C_{2k}^k . \quad (21)$$

If we put

$$\frac{1}{s+1} C_{2s}^s = T_s,$$

then Formula (21) takes the form

$$T_0 T_{k-1} + T_1 T_{k-2} + \cdots + T_{k-1} T_0 = T_k. \tag{22}$$

Another relation for the C_n^m is obtained by separating into classes the good permutations of m d's and k h's. Let l be a given number, $1 \leqslant l \leqslant m$. We put in the sth class all good permutations with just s d's among the first l elements. Then the first l elements contain $l - s$ h's. Since the number of h's cannot be less than the number of d's, we must have $0 \leqslant 2s \leqslant l$.

We compute the number of permutations in the sth class. Each of these permutations consists of two parts. The first part is formed of the first l letters and the second part is formed of the remaining $m + k - l$ letters. Of the first l letters, $l - s$ are h's and s are d's. Since the permutation is good, its part consisting of the first l letters is also good. The number of good permutations of $l - s$ h's and s d's is $[(l - 2s + 1)/(l - s + 1)]C_l^s$.

Observe that the first part of a permutation in the sth class has a surplus of $(l - s) - s = l - 2s$ h's. This means that the second part of the permutation determines a good permutation of $l - 2s$ h's followed by $m + k - l$ elements of which $k - l + s$ are h's and $m - s$ are d's. The number of such permutations can be computed from Formula (17) with q, k, and m replaced by $l - 2s$, $k - l + s$, and $m - s$, respectively, and turns out to be equal to $C_{m+k-l}^{m-s} - C_{m+k-l}^{m+s-l-1}$. By the rule of product, the number of permutations in the sth class is

$$\frac{l - 2s + 1}{l - s + 1} C_l^s [C_{m+k-l}^{m-s} - C_{m+k-l}^{m+s-l-1}].$$

Since the number of good permutations of m d's and k h's is $[(k - m + 1)/(k + 1)]C_{m+k}^m$ and the number of classes is $E(l/2)$ (as usual, $E(l/2)$ denotes the integral part of $l/2$), it follows that

$$\sum_{s=0}^{E(l/2)} \frac{l - 2s + 1}{l - s + 1} C_l^s [C_{m+k-l}^{m-s} - C_{m+k-l}^{m+s-l-1}] = \frac{k - m + 1}{k + 1} C_{m+k}^m. \tag{23}$$

(Here we assign the value zero to all C_r^p with $p < 0$.)

By using different ways of separating permutations into classes, the reader will find it easy to establish analogous relations.

IV

Distributions and Partitions

In this chapter we consider problems which involve the separation of a series of elements into a series of classes. Our task is to compute the number of ways in which this can be done. There is a great deal of variety in this group of problems. Sometimes the order of the elements in the various classes is essential (for example, when signaling with flags it is important not only to choose the right flags for a particular mast but also to arrange them in the right order on the mast), and sometimes it is irrelevant (for example, a domino player is concerned with the set of dominos at his disposal and not with the order in which he picked them from the pile). Sometimes the order of the classes plays a role (for example, participants in a domino game are seated in a definite order and it is important to know who gets a particular set of dominos), and sometimes it plays none (for example, when I put snapshots, which I am about to mail, into identical envelopes, the content of each envelope is important but the order of the envelopes is not; the people at the post office are bound to shuffle them thoroughly).

There may be other differences: We may or may not distinguish between elements; we may or may not distinguish between classes of elements; we may or may not allow the empty class (the class without elements). All these possibilities account for the wide range of these problems.

A Domino Game

Twenty eight dominos are divided equally among 4 players. In how many ways can this be done?

The division of the dominos can be effected in the following manner: First all the dominos are arranged in a row. Then the first player takes the first 7 dominos, the second player takes the next 7 dominos, the third player takes the following 7 dominos, and the fourth player takes the remaining 7 dominos. It is clear that in this way we can realize all possible ways of dividing the dominos.

At first glance it might appear that the number of ways in which the dominos can be distributed is equal to 28!, the number of permutations of 28 objects. This, however, is not the case. After all, the first player does not care whether the first domino he picks up is the domino 6 : 6 or the domino 3 : 4; what interests him is what 7 dominos he ends up with. This means that an arbitrary permutation of the first 7 dominos does not affect the distribution of the dominos. The same is true of a permutation of the second group of 7 dominos, of a permutation of the third group of 7 dominos, and of a permutation of the last group of 7 dominos. By the rule of product there are $(7!)^4$ permutations of the dominos which do not affect the distribution of the dominos. It follows that there are $28!/(7!)^4$ ways of distributing the dominos. This number is approximately equal to $4.7 \cdot 10^{15}$.

This result can be obtained in a different way: The first player chooses 7 out of 28 dominos. Since he is not concerned with the order of the dominos, he has C_{28}^7 choices. Then the second player chooses 7 out of the remaining 21 dominos and so has C_{21}^7 choices. The third player chooses 7 out of 14 dominos and so has C_{14}^7 choices. The last player has C_7^7, that is, one choice. By the rule of product the number of possibilities is

$$C_{28}^7 C_{21}^7 C_{14}^7 C_7^7 = \frac{28!}{21!\,7!} \cdot \frac{21!}{14!\,7!} \cdot \frac{14!}{7!\,7!} = \frac{28!}{(7!)^4}.$$

If we apply the same type of argument to the game of preference (where there are 32 cards, each of the 3 players gets 10 cards, and 2 cards are set aside), then we see that the number of deals is

$$\frac{32!}{10!\,10!\,10!\,2!} = 2{,}753{,}294{,}408{,}504{,}640.$$

Some readers may think it a waste of time to study card games. To these people we wish to point out that combinatorics and probability have their origin in the study of games of chance. The game of dice, card games, and the game of heads or tails gave rise to problems which outstanding mathematicians like Pascal, Bernoulli, Euler, and Chebyshev used to refine the ideas and methods of combinatorics and probability. Also, many of the ideas of game theory (a branch of mathematics with many applications in economics and military strategy) were the result of the study of very simple card games.

Putting Things in Boxes

The domino and preference problems just studied belong to the class of combinatorial problems which involve putting things in boxes. The general formulation of such problems is as follows:

Given n different objects and k boxes. n_1 of these objects are to be put in the first box, n_2 in the second box,..., n_k in the kth box, with $n_1 + n_2 + \cdots + n_k = n$. In how many ways can this be accomplished?

In the domino problem the players were the counterpart of the boxes, and the dominos were the counterpart of the objects. Reasoning as before, we conclude that the answer to the general problem is

$$P(n_1, n_2, ..., n_k) = \frac{n!}{n_1!\, n_2! \cdots n_k!} \,. \tag{1}$$

We came across this formula earlier when we solved the following, seemingly different, problem.

We are given objects of k types. What is the number of permutations of n_1 objects of the first type, n_2 objects of the second type,..., n_k objects of the kth type?

Here, too, the answer is given by the formula

$$P(n_1, n_2, ..., n_k) = \frac{n!}{n_1!\, n_2! \cdots n_k!} \,,$$

where $n = n_1 + n_2 + \cdots + n_k$ (see p. 24). In order to establish the connection between the two problems, we number the n positions which can be occupied by our objects. Each permutation of the n position numbers determines a division of these numbers into k classes: the first class consists of the numbers of the positions occupied by the elements of the first type, the second class consists of the numbers of the positions occupied by the elements of the second type, and so on. This connection between permutations with repetitions and the placing of the numbers of the positions in "boxes" explains the fact that the solutions of both problems are given by the same formula.

A Bouquet

In the problem of placing objects in boxes we supposed the number of objects placed in a given box known (for example, we supposed the number of dominos selected by each player known). In most distribution problems these numbers are not given.

Two children collected 10 camomiles, 15 cornflowers, and 14 forget-me-nots. In how many ways can they divide the flowers?

It is clear that the camomiles can be divided in 11 ways (the first child may choose no camomile, 1 camomile, 2 camomiles,..., all 10 camomiles), the cornflowers in 16 ways, and the forget-me-nots in 15 ways. Since flowers of each kind are chosen independently, it follows by the rule of product that there are $11 \cdot 16 \cdot 15 = 2640$ ways of dividing the flowers.

Some of these ways of dividing the flowers are downright unfair; surely, it is not fair if one of the children gets no flowers at all. Therefore, we impose the restriction that each child must get at least 3 flowers of each kind. Then the camomiles can be divided in 5 ways (the first child can get 3, 4, 5, 6, or 7 camomiles), the cornflowers in 10 ways, and the forget-me-nots in 9 ways. This gives $5 \cdot 10 \cdot 9 = 450$ ways of dividing the flowers.

In general, if there are n_1 objects of one kind, n_2 objects of another kind,..., n_k objects of a kth kind, then these objects can be divided between two people in

$$(n_1 + 1)(n_2 + 1) \cdots (n_k + 1) \tag{2}$$

ways. In particular, if all the objects are different and there are k of them, then $n_1 = n_2 = \cdots = n_k = 1$, and this means that the objects can be divided in 2^k ways.

If we impose the restriction that each of the two people is to get at least s_1 objects of the first kind, s_2 objects of the second kind,..., s_k objects of the kth kind, then the number of ways in which the division can be carried out is

$$(n_1 - 2s_1 + 1)(n_2 - 2s_2 + 1) \cdots (n_k - 2s_k + 1). \tag{3}$$

We leave it to the reader to prove these assertions.

The Number of Divisors

Formula (2) enables us to solve the following problem in number theory:
Find the number of divisors of a natural number N.

Let $N = p_1^{n_1} p_2^{n_2} \cdots p_k^{n_k}$, where $p_1, ..., p_k$ are distinct primes (for example, $360 = 2^3 \cdot 3^2 \cdot 5$). If we write N as a product of two factors N_1 and N_2, then the primes are divided between N_1 and N_2. Specifically, if the prime p_j enters N_1 to the power m_j, $j = 1, ..., k$, then

$$N = (p_1^{m_1} \cdots p_k^{m_k})(p_1^{n_1 - m_1} \cdots p_1^{n_k - m_k}).$$

This means that, from the combinatorial point of view, writing N as a product of two factors amounts to dividing n_1 objects of one kind, n_2 objects of another kind,..., n_k of objects of a kth kind into two parts. Formula (2) shows that this can be done in $(n_1 + 1) \cdots (n_k + 1)$ ways. Hence the number of divisors of a natural number $N = p_1^{n_1} \cdots p_k^{n_k}$ is $(n_1 + 1) \cdots (n_k + 1)$. This number is denoted by the symbol $\tau(N)$.

Picking Apples

Three children picked 40 apples. In how many ways can the children divide the apples if all the apples are supposed alike (that is, our concern is only with the number of apples obtained by each child) ?

To solve this problem we proceed as follows: We add 2 identical pears to the apples and then permute the 40 apples and the 2 pears. By the formula for permutations with repetitions we see that the number of such permutations is

$$P(40, 2) = C_{42}^2 = \frac{42!}{40!\, 2!} = 861.$$

Each of these permutations determines a division of the apples: The apples preceding the first pear go to the first child, the apples between the two pears go to the second child, and the apples following the second pear go to the third child. It is clear that different permutations determine different ways of dividing the apples. It follows that the apples can be divided in 861 ways. It may happen that one (or even two) of the children won't get any apples. For example, if one of the pears comes first in the permutation, then the first child gets no apples, and if one of the pears comes last in the permutation, then the third child gets no apples. If no pear is in the beginning or at the end of the permutation but both pears occur together, then the second child gets no apples. We leave it to the reader to decide what happens if both pears occur together in the beginning of the permutation or at the end of the permutation.

A similar argument shows that n like objects can be divided among k people in

$$P(n, k - 1) = C_{n+k-1}^n = C_{n+k-1}^{k-1} \tag{4}$$

ways.

To assure a measure of fairness, we now require that each person receive at least r objects. Then there are $n - kr$ objects which can be distributed in an arbitrary manner. We know that this can be done in $C_{n-kr+k-1}^{k-1} = C_{n-k(r-1)-1}^{k-1}$ ways.

In particular, *if each of the k people is to receive at least 1 object, then the problem has* C_{n-1}^{k-1} *solutions.*

The last result can be arrived at differently. Let us line up the n given objects in a row. There are $n - 1$ spaces between the successive objects. If we place dividers in $k - 1$ of these spaces, then the row of objects is partitioned into k nonempty parts. The first part goes to the first person, the second part goes to the second person, and so on. Since there are C_{n-1}^{k-1} ways of placing $k-1$ dividers in $n-1$ spaces, it follows that the number of distributions is C_{n-1}^{k-1}.

Picking Mushrooms

In order to find the number of ways of distributing objects of different kinds we must compute the number of ways of distributing objects of each kind and then multiply these numbers. By way of example we solve the following problem:

In how many ways can one divide 10 meadow mushrooms, 15 button mushrooms, and 8 truffles among 4 children?

Using the results in the previous section we see that the number in question is

$$C_{13}^3 C_{18}^3 C_{11}^3 = 41,771,040.$$

If each child is to get at least one mushroom of a kind, then the corresponding number is

$$C_9^3 C_{14}^3 C_7^3 = 1,070,160.$$

If n different objects are divided without any restrictions among k people, then each object can be handed out in k ways (since there are k possible recipients). But this means that the problem has k^n solutions.

For example, 8 different pastries can be divided among 5 people in $5^8 = 390,625$ ways.

Sending Photographs

I wish to send a friend 8 different photographs. In how many ways can this be done if the photographs are to be placed in 5 different envelopes?

This problem is similar to the last problem in the previous section. We would therefore be inclined to say that the answer to the present problem is also $5^8 = 390,625$. However, one could hardly be expected to mail empty envelopes, so it is natural to require that each envelope contains at least 1 photograph. To compute the answer, we make use of the principle of inclusion and exclusion (the answer C_{8-1}^{5-1} is incorrect since the photographs are different).

First we compute the number of distributions of the photographs in which r specified envelopes are empty (and the remaining envelopes are empty or not). This number is equal to the number of ways in which 8 different photographs can be placed without restriction in $5-r$ envelopes. The latter number is $(5-r)^8$ (see above). On the other hand, there are C_5^r ways of choosing r out of 5 envelopes. By the rule of product the number of distributions in which r

envelopes are known to be empty is $C_5^r(5-r)^8$. By the formula of inclusion and exclusion, the number of distributions without empty envelopes is

$$5^8 - C_5^1 \cdot 4^8 + C_5^2 \cdot 3^8 - C_5^3 \cdot 2^8 + C_5^4 \cdot 1^8 = 126,000.$$

Similarly, disallowing empty envelpes, we can send n different photographs in k different envelopes in

$$k^n - C_k^1(k-1)^n + C_k^2(k-2)^n - \cdots + (-1)^{k-1}C_k^{k-1} \cdot 1^n \qquad (5)$$

ways.

We leave it to the reader to analyze the following problem:

Given n_1 objects of the first kind, n_2 objects of the second kind,..., n_s objects of the sth kind. In how many ways can these objects be divided among k people if each person is to get at least 1 object?

The answer to this problem is given by the formula

$$C_{n_1+k-1}^{k-1}C_{n_2+k-1}^{k-1} \cdots C_{n_s+k-1}^{k-1} - C_k^1 C_{n_1+k-2}^{k-2}C_{n_2+k-2}^{k-2} \cdots C_{n_s+k-2}^{k-2}$$

$$\qquad (6)$$

$$+ C_k^2 C_{n_1+k-3}^{k-3}C_{n_2+k-3}^{k-3} \cdots C_{n_s+k-3}^{k-3} - \cdots + (-1)^{k-1}C_k^{k-1}.$$

For example, the number of ways in which 8 apples, 10 pears, and 7 oranges can be divided among 4 children if each child is to get at least 1 fruit is

$$C_{11}^3 C_{13}^3 C_{10}^3 - C_4^1 C_{10}^2 C_{12}^2 C_9^2 + C_4^2 C_9^1 C_{11}^1 C_8^1 - C_4^3 = 5,464,800.$$

Flags on Masts

In the distribution problems discussed so far the order of the elements in a box was irrelevant. However, in some of these problems this order must be taken into consideration.

n different signal flags are displayed on k masts. The meaning of a signal depends on the order of the flags. In how many ways can the flags be displayed if all of the flags must be used but some of the masts may be flagless?

Each display of the flags can be realized in two stages: First we select a suitable permutation of the flags and then place appropriate numbers of the flags on the masts in the order determined by the permutation. Since n different flags can be permuted in $n!$ ways and n identical flags can be distributed on k masts in C_{n+k-1}^{k-1} ways, it follows by the rule of product that the number of the displays in question is

$$n! \, C_{n+k-1}^{k-1} = \frac{(n+k-1)!}{(k-1)!} = A_{n+k-1}^n . \qquad (7)$$

Quite generally, the number of ways of placing *n different objects in k different boxes with the objects in each box arranged in a definite order is* A_{n+k-1}^{k-1} .

The same result can be obtained in a different manner: We add to the *n* objects $k - 1$ identical spheres and consider all the permutations of the resulting set of $n + k - 1$ objects. The distribution determined by one of these permutations is the distribution in which the objects in the first box are the objects in the permutation located before the first sphere (if the first element in our permutation is a sphere, then the first box will be empty), the objects in the second box are the objects in the permutation located between the first and second spheres,..., the objects in the *k*th box are the objects in the permutation following the last sphere. It is clear that in this way we obtain all of the required distributions. But then their number is equal to the number of permutations of *n* different objects and *k* identical spheres, that is

$$P(\underbrace{1, 1,..., 1}_{n \text{ times}}, k - 1) = \frac{(n + k - 1)!}{1! \cdots 1! (k - 1)!} = A_{n+k-1}^n .$$

Similar reasoning applies to the case when there is at least one flag on each mast (or at least one object in each box) except that now the factor C_{n+k-1}^{k-1} must be replaced by the factor C_{n-1}^{k-1}; that this is so is implied by the formula derived on p. 68. Another way of arriving at this answer is to count the number of ways of placing dividers in $k - 1$ of the $n - 1$ spaces between *n* objects.

Count of All Signals

So far we assumed that in transmitting a signal we invariably used all the flags. However, some signals may not require the use of all the flags. This suggests the problem of computing *the number of signals which can be transmitted by using some or all of n flags.* As before we allow flagless masts.

We separate the signals into classes according to the number of flags involved in the transmission of a particular signal.

By Formula (7) the number of signals transmitted by means of *s* flags is A_{s+k-1}^s (there are *k* masts). Since *s* flags can be chosen out of *n* flags in C_n^s ways, the number of signals in the *s*th class is $C_n^s A_{s+k-1}^s$. It follows that the number of signals in all the classes is

$$C_n^0 A_{k-1}^0 + C_n^1 A_k^1 + C_n^2 A_{k+1}^2 + \cdots + C_n^n A_{n+k-1}^n . \tag{8}$$

For example, with 6 different flags on 3 masts we can transmit

$$1 + C_6^1 A_3^1 + C_6^2 A_4^2 + C_6^3 A_5^3 + C_6^4 A_6^4 + C_6^5 A_7^5 + C_6^6 A_8^6 = 42{,}079$$

signals.

If we disallow empty masts then in place of Formula (8) we obtain formula

$$C_n^k C_{k-1}^{k-1} k! + C_n^{k+1} C_k^{k-1}(k+1)! + C_n^{k+2} C_{k+1}^{k-1}(k+2)! + \cdots + C_n^n C_{n-1}^{k-1} n!. \qquad (9)$$

Various Statistics

The problem of placing things in boxes is of great importance in statistical physics. Statistical physics studies the distribution of physical particles in relation to certain properties. An example of a problem in statistical physics is the determination of the fraction of molecules of a gas at a given temperature which have a certain velocity. The possible states (known as phase states) can be thought of as tiny cells and each particle belongs to one of these cells. (In what follows, k denotes the number of cells and n the number of particles in a sample.)

The statistics which applies in a particular situation depends on the nature of the particles involved. The classical statistical physics of Maxwell and Boltzmann regards all particles as different from one another and is applicable, among others, to molecules of a gas. We know that n different particles can be placed in k cells in k^n ways. If for a given energy all of these k^n ways of placing the particles are assigned the same probability, then we speak of *Maxwell–Boltzmann statistics*.

Not all particles are governed by Maxwell–Boltzmann statistics. Photons, atomic nuclei, and atoms containing an even number of particles are governed by the statistics due to Einstein and the Indian scientist Bose. In *Bose–Einstein statistics* particles are indistinguishable, so that we are not concerned with the identity of the individual particles but only with the number of particles in each cell. This means that the count of admissible distributions is the same as the count of distributions of apples discussed on p. 68 and is therefore equal to $C_{n+k-1}^{k-1} = C_{n+k-1}^n$. In Bose–Einstein statistics all of these distributions are assigned equal probabilities.

Particles such as electrons, neutrons, and protons are not governed by either of the above statistics. In the case of these particles, each cell contains no more than one particle, and each distribution satisfying this condition is assigned the same probability. The number of different distributions is C_k^n. This statistics is called the *Dirac–Fermi statistics*.

Partitions of Numbers

In most of the distribution problems discussed earlier the objects were supposed distinguishable. Now we will treat distribution problems where the objects are supposed indistinguishable. This allows us to replace distributions of N objects with partitions of the natural number N into summands of natural numbers, that is, with ways of writing N as a sum of natural numbers.

There are many different problems of this type. We may pay attention to the order of the summands or disregard it. We may restrict our attention to partitions with an odd number of summands, or to partitions with an even number of summands. We may allow partitions with arbitrary summands or partitions with different summands only, and so on. The basic technique for the solution of partition problems is to reduce the solution of the original partition problem to the solution of a partition problem involving a smaller number or to a partition problem involving a smaller number of summands.

Mailing of Printed Matter

Postage on some printed matter is 18 cents. We have at our disposal (unlimited amounts of) 4-cent stamps, 6-cent stamps, and 10-cent stamps. In how many ways can we paste on the required 18 cents worth of stamps if two ways which differ in the order of the stamps are regarded as different?

Let $f(N)$ denote the number of arrangements of 4-, 6-, and 10-cent stamps which add up to N cents. We claim that

$$f(N) = f(N - 4) + f(N - 6) + f(N - 10). \tag{10}$$

Consider an arrangement of the stamps in which the last stamp is a 4-cent stamp. Then the value of the remaining stamps is $N - 4$ cents. Conversely, addition of a 4-cent stamp to an arrangement of stamps worth $N - 4$ cents yields an arrangement of stamps worth N cents. Also, different arrangements of stamps worth $N - 4$ cents yield different arrangements of stamps worth N cents. It follows that the number of the required arrangements terminating in a 4-cent stamp is $f(N - 4)$.

The same type of argument shows that the number of the required arrangements terminating in a 6-cent stamp is $f(N - 6)$, and the number of the required arrangements terminating in a 10-cent stamp is $f(N - 10)$. Since each of our arrangements terminates in a 4-, 6-, or 10-cent stamp, it follows by the rule of sum that $f(N)$ is indeed equal to the sum in (10).

Relation (10) enables us to reduce the problem of computing the number of arrangements of stamps which add up to N cents to similar problems in which the stamps add up to less than N cents. For small values of N it is easy to compute $f(N)$ directly. Specifically,

$$f(0) = 1, \quad f(1) = f(2) = f(3) = 0, \quad f(4) = 1, \quad f(5) = 0,$$
$$f(6) = 1, \quad f(7) = 0, \quad f(8) = 1, \quad f(9) = 0.$$

(The equality $f(0) = 1$ reflects the fact that the only way to get 0 cents is not to paste on any stamps. The equalities $f(1) = f(2) = f(3) = f(5) = f(7) = f(9) = 0$ reflect the fact that using 4-, 6-, and 10-cent stamps we cannot get arrangements of stamps which add up to 1, 2, 3, 5, 7, or 9 cents.) Using these values of $f(N)$ for $N = 0, 1, 2, 3, 4, 5, 6, 7, 8, 9$, we see that

$$f(10) = f(6) + f(4) + f(0) = 3.$$

Next we find that

$$f(11) = f(7) + f(5) + f(1) = 0,$$
$$f(12) = f(8) + f(6) + f(2) = 2$$

and so on. Eventually we obtain the value $f(18) = 8$. The eight arrangements of the stamps are:

10, 4, 4; 4, 10, 4; 4, 4, 10; 6, 4, 4, 4; 4, 6, 4, 4; 4, 4, 6, 4; 4, 4, 4, 6; 6, 6, 6.

We note that the values of $f(N)$ for $N = 1, 2, 3, 4, 5, 6, 7, 8, 9$ can be obtained without direct computation. Observe that a nonnegative number of stamps cannot add up to a negative number of cents, so that $f(N) = 0$ for $N < 0$. We saw that $f(0) = 1$. Therefore

$$f(1) = f(-3) + f(-5) + f(-9) = 0.$$

Similarly, $f(2) = 0, f(3) = 0$. For $N = 4$ we have

$$f(4) = f(0) + f(-2) + f(-6) = 1,$$

and so on.

The General Postage Problem

The preceding problem is a special case of the following problem.

Given (unlimited numbers of) stamps of k different denominations n_1, n_2,..., n_k. Find the number of arrangements of stamps which add up to N cents if two arrangements which differ in the order of the stamps are regarded as different.

In this case $f(N)$ satisfies the relation

$$f(N) = f(N - n_1) + f(N - n_2) + \cdots + f(N - n_k). \tag{11}$$

We have $f(N) = 0$ for $N < 0$ and $f(0) = 1$. To compute $f(N)$ for a preassigned value of N we use (11) to compute successively the numbers $f(1), f(2),..., f(N-1)$. Then a final application of (11) yields the value of $f(N)$.

We consider the special case of this problem when $n_1 = 1, n_2 = 2,..., n_k = k$. The resulting problem is the same as the problem of computing the number of partitions of the number N into summands selected from among the numbers $1, 2,..., k$ with the provision that two partitions which differ in the order of their summands are to be viewed as different. We denote the number of such partitions by $\varphi(k; N)$. Relation (11) implies that

$$\varphi(k; N) = \varphi(k; N - 1) + \varphi(k; N - 2) + \cdots + \varphi(k; N - k). \qquad (12)$$

Also,

$$\varphi(k; 0) = 1, \quad \text{and} \quad \varphi(k; N) = 0, \quad \text{for} \quad N < 0.$$

The computation of $\varphi(k; N)$ is simplified by noting that

$$\varphi(k; N - 1) = \varphi(k; N - 2) + \cdots + \varphi(k; N - k) + \varphi(k; N - k - 1).$$

Using this relation we see that

$$\varphi(k; N) = 2\varphi(k; N - 1) - \varphi(k; N - k - 1). \qquad (13)$$

It is clear that the summands cannot exceed N. Hence $\varphi(N; N)$ is the number of partitions of N into positive integral summands (including the "partition" $N = N$) with order taken into consideration. The number of partitions involving s summands is C_{N-1}^{s-1} (see p. 68). Therefore

$$\varphi(N, N) = C_{N-1}^0 + C_{N-1}^1 + \cdots + C_{N-1}^{N-1} = 2^{N-1}.$$

We have shown that when order is taken into consideration, then the number of partitions of a natural number N is 2^{N-1}. For example, there are $2^{5-1} = 16$ (ordered) partitions of the number 5:

$$
\begin{array}{lll}
5 = 5 & 5 = 3 + 1 + 1 & 5 = 1 + 2 + 2 \\
5 = 4 + 1 & 5 = 1 + 3 + 1 & 5 = 2 + 1 + 1 + 1 \\
5 = 1 + 4 & 5 = 1 + 1 + 3 & 5 = 1 + 2 + 1 + 1 \\
5 = 2 + 3 & 5 = 2 + 2 + 1 & 5 = 1 + 1 + 2 + 1 \\
5 = 3 + 2 & 5 = 2 + 1 + 2 & 5 = 1 + 1 + 1 + 2 \\
& & 5 = 1 + 1 + 1 + 1 + 1.
\end{array}
$$

Combinatorial Problems in Information Theory

In information theory we find it necessary to solve the following problem which is very similar to the main problem in the previous section: A message is to be transmitted using symbols of various types. Transmission of a signal of the first type takes t_1 time units, transmission of a signal of the second type takes t_2 time units,..., transmission of a signal of the kth type takes t_k time units. *How many different messages can be transmitted in T time units?* Here we have in mind only "maximal" messages, that is messages to which no signal can be added without exceeding the time limit T.

Let $f(T)$ denote the number of messages whose "transmission time" is T. Reasoning as in the problem about stamps we see that

$$f(T) = f(T - t_1) + \cdots + f(T - t_k). \tag{14}$$

Also, $f(T) = 0$ for $T < 0$ and $f(0) = 1$.

Problem of the University Applicant

A university applicant is required to pass 4 examinations. The numerical passing grades are 3, 4, and 5. The minimal cumulative score for admission to the university is 17. In how many ways can the applicant accumulate the necessary 17 or more points?

This problem is similar to the problem about stamps except that the number of "stamps" which "add up to 17 points" is prescribed. Let $F(k; N)$ denote the number of ways in which it is possible to accumulate N points after passing k examinations. Reasoning as in the case of Formula (11) on p. 74 we find that

$$F(k; N) = F(k - 1; N - 3) + F(k - 1; N - 4) + F(k - 1; N - 5).$$

This relation and the fact that $F(2; 11) = 0$ and $F(2; 10) = 1$ yield

$$\begin{aligned}
F(4; 17) &= F(3; 14) + F(3; 13) + F(3; 12) \\
&= F(2; 11) + 2F(2; 10) + 3F(2; 9) + 2F(2; 8) + F(2; 7) \\
&= 2 + 3F(2; 9) + 2F(2; 8) + F(2; 7).
\end{aligned}$$

In turn, this yields
$$F(4; 17) = 2 + 3F(1; 6) + 5F(1; 5) + 6F(1; 4) + 3F(1;3) + F(1; 2).$$
Since $F(1; 6) = F(1; 2) = 0$, $(F(1; 6) = 0$ because the most one can earn in 1 exam is a 5. On the other hand, $F(1; 2) = 0$ because the least grade which counts is a 3), and $F(1; 5) = F(1; 4) = F(1; 3) = 1$, it follows that $F(4; 17) = 16$. Similarly, $F(4; 18) = 10$, $F(4; 19) = 4$, and $F(4; 20) = 1$. This means that there are $16 + 10 + 4 + 1 = 31$ ways of passing the examinations.

There is another way of ariving at these results. It is easy to see that there are two essentially different ways of accumulating 17 points: one way is to earn

2 grades of 5, 1 grade of 4, and 1 grade of 3, and the other is to earn 1 grade of 5 and 3 grades of 4. These sets of grades may be associated with the 4 examinations in various ways. Since

$$P(2, 1, 1) + P(1, 3) = \frac{4!}{2!\,1!\,1!} + \frac{4!}{3!\,1!} = 16,$$

it follows that 17 points can be earned in 16 ways. In this way we can also compute the number of ways of earning 18, 19, and 20 points, respectively.

In general, let $F(m; N)$ denote the number of partitions (with order taken into consideration) of N into m summands each of which is equal to one of the numbers n_1, n_2,..., n_k. Arguing as in the case of relation (11), we see that

$$F(m; N) = F(m - 1; N - n_1) + \cdots + F(m - 1; N - n_k). \qquad (15)$$

In particular, if $n_1 = 1, n_2 = 2,..., n_k = k$, then we obtain the number of (ordered) partitions of N into m summands each of which is equal to one of the numbers 1, 2,..., k. If we denote this number by $F(m; N; k)$, then we have

$$F(m; N; k) = F(m - 1; N - 1; k)$$
$$+ F(m - 1; N - 2; k) + \cdots + F(m - 1; N - k; k). \qquad (16)$$

By imitating the argument on p. 75 we obtain the following simpler version of (16):

$$F(m; N; k) = F(m; N - 1; k)$$
$$+ F(m - 1; N - 1; k) - F(m - 1; N - k - 1; k). \qquad (17)$$

Now we consider partition problems in which we identify partitions differing only in the order of the summands.

Paying for a Purchase

A purse contains single coins of the following denominations: 1, 2, 3, 5, 10, 15, 20, and 50 cents.* In how many ways is it possible to pay with these coins for a 73 cent purchase?

Here the order of the coins is irrelevant. What matters is which coins are used to make up the required total. We denote by

$$F(n_1, n_2,..., n_m; N)$$

---

* Ignoring the realities of our system of coins is both harmless and convenient (Translators).

the number of ways of paying N cents with coins of different denominations selected from among coins whose denominations are n_1, n_2,..., n_m cents.

We separate all the payments into two classes: the class of payments which include an n_m-cent coin and the class of payments which do not include an n_m-cent coin. If a payment includes an n_m-cent coin, then the sum of $N - n_m$ cents must be paid with coins whose denominations are restricted to n_1, n_2,..., n_{m-1} cents, and this can be done in $F(n_1, n_2,..., n_{m-1}; N - n_m)$ ways. If a payment does not include an n_m-cent coin, then the full amount of N cents must be paid with coins whose denominations are restricted to n_1, n_2,..., n_{m-1} cents, and this can be done in $F(n_1, n_2,..., n_{m-1}; N)$ ways. Hence

$$F(n_1, n_2,..., n_m; N) = F(n_1, n_2,..., n_{m-1}; N - n_m) + F(n_1, n_2,..., n_{m-1}; N). \quad (18)$$

In the original problem we were required to select coins from a total of m coins. Relation (18) reduces this total to $m - 1$. Repeated use of (18) enables us to reduce this total to $m - 2$, and so on. Eventually we reach the point when there is nothing to pay or when the total of coins to select from is down to one. Each of the last two problems has a unique solution. Also, many summands vanish in the process of computation; in fact, if $n_1 + n_2 + \cdots + n_m < N$ then $F(n_1, n_2,..., n_m; N) = 0$ (since the coins add up to less then the required payment), and if $n_m > N$, then (18) reduces to

$$F(n_1, n_2,..., n_m; N) = F(n_1, n_2,..., n_{m-1}; N)$$

(since n_m must not be included in a payment).

We apply the method just outlined to the solution of our problem. By Eq. (18) we have

$$\begin{aligned} F(1, 2, 3, 5, 10, 15, 20, 50; 73) &= F(1, 2, 3, 5, 10, 15, 20; 23) \\ &\quad + F(1, 2, 3, 5, 10, 15, 20; 73) \\ &= F(1, 2, 3, 5, 10, 15, 20; 23) \end{aligned}$$

$(1+2+3+5+10+15+20 < 73$ implies that $F(1, 2, 3, 5, 10, 15, 20; 73) = 0)$.

Again,

$$F(1, 2, 3, 5, 10, 15, 20; 23) = F(1, 2, 3, 5, 10, 15; 3) + F(1, 2, 3, 5, 10, 15; 23).$$

The first of these summands is

$$\begin{aligned} F(1, 2, 3, 5, 10, 15; 3) &= F(1, 2, 3; 3) \\ &= F(1, 2; 0) + F(1, 2; 3) \\ &= 1 + F(1; 3) + F(1; 1) = 2. \end{aligned}$$

Now,

$$F(1, 2, 3, 5, 10, 15; 23) = F(1, 2, 3, 5, 10; 8) + F(1, 2, 3, 5, 10; 23)$$
$$= F(1, 2, 3, 5, 10; 8)$$

$(1 + 2 + 3 + 5 + 10 < 23$ implies that $F(1, 2, 3, 5, 10; 23) = 0)$, and $F(1, 2, 3, 5; 8) = F(1, 2, 3; 3) = 2$.

All in all, we have

$$F(1, 2, 3, 5, 10, 15, 20, 50; 73) = 4.$$

Indeed, we can pay the required 73 cents with a 50-, a 20-, and a 3-cent coin; with a 50-, a 20-, a 2-, and a 1-cent coin; with a 50-, a 15-, a 5-, and a 3-cent coin; and with a 50-, a 15-, a 5-, a 2-, and a 1-cent coin.

Purchase of Candy

A store sells various kinds of candy: 3 kinds sell for 2 cents a piece and 2 kinds sell for 3 cents a piece. In how many ways can one buy 8 cents worth of candy if no two pieces of candy are to be of the same kind?

The required computations are:

$$F(2, 2, 2, 3, 3; 8) = F(2, 2, 2, 3; 5) + F(2, 2, 2, 3; 8)$$
$$= F(2, 2, 2; 2) + 2F(2, 2, 2; 5) + F(2, 2, 2; 8)$$
$$= F(2, 2, 2; 2) = F(2, 2; 0) + F(2, 2; 2)$$
$$= 1 + F(2; 0) + F(2; 2) = 3.$$

Each of the 3 ways of buying candy consists in combining two 3-cent pieces of different kinds with a 2-cent piece.

Consider next the following problem:

A purse contains 3 two-cent coins and 2 three-cent coins: In how many ways can one pay 8 cents using these coins?

At first sight this problem seems to be identical with the previous problem. However, everything depends on whether or not we distinguish between coins of the same denomination: If we do, then our present problem is indeed identical with the previous problem, and there are 3 ways of paying; if we do not, then there is only one way of paying, namely by combining 2 three-cent coins and 1 two-cent coin.

These examples show that the nature of a payment problem depends on whether or not we distinguish between coins of the same denomination. The method discussed in the previous section applies only when the various coins are viewed as different regardless of their denominations. Now we will show what method to use when we don't distinguish between coins of the same denomination.

A purse contains 10 two-cent coins and 5 three-cent coins. In how many ways can we pay the sum of 22 cents if we do not distinguish between coins of the same denomination?

We denote the number of solutions by the symbol $\Phi(10 \cdot 2, 5 \cdot 3; 22)$ ($10 \cdot 2$ indicates that there are 10 two-cent coins, and $5 \cdot 3$ indicates that there are 5 three-cent coins). We separate the ways of effecting payment into classes in accordance with the number of 3-cent coins used in a payment. For example, if 2 three-cent coins are used, then the remaining 16 cents are paid with 2-cent coins, and if all 5 three-cent coins are used then it remains to pay 7 cents. If no 3-cent coins are used in a payment, then the full sum of 22 cents must be paid with 2-cent coins. This leads to the equality

$$\Phi(10 \cdot 2, 5 \cdot 3; 22) = \Phi(10 \cdot 2; 22)$$
$$+ \Phi(10 \cdot 2; 19) + \Phi(10 \cdot 2; 16)$$
$$+ \Phi(10 \cdot 2; 13) + \Phi(10 \cdot 2; 10) + \Phi(10 \cdot 2; 7). \quad (19)$$

(The 6 summands on the right-hand side of (19) correspond to the 6 classes of payments.) Observe that 22 cents cannot be paid with 10 two-cent coins; that an odd sum can't be paid with 2-cent coins; and that an even sum can be paid with 2-cent coins in just one way. With these facts in mind we see from (19) that

$$\Phi(10 \cdot 2; 5 \cdot 3; 22) = 2.$$

The 2 ways are

$$22 = 8 \cdot 2 + 2 \cdot 3 = 5 \cdot 2 + 4 \cdot 3.$$

Getting Change

Consider the following problem:

In how many ways is it possible to change a 10-cent coin into 1-, 2-, 3-, and 5-cent coins?

This problem is similar to the problem at the end of the last section except that now no restrictions are imposed on the number of coins of various denominations. To indicate this fact we denote the number of solutions by the new symbol $\Phi(1, 2, 3, 5; 10)$. Reasoning as in the preceding section, we obtain the relation

$$\Phi(1, 2, 3, 5; 10) = \Phi(1, 2, 3; 10) + \Phi(1, 2, 3; 5) + \Phi(1, 2, 3; 0) \quad (20)$$

(the 3 summands on the right-hand side of (20) reflect the fact that the change may contain no 5-cent coin, 1 five-cent coin, or 2 five-cent coins). It is clear that $\Phi(1, 2, 3; 0) = 1$ (there is just one way of changing a 0 coin).

To compute $\Phi(1, 2, 3; 5)$ we separate the ways of changing a 5-cent coin into 1-, 2-, and 3-cent coins into classes in accordance with the number of 3-cent coins in the change. This yields the relation

$$\Phi(1, 2, 3; 5) = \Phi(1, 2; 5) + \Phi(1, 2; 2)$$

(the first summond corresponds to the case when there are no 3-cent coins in the change, and the second summand corresponds to the case when there is just 1 three-cent coin in the change).

Continuing these computations we obtain the relation

$$\Phi(1, 2, 3; 5) = \Phi(1; 5) + \Phi(1; 3) + \Phi(1; 1) + \Phi(1; 2) + \Phi(1; 0).$$

Since the value of each summand in this sum is 1 (the change consists of pennies only!), it follows that $\Phi(1, 2, 3; 5) = 5$. A similar computation shows that $\Phi(1, 2, 3; 10) = 14$. This means that there are $14 + 5 + 1 = 20$ ways of changing our 10-cent coin.

In place of (20) we could take the relation

$$\Phi(1, 2, 3, 5; 10) = \Phi(1, 2, 3; 10) + \Phi(1, 2, 3, 5; 5).$$

This relation states that the ways of changing our 10-cent coin can be separated into two classes: the class of ways in which no 5-cent coin is used and the class of ways in which at least one 5-cent coin is used.

In general, if the change adds up to N cents and consists of coins of denominations $n_1, n_2, ..., n_k$ cents, then the analog of Eq. (20) is

$$\Phi(n_1, ..., n_{k-1}, n_k; N) = \Phi(n_1, ..., n_{k-1}; N) + \Phi(n_1, ..., n_{k-1}, n_k; N - n_k). \quad (21)$$

Relation (21) shows that if no n_k-cent coin is included in the change, then the full sum N is made up of coins of denominations $n_1, n_2, ..., n_{k-1}$, and if at least one n_k-cent coin is used, then the remainder $N - n_k$ may include coins of all the denominations $n_1, n_2, ..., n_{k-1}, n_k$.

If, as on p. 77 the change must consist of different coins, then (21) is replaced with the following analog of Formula (18):

$$F(n_1, ..., n_{k-1}, n_k; N) = F(n_1, ..., n_{k-1}; N) + F(n_1, ..., n_{k-1}; N - n_k). \quad (22)$$

(Unordered) Partitions of an Integer

We consider a special case of the problem about change when all the coins from 1 to n cents are admissible. In other words, we consider the following problem:

Find the number of partitions of a natural number N into summands selected from among the numbers 1, 2,..., n (order of the summands is immaterial).

We denote the number of such partitions by Π_N^n; in particular, we define $\Pi_0^n = 1$. Observe that the number of all partitions of N is Π_N^N (no partition of N can include a summand $> N$). This means that if $n > N$, then $\Pi_N^n = \Pi_N^N$. Also, $\Pi_N^1 = 1$. If $1 < n \leqslant N$, then

$$\Pi_N^n = \Pi_N^{n-1} + \Pi_{N-n}^n . \tag{23}$$

In fact, if n is not used as a summand, then N is partitioned into summands selected from among the numbers 1, 2,..., $n - 1$, and the number of such partitions is Π_N^{n-1}. On the other hand, if n is used as a summand, then $N - n$ is partitioned into summands selected from among the numbers 1, 2,..., n, and the number of such partitions is Π_{N-n}^n.

Now we require that *all the summands in a partition must be different*. In this case we denote the number of partitions by Φ_N^n and put $\Phi_0^n = 1$. Observe that if $n > N$, then $\Phi_N^n = \Phi_N^N$ (see above). Also, $\Phi_1^1 = 1$ and $\Phi_N^1 = 0$ for $N > 1$. If $1 < n \leqslant N$ then we claim that

$$\Phi_N^n = \Phi_N^{n-1} + \Phi_{N-n}^{n-1} \tag{24}$$

(the number n can appear as a summand only once). We leave the proof of (24) as an exercise for the reader.

Using (24) and the above few facts we can evaluate Φ_N^n for all values of n and N. In evaluating Π_N^n it is convenient to replace (23) with

$$\Pi_N^n = \Pi_N^{n-1} + \Pi_{N-n}^{n-1} + \Pi_{N-2n}^{n-1} + \cdots + \Pi_{N-kn}^n , \qquad 0 \leqslant N - kn < n, \tag{25}$$

obtained by repeated application of (23). If in (25) $N - kn = 0$, then $\Pi_{N-kn}^n = \Pi_0^n = 1$. If $0 < N - kn < n$, then $\Pi_{N-kn}^n = \Pi_{N-kn}^{N-kn}$. Using (25) we evaluate succesively Π_N^2 for all required N, then Π_N^3, and so on.

Use of Graphs

The original methods of proving theorems about partitions of an integer were very complicated. Just as in other areas of mathematics, so here too, the introduction of geometric considerations simplified proofs and made them more intuitive.

Every partition of an integer can be represented by means of a graph consisting of rows of points. Each row of the graph has as many points as there are 1's in the corresponding summand of the partition. For example, the partition $7 = 1 + 1 + 2 + 3$ is represented by the graph in Fig. 10.

FIG. 10

Since we disregard the order of the summands in a partition, we can arrange the rows of its graph in such a way that the number of points in each successive row does not decrease. Also, we can line up the rows so that the first points of the rows form a vertical column. Such graphs will be called *normal*.

Graphs enable us to give easy proofs of many properties of partitions. Consider, for example, the assertion that *the number of partitions of an integer N into at most m summands is the same as the number of partitions of the integer $N + m$ into m summands.* The graph of a partition of N into at most m summands consists of N points arranged in at most m rows. If we add to such a graph a column of m points (see Fig. 11 which illustrates this transformation for the case $N = 5$,

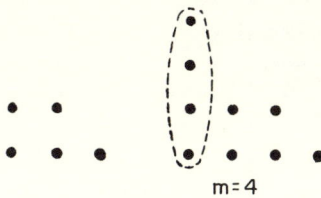

m = 4

FIG. 11

$m = 4$), then we obtain a graph consisting of $N + m$ points arranged in m rows. Conversely, if we subtract the first column from an m-rowed graph consisting of $N + m$ points, then we obtain a graph consisting of N points arranged in at most m rows. This establishes a one-to-one correspondence between the two classes of graphs. But then each class contains the same number of graphs. Our assertion follows.

The following result (due to Euler) is somewhat harder to prove:

The number of partitions of N into at most m summands is equal to the number of partitions of $N + m(m + 1)/2$ into m unequal summands.

Every partition of N into at most m summands is represented by means of a graph consisting of N points arranged in at most m rows. We add to such a graph an isosceles right triangle consisting of m rows and put the new graph in normal form (Fig. 12 illustrates this transformation for the case $N = 6$,

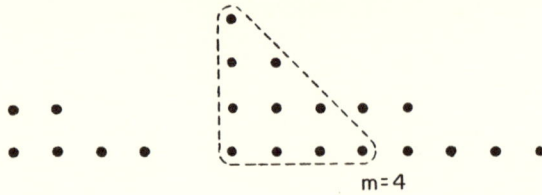

FIG. 12

$m = 4$). Since the triangle contains $m(m + 1)/2$ points, the resulting graph contains $N + m(m + 1)/2$ points arranged in m rows of increasing length (the length of a row is its number of points); this last assertion follows from the fact that our final graph is the result of adding a graph with increasing row length (the right isosceles triangle) to a graph with nondecreasing row length (the original graph). Conversely, deletion of a right isosceles m-rowed triangle from a graph of a partition of $N + m(m + 1)/2$ into unequal summands results in a graph of a partition of N into at most m summands.

It is clear that we have established a one-to-one correspondence between the graphs of the partitions of N into at most m rows and the graphs of the partitions of $N + m(m + 1)/2$ into m unequal rows. This shows that the two classes contain the same number of graphs. This completes the proof.

Dual Graphs

By rotating a graph through 90° and putting the result in normal form we obtain a new graph, the *dual* of the original graph. This transformation is illustrated in Fig. 13.

FIG. 13

It is clear that by repeating this transformation we obtain the original graph. Hence all graphs can be arranged in pairs of graphs dual to each other (with some graphs self dual, that is, equal to their duals; see Fig. 14).

FIG. 14

It is clear that the rows and columns of a graph are, respectively, the columns and rows of its dual.

Using duality of graphs we can compare partitions in which the *magnitude* of the summands is subject to restrictions with partitions in which the *number* of summands is subject to restrictions. For example, consider the assertion:

The number of partitions of N in which the summands do not exceed n is equal to the number of partitions of N in which the number of summands does not exceed n.

For proof, note that a graph of a partition of N in which the summands do not exceed n consists of N points arranged in rows at most n long. This means that the graph has at most n columns. But then the dual graph has at most n rows, that is, it corresponds to a partition of N into at most n summands.

A similar argument shows that the number of partitions of N into n summands is equal to the number of partitions of N into summands not larger than n of which at least one is equal to n.

Next we consider partitions of N into even summands. Such partitions are represented by graphs whose rows contain an even number of points. But then in the dual graphs the number of rows of equal length is invariably even (see Fig. 15). This proves that

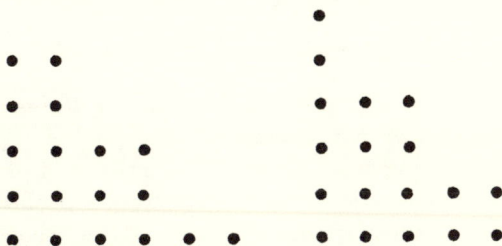

FIG. 15

The number of partitions of N into even summands is equal to the number of partitions of N in which each summand enters an even number of times.

A similar argument shows that

The number of partitions of N into odd summands is equal to the number of partitions in which every summand other than the largest one enters an even number of times and the largest summand enters an odd number of times.

Euler's Formula*

In connection with certain partition problems Euler studied the infinite product

$$A = (1 - x)(1 - x^2)(1 - x^3) \cdots (1 - x^n) \cdots . \tag{26}$$

If we remove the parentheses in the first 22 factors of this product, then we obtain the expression

$$A = [1 - x - x^2 + x^5 + x^7 - x^{12} - x^{15} + x^{22} + \cdots]$$
$$\times (1 - x^{23})(1 - x^{24}) \cdots (1 - x^n) \cdots,$$

where dots replace all summands in which the exponent of x exceeds 22. The reason for this is that multiplication of the expression in square brackets by $1 - x^{23}$, $1 - x^{24}$,..., does not affect the powers of x whose exponent does not exceed 22 but does affect the higher powers of x. Thus removal of all the parentheses in (26) yields an infinite series whose first few terms are

$$1 - x - x^2 + x^5 + x^7 - x^{12} - x^{15} + x^{22} + \cdots. \tag{27}$$

The expression in (27) suggests that two negative terms are followed by two positive terms which are again followed by two negative terms, and so on. On the other hand, the law of formation of the exponents of the successive terms of the series is far more elusive. After a great deal of experimentation Euler found that

If we write the infinite product

$$(1 - x)(1 - x^2)(1 - x^3) \cdots (1 - x^n) \cdots$$

as an infinite series, then the only nonzero terms are the terms of the form $(-1)^k x^{(3k^2 \pm k)/2}$ *where k is a natural number.*

Euler's result is very important not only in the theory of partitions but also in the theory of elliptic functions and in other areas of mathematical analysis. Most proofs of this theorem are rather difficult. We are about to give a very simple geometric proof of this theorem. Before we do so we will find it convenient to formulate the theorem in terms of the theory of partitions.

Upon removal of the parentheses in (26) and prior to any reductions in the resulting series, the number of occurrences of $\pm x^N$ is equal to the number of partitions of N into sums with different summands. Furthermore, the number of occurrences of x^N is equal to the number of such partitions with an even number of summands, and the number of occurrences of $-x^N$ is equal to the number of

---

* This section can be omitted in a first reading.

such partitions with an odd number of summands. For example, the partition $12 = 5 + 4 + 2 + 1$ corresponds to the occurrence of x^{12} thought of as the product $(-x^5)(-x^4)(-x^2)(-x)$, and the partition $12 = 5 + 4 + 3$ corresponds to the occurrence of $-x^{12}$ thought of as the product $(-x^5)(-x^4)(-x^3)$. It follows that the coefficient of x^N in (27) is equal to the difference between the number of partitions of N into an even number of distinct summands and the number of partitions of N into an odd number of distinct summands. Expressed in terms of partitions, Euler's theorem states that

If N is not of the form $N = (3k^2 \pm k)/2$, then the difference between the number of partitions of N with an even number of distinct summands and the number of partitions of N with an odd number of distinct summands is zero. If $N = (3k^2 \pm k)/2$, then the corresponding difference is $(-1)^k$.

Specifically, if k is even, then this difference is 1, and if k is odd, then this difference is -1.

The proof is based on a method for converting a graph with an even number of rows into a graph with the same number of points and an odd number of rows, and conversely. Since the summands in any one of our partitions are distinct, its graph can be thought of as a tower of trapezoids. We denote the number of points in the top row of the graph by m, and the number of rows in the bottom trapezoid by n. For the graph in Fig. 16 we have $m = 2$ and $n = 3$.

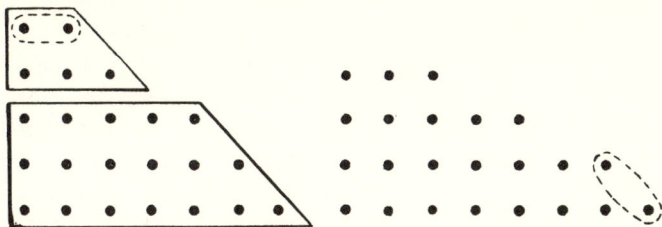

FIG. 16

Suppose the graph consists of two or more trapezoids and that $m \leqslant n$. Then we delete the top row and add one point to each of the last m rows of the bottom trapezoid. The resulting graph has the same number of points as the original graph, it has rows of different lengths, but the parity of its number of rows is different from the parity of the number of rows of the original graph (see Fig. 16). The same kind of transformation can be applied to a graph consisting of one trapezoid provided that $m \leqslant n - 1$; this case is illustrated in Fig. 17a.

Now suppose the graph consists of two or more trapezoids and that $m > n$. Then we remove one point from each of the rows of the bottom trapezoid and place these n points as a new row on top of our graph (this can be done because the new top row has n points and the old top row has $m > n$ points). The trans-

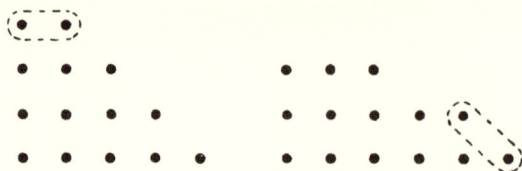

Fig. 17a

formed graph has the same number of points as the original graph, it has rows
of different lengths, but the parity of its number of rows is different from the
parity of the number of rows of the original graph; this case is illustrated in
Fig. 17b. The same transformation can be applied to a graph with one trapezoid

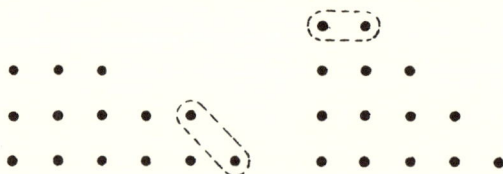

Fig. 17b

provided that $n \leqslant m - 2$. Inspection of Figs. 17a and 17b shows *there is a
one-to-one correspondence between the transformations of the first type and the
transformations of the second type which links mutually inverse transformations;
if a graph admits one of a pair of corresponding transformations, then its image
admits the other transformation of the pair and the final image coincides with the
original graph.* In other words, there are *as many graphs of partitions of N which
admit one of our transformations and have an even number of rows as there are
graphs of partitions of N which admit one of our transformations and have an odd
number of rows.* The graphs which do not admit any of our transformations
consist of single trapezoids for which $m = n$ or $m = n + 1$; in the first case N,
the number of points in the graph, is $(3n^2 - n)/2$, and in the second case N is
$(3n^2 + n)/2$ (see Fig. 18). Finally, there is at most one graph with a prescribed
number N of points which does not admit any of our transformations.

Our discussion shows that if N is not of the form $(3n^2 \pm n)/2$, then it has as
many partitions with different summands and an even number of rows as it

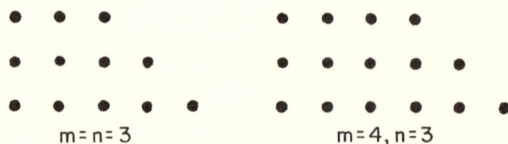

m = n = 3   m = 4, n = 3

Fig. 18

has partitions with different summands and an odd number of rows. If $N = (3n^2 \pm n)/2$ and n is even, then there is just one graph with N points which admits none of our transformations and has an even number of rows; in that case the number of partitions of N into an even number of distinct summands exceeds by 1 the number of partitions of N into an odd number of distinct summands. If $N = (3n^2 \pm n)/2$ and n is odd, then the number of partitions of N into an odd number of distinct summands exceeds by 1 the number of partitions of N into an even number of distinct summands. This completes the proof.

V

Combinatorics on a Chessboard

Walking in a Town

Figure 19 shows the plan of a town (such as, for example, the Australian capital Canberra). *The town consists of* $n \times k$ *rectangular blocks separated by* $n - 1$ *"horizontal" streets and* $k - 1$ *"vertical" streets. A traveler wishes to get from A to B along the shortest route. In how many ways can he do so?*

Fig. 19

It is clear that regardless of his choice of route, the traveler must invariably go "from left to right" or "up" and so must pass through $k + n$ intersections (including A but excluding B). Consider a route followed by our traveler. We mark an intersection belonging to the route with a 1 or a 0 according as the segment of the route following the intersection is vertical or horizontal. In this

way, each route determines a unique sequence of k zeros and n ones, and, conversely, each sequence of k zeros and n ones determines a unique optimal route from A to B; the route shown in Fig. 19 corresponds to the sequence 0110001100. Since the number of sequences of k zeros and n ones is

$$P(k, n) = C_{n+k}^n = \frac{(n + k)!}{n!\, k!} \,, \tag{1}$$

it follows that this is also the required number of routes.

The Arithmetical Square

Our traveler moves in the manner of a rook on a chessboard. Now consider an infinite chessboard bounded by two mutually perpendicular rays pointing down and to the right. Starting at the vertical edge of the board, we number the columns 0, 1, 2,.... . Starting at the horizontal edge of the board we number the rows 0, 1, 2,.... . The square at the intersection of the kth column and nth row is assigned the coordinates (k, n). In view of the outcome of the previous problem, we see that if a rook located originally at the square $(0, 0)$ moves only down and to the right, then it can get to the square (k, n) in C_{n+k}^k ways (the rook must make

k moves to the right and n moves down). If we write the number C_{n+k}^k in the square (k, n), then we obtain the following table:

TABLE 3

1	1	1	1	1	1	...
1	2	3	4	5	6	...
1	3	6	10	15	21	...
1	4	10	20	35	56	...
1	5	15	35	70	126	...
1	6	21	56	126	252	...
⋮	⋮	⋮	⋮	⋮	⋮	...

This table is called the *arithmetical square*. Careful study of this table shows that *any number* (greater than 1) *in the arithmetical square is the sum of the number above it and the number to the left of it.* For example, 4 is above, and 6 is to the left of 10 = 4 + 6.

This rule follows readily from the equality $C_n^k = C_{n-1}^k + C_{n-1}^{k-1}$ proved earlier (see p. 35). However, we can also prove it directly using the following argument: The rook can reach the square (k, n) from the square $(k - 1, n)$ or from the square $(k, n - 1)$. By the rule of sum, it follows that the number of ways of reaching the square (k, n) is equal to the sum of the number of ways of reaching the square $(k - 1, n)$ and the number of ways of reaching the square $(k, n - 1)$, which is what we wished to prove.

The relation $C_{n+k}^k = C_{n+k}^n$ shows that the arithmetical square is symmetric with respect to the diagonal passing through the corner (this diagonal is referred to in the sequel as the *principal diagonal*). To see this "geometrically," observe that the number of ways of reaching the intersection of the nth column and the kth row is the same as the number of ways of reaching the intersection of the kth column and the nth row.

Figurate Numbers

In the previous section we expressed each entry in Table 3 as a sum of an entry in the preceding row and an entry in the preceding column. Each entry in this table can also be expressed as the sum of entries in the preceding row alone. In fact, the equality

$$C_{n+k}^k = C_{n+k-1}^k + C_{n+k-2}^{k-1} + \cdots + C_{n-1}^0$$

(Formula 15, p. 36) shows that every entry in our table is the sum of the entries in the previous row beginning with the first and ending with the one above the entry in question. This shows how to compute the entries in our table row by row.

This method of computing the entries in Table 3 is connected with the study of figurate numbers, which goes back to Pythagoras and Nicomachus. If we represent the numbers 1, 2, 3,... as rows of one, two, three, and so on, dots, and group the rows into triangles, then (see Fig. 20) the number of dots in each

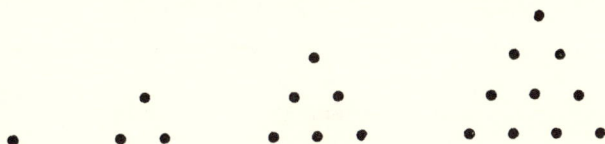

FIG. 20

triangle is equal to the corresponding entry in the second* row of Table 3. This explains why the numbers 1, 3, 6, 10, 15, 21,..., are called *triangular numbers*. The kth triangular number is

$$C^2_{k+1} = \frac{(k+1)k}{2}.$$

The triangles in Fig. 20 can be stacked into pyramids. The number of points in each pyramid is equal to the corresponding number in the third row of Table 3. This explains why the numbers 1, 4, 10, 20, 35,..., are called *pyramidal numbers*. The kth pyramidal number is

$$C^3_{k+2} = \frac{(k+2)(k+1)k}{1 \cdot 2 \cdot 3}.$$

In order to continue with this interpretation of the numbers in Table 3, it is necessary to consider pyramids in spaces of more than three dimensions.

The study of figurate numbers attracted the attention of mathematicians for centuries and was at one time an important part of the theory of numbers.

Arithmetical Triangle

Now we consider a chessboard in the form of a half plane and place a checker in square A of the zeroth row (Fig. 21). Moving in accordance with the rules of

---

* We remind the reader that the rows in Table 3 are numbered 0, 1, 2,... .

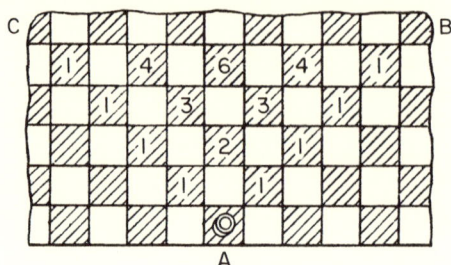

FIG. 21

the game (that is, the game of checkers) the checker can reach each of the (shaded) squares in the region bounded by the lines *AB* and *AC* (Fig. 21). If we write in each square the number of paths by which the square in question can be reached by the checker, then we see that, except for position, these numbers are the same as the numbers in the arithmetical square. And small wonder: If we rotate the board through 45°, then our checker moves along vertical and horizontal lines, that is, in the manner of a rook. It is customary to represent the numbers in Fig. 21 in the form of a triangle (Table 4). Then every number

TABLE 4

```
          1
        1   1
      1   2   1
    1   3   3   1
  1   4   6   4   1
. . . . . . . . . . . .
```

(greater than 1) is the sum of the two numbers which are in the row above this number and to either side of it. This triangle is frequently called the *Pascal triangle*. However, Tartaglia* (1500–1557), who preceded Pascal (1623–1662), was familiar with it. Also, long before Tartaglia, this triangle turned up in the works of the Arab mathematicians Nasir Eddin and Omar Khayyam. In view of these facts, it is perhaps best to call this triangle the *arithmetical triangle*.

---

* Tartaglia was a remarkable mathematician. In addition to discovering the arithmetical triangle, he also discovered the formula for the solution of cubic equations. Tartaglia told the formula to another Italian mathematician, D. Cardano, who swore to keep it secret but promptly published it in his textbook of algebra with the result that the formula for the solution of cubic equations is called "Cardano's formula."

The arithmetical triangle can be written in the following form:

TABLE 5

1	0	0	0	0	0	$\cdots$
1	1	0	0	0	0	$\cdots$
1	2	1	0	0	0	$\cdots$
1	3	3	1	0	0	$\cdots$
1	4	6	4	1	0	$\cdots$
1	5	10	10	5	1	$\cdots$
.	.	.	.	.	.	.

Here at the intersection of the kth column and the nth row we have the number C_n^k (we remind the reader that the number of the top row and of the leftmost column is zero). Each number in Table 5 (not in the top row or leftmost column) is the sum of the number directly above it and the number directly above and to the left of it. For example, in the case of the number 4 in the fourth row the number directly above it is 1 and the number directly above and to the left of it is 3, and $4 = 1 + 3$. We note that *all the entries in Table 5 above the principal diagonal are zeros, and all the entries in the zero column are ones.* Also, the entries in the nth row, that is, the numbers C_n^k for fixed n, are the coefficients in the expansion of $(1 + x)^n$ in powers of x. Since $1 + x$ is a binomial, the numbers C_n^k for fixed n are called *binomial coefficients*. We will come back to this issue in Chap. VII.

The Extended Arithmetical Triangle

The arithmetical triangle takes up only part of the plane. We now extend it to the whole plane in such a way that, just as before, *every element is equal to the sum of the element directly above it and the element directly above and to the left of it.* Also, since the zeroth column in the arithmetical triangle consists of ones, it is natural to fill the zeroth column in the extended arithmetical triangle with ones.

Application of our "rule of formation" to the zeroth column shows that the column directly to the left of it must consist of zeros. But then all the columns to the left of the zeroth column must consist of zeros. Now consider the rows above the zeroth row. The first element of the zeroth row is 0. Since the number directly above and to the left of it is 1, we must write directly above it the number -1 $(1 + (-1) = 0)$. But then to get 0 in the second position in the zeroth row, we must write 1 directly above it. Continuing in this way, we see that the row directly above the zeroth row consists of an alternating sequence of 1's and -1's. By now it is clear how to fill each successive new row.

The result of our extension work is the following table:

TABLE 6

··· 0	1	−5	15	−35	70	−126	···
··· 0	1	−4	10	−20	35	−56	···
··· 0	1	−3	6	−10	15	−21	···
··· 0	1	−2	3	−4	5	−6	···
··· 0	1	−1	1	−1	1	−1	···
··· 0	1	0	0	0	0	0	···
··· 0	1	1	0	0	0	0	···
··· 0	1	2	1	0	0	0	···
··· 0	1	3	3	1	0	0	···
··· 0	1	4	6	4	1	0	···
··· 0	1	5	10	10	5	1	···

A look at the (nonzero) part of our table above the zeroth row shows that, apart from sign, its elements coincide with the appropriate elements of the arithmetical square on p. 92. Specifically, the element of the extended arithmetical triangle at the intersection of the kth column and $(-n)$th row is $(-1)^k C^k_{n+k-1}$. That this is true for all values of k and $-n$ follows from the equality

$$(-1)^k C^k_{n+k-1} + (-1)^{k-1} C^{k-1}_{n+k-2} = (-1)^k [C^k_{n+k-1} - C^{k-1}_{n+k-2}] = (-1)^k C^k_{n+k-2}$$

(see Formula (11) on p. 35). This equality shows that in the table composed of the numbers $(-1)^k C^k_{n+k-1}$, the kth element of the $(-n+1)$th row is the sum of the elements of the $(-n)$th row occupying the positions k and $k-1$. This means that the law of formation of the table composed of the numbers $(-1)^k C^k_{n+k-1}$ is the same as the law of formation of the extended arithmetical triangle. Moreover both tables have the same zeroth column and (-1)th row. It follows that the two tables coincide.

In the original arithmetical triangle (Table 5), the number at the intersection of the kth column and nth row is C^k_n. In the extended arithmetical triangle, the number at the intersection of the kth column and $(-n)$th row is $(-1)^k C^k_{n+k-1}$. This suggests that we define the symbol C^k_n for negative subscripts by means of the equation

$$C^k_{-n} = (-1)^k C^k_{n+k-1}.$$ (2)

Table 6 shows that the extension of C^k_n to negative values of k is trivial: $C^k_n = 0$ for $k < 0$ (see also p. 158). Furthermore, $C^k_n = 0$ for $0 \leqslant n < k$.

Chess King

Consider a chess king placed in the upper left-hand corner of an infinite chessboard and allowed to move (one square at a time) either forward or forward and to the right. If we write in each square the number of ways in which this "one-sided chess king" can reach the square in question, then we obtain the arithmetical triangle.

Now we replace the "one-sided king" with the usual chess king but keep the restriction that the king must invariably move forward, that is to the next new row. To enable the king to take advantage of his new possibilities, we place him on a board bounded by a single straight line. Figure 22 shows a board of this type. The numbers in the squares indicate in how many ways the king can reach the square in question if he starts from the square marked with a crown.

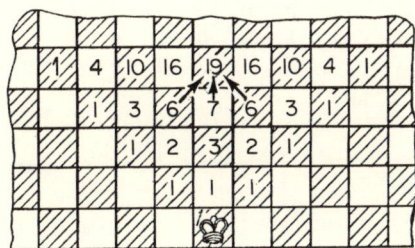

Fig. 22

To see how this table is put together, we assume that we know in how many ways the royal wanderer can reach each of the squares in the $(n-1)$th row and try to compute the number of ways in which he can reach various squares in the nth row. It is clear that the king can reach a square in the nth row from the square directly below it, from the square directly below and to the left of it, and from the square directly below and to the right of it (see Fig. 22). By the rule of sum it follows that:

The number of ways in which the chess king can reach a square in the nth row is equal to the sum of the number of ways in which he can reach the three nearest squares in the $(n-1)$th row; here we assume that the chess king can reach the initial square in just one way (by just staying put) and that he cannot reach the remaining squares in the zeroth row.

The Generalized Arithmetical Triangle

If we move the numbers in the triangle in Fig. 22 to the left so that the left diagonal of 1's assumes a vertical position, and flip the resulting table about the horizontal boundary line, then we obtain Table 7 (apart from the two columns of zeros on the left). The law of formation for this table can be stated as follows:

Each number is the sum of three numbers in the preceding row, namely, the number directly above it and its two immediate neighbors to the left.

For example, the number 16 in the fourth row is the sum of the numbers 3, 6, and 7 in the third row.

It is clear how to generalize the arithmetical triangle. In place of the number 3, we take any natural number m. We fill the zeroth row with a one followed by zeros. In filling the first row we require that any number in that row be the sum of m numbers in the zeroth row, namely, the number just above it and the $m - 1$ immediate left neighbors of that number (with zeros making up any

TABLE 7

0	0	1	0	0	0	0	0	0	0	0
0	0	1	1	1	0	0	0	0	0	0
0	0	1	2	3	2	1	0	0	0	0
0	0	1	3	6	7	6	3	1	0	0
0	0	1	4	10	16	19	16	10	4	1

shortage). The remaining rows are filled in an analogous manner. The result is a table in which each number is the sum of m numbers in the preceding row, namely, the number directly above it and the immediate $m - 1$ left neighbors of that number. For $m = 2$, we obtain the arithmetical triangle, and for $m = 3$, we obtain Table 7.

To distinguish between the arithmetical triangles coresponding to different values of m, we speak of *m-arithmetical triangles*. We denote the number in an m-arithmetical triangle located at the intersection of the kth column and nth row by $C_m(k, n)$. The definition of an m-arithmetical triangle implies that the numbers $C_m(k, n)$ satisfy the relation

$$C_m(k, n) = C_m(k, n-1) + C_m(k-1, n-1) + \cdots + C_m(k-m+1, n-1). \qquad (3)$$

Also, the following relations hold:

$$C_m(k, 1) = \begin{cases} 1, & \text{if } 0 \leqslant k \leqslant m-1, \\ 0, & \text{if } k \geqslant m. \end{cases}$$

Generalized Arithmetical Triangles and Numbers to the Base m

There is a connection between the numbers $C_m(k, n)$ and numbers to the base m. Specifically, *there are $C_m(k, n)$ n-digit numbers to the base m with digit sum k.* (In counting

n-digit numbers we must include numbers consisting of n digits and starting with one or more zeros; for example, 001,215 is to be regarded as a 6-digit number with digit sum 9.)

We denote by $B_m(k, n)$ the number of n-digit numbers to the base m with digit sum k and prove that the numbers $B_m(k, n)$ satisfy the same relation which is satisfied by the numbers $C_m(k, n)$, namely relation (3). The last digit of a number to the base m is one of the digits $0, 1,..., m - 1$. It follows that the possible values of the digit sum of an $(n - 1)$-digit number obtained by erasing the last digit of an n-digit number are $k, k - 1,..., k - m + 1$. But then, in view of the rule of sum, we have

$$B_m(k, n) = B_m(k, n - 1) + \cdots + B_m(k - m + 1, n - 1). \tag{4}$$

Also, it is clear that $B_m(k, 1)$ is 1 if $0 \leqslant k \leqslant m - 1$ and 0 otherwise (if $0 \leqslant k \leqslant m - 1$, then there is just one 1-digit number to the base m with digit sum k, and for $k \geqslant m$ there are no 1-digit numbers to the base m with digit sum k). This means that the top row of the table of the $B_m(k, n)$ coincides with the top row of the table of the $C_m(k, n)$. Also, the law of formation (3) governing the $C_m(k, n)$ is the same as the law of formation (4) governing the $B_m(k, n)$. It follows that $B_m(k, n) = C_m(k, n)$ for all k and n.

Some Properties of the Numbers $C_m(k, n)$

We shall now state and prove a few properties of the numbers $C_m(k, n)$. Some of these properties reflect certain properties of the numbers C_k^n, and some establish a direct connection between these two classes of numbers. The existence of such connections is not surprising if we remember that the structure of the arithmetical triangle implies the relation $C_2(k, n) = C_k^n$.

The numbers $C_m(k, n)$ are different from zero only if $0 \leqslant k \leqslant n(m - 1)$; this follows from the fact that each row of nonzero entries in an m-arithmetical triangle contains $m - 1$ more entries than its predecessor.

The numbers $C_m(k, n)$ have the following symmetry property:

$$C_m(k, n) = C_m(n(m - 1) - k, n). \tag{5}$$

To prove (5), we associate with each n-digit number to the base m its "complement" obtained by replacing each digit in the original number with the difference between $m - 1$ and the digit in question; for example, if $m = 7$, then the complement of 3,140,216 is 3,526,450. It is clear that if the digit sum of the original number is k, then the digit sum of the complement is $n(m - 1) - k$. It follows that there are as many n-digit numbers with digit sum k as there are n-digit numbers with digit sum $n(m - 1) - k$, which is what is claimed in (5).

Since there are m^n n-digit numbers to the base m (see p. 3), it follows that

$$C_m(0, n) + C_m(1, n) + \cdots + C_m(n(m - 1), n) = m^n. \tag{6}$$

Now we prove that

$$C_m(0, l) C_m(k, n - l) + C_m(1, l) C_m(k - 1, n - l) + \cdots + C_m(k, l) C_m(0, n - l) = C_m(k, n), \tag{7}$$

where $0 \leqslant l \leqslant n$. We separate all n-digit numbers with digit sum k into classes. The numbers in the sth class are the numbers for which the sum of the first l digits is s. The sum of the last $n - l$ digits of a number in the sth class is $k - s$. By the rule of product, the sth class contains $C_m(s, l) \, C_m(k - s, n - l)$ numbers. Since there are $C_m(k, n)$ n-digit numbers with digit sum k, relation (7) follows.

For $l = 1$, relation (7) reduces to relation (3); this is so because $C_m(k, 1) = 1$ for $0 \leqslant k \leqslant m - 1$ and $C_m(k, 1) = 0$ for $k \geqslant m$.

Finally we prove the relation

$$C_n^0 C_{m-1}(k - n, n) + C_n^1 C_{m-1}(k - n + 1, n - 1) +$$

$$\cdots + C_n^s C_{m-1}(k - n + s, n - s) + \cdots + C_n^n C_{m-1}(k, 0) = C_m(k, n). \qquad (8)$$

We separate all n-digit numbers to the base m with digit sum k into classes. We put in the sth class, $0 \leqslant s \leqslant n$, all numbers with exactly s zeros.

Now we compute the number of elements in the sth class. Each number in this class can be selected in two stages. First we select the positions of the s zeros. Since each number has n digits, this can be done in C_n^s ways. Then we ignore the s zeros in each of our numbers and decrease each of the remaining digits by 1. We end up with an $(n - s)$-digit number with digits belonging to the sequence $0, 1, \ldots, m - 2$, and digit sum $k - (n - s) = k - n + s$. This class of numbers coincides with the class of $(n - s)$-digit numbers to the base $m - 1$ and digit sum $k - n + s$. There are $C_{m-1}(k - n + s, n - s)$ such numbers. By the rule of product, we see that there are $C_n^s \, C_{m-1}(k - n + s, n - s)$ numbers in the sth class. Since there are $C_m(k, n)$ n-digit numbers to the base m with digit sum k, the rule of sum implies the validity of (8).

Starting with the relation $C_2(k, n) = C_k^n$ and making repeated use of (8), we can express $C_m(k, n)$ in terms of the binomial coefficients. For example, using (8) once, we obtain the relation

$$C_3(k, n) = C_n^0 C_{k-n}^n + C_n^1 C_{k-n+1}^{n-1} + \cdots + C_n^n C_k^0 .$$

Checker in a Corner

We place a checker in the corner of an infinite checkerboard bounded by two mutually perpendicular rays (Fig. 23).* In each square we write the number of ways in which the checker can reach the square in question. The result is different from that obtained on p. 94, where the board was bounded by a single line. The difference is due to the fact that now the checker cannot cross the vertical boundary. In general, the existence of the vertical boundary reduces the number of ways in which the checker can reach a particular square. For example, in the present case each square adjoining what is now the vertical boundary line can be reached from only one neighboring square, whereas in the case discussed on p. 94 each of these squares could be reached from two neighboring squares. In the case discussed on p. 94, we saw that the number appearing in a black square was the sum of the two numbers appearing in the neighboring black

---

* The extra column on the left of the board in Fig. 23 is used in the sequel.

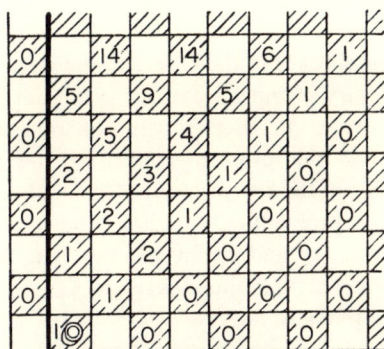

FIG. 23

squares in the row just below the black square in question. For this rule to hold in the present case, we must add a column to the left of the vertical boundary and write a zero in each of its black squares (this agrees with the fact that the extra squares cannot be reached by our checker).

We wish to compute the number of ways in which the checker can reach a particular square in Fig. 23. Let a zero denote a move to the left and a one a move to the right. Then each path is denoted by a sequence of zeros and ones. Also, the number of zeros and ones in a sequence is uniquely determined by the terminal square of the path associated with the sequence. For example, any path with 4 zeros and 6 ones leads to the square at the intersection of the second column and tenth row (when numbering the rows and columns we start, as before, with the number 0).

It is clear that only certain sequences of zeros and ones are admissible as descriptions of paths which can be reached by our checker. For example, no sequence can begin with a zero, since this would immediately force the checker off the board. The admissible sequences can be characterized as sequences in which the number of ones preceding any entry is at least as large as the number of zeros preceding this entry. Another way of saying the same thing is that at any move of a given path the number of moves to the right preceding this move is at least as large as the number of moves to the left preceding this move; violation of this condition would force our checker off the board.

We see that our problem reduces to the problem of computing the number of sequences of k zeros and m ones satisfying the following condition: the number of ones preceding any entry is at least as large as the number of zeros preceding this entry. This problem was solved on p. 59 (there in place of zeros and ones we used the letters d and h), where it was found that the number of admissible sequences was $[(m - k + 1)/(m + 1)]C_{m+k}^k$. This then is the number which belongs in the square at the intersection of the $(m - k)$th column and $(m + k)$th row.

Rather than place our checker in the corner we now place it in the qth square of the zeroth row (contrary to the rules of the game this may turn out to be a white square). This provides the checker with a margin of q moves to the left. This case corresponds to the problem considered on p. 60 where, at the beginning of the sale of tickets, the cashier had a supply of q half-dollar coins. The solution of the latter problem tells us that, if our checker reaches a certain square after k moves to the left and m moves to the right, $0 \leqslant k \leqslant m + q$, then the number of ways of reaching this square is $C_{m+k}^k - C_{m+k}^{k-q+1}.$* Figure 24 shows the table corresponding to the case $q = 3$.

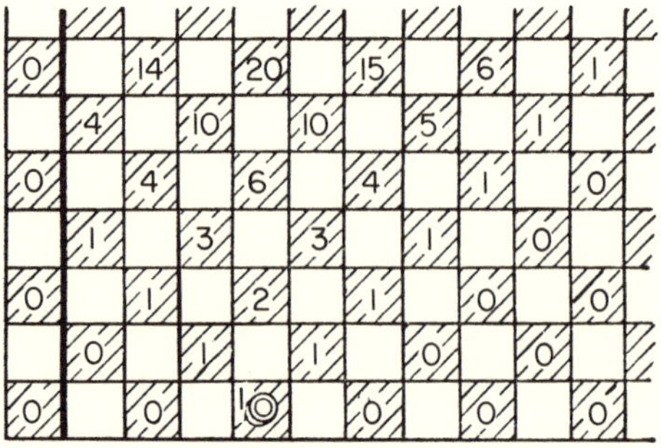

FIG. 24

The Arithmetical Pentagon

If we rotate the board in Fig. 23 through a 45° angle, then the checker moves along vertical and horizontal lines, and the boundary forms a 45° angle with these lines. This shows that the problem of the checker is equivalent to the following problem:

A rook is placed at the corner of a chessboard. In how many ways can the rook reach the square (k, m) if it moves along the shortest possible paths and does not cross the diagonal of the board (but is allowed to reach it)?

Let the rook be initially in the upper left-hand corner of the board. If (k, m) denotes the square at the intersection of the kth column and mth row ($k, m = 0, 1, 2,...$), then the results of the previous section show that, for $k \leqslant m$, the required number of ways is $[(m - k + 1)/(m + 1)]C_{m+k}^k$, and, for

---

* This number belongs in the square at the intersection of the $(m - k + q)$th column and $(m + k)$th row (Translators).

$k > m$, this number is zero. If the diagonal is moved q squares to the right, then for $0 \leqslant k \leqslant m + q$ the required number of ways is $C_{m+k}^{k} - C_{m+k}^{k-q-1}$, and for $k > m + q$ it is zero.

With the diagonal moved to the right and the board finite, the nonzero numbers in the corresponding table fill a pentagon (Fig. 25). This pentagon is

1	1	1	1	0	0	0
1	2	3	4	4	0	0
1	3	6	10	14	14	0
1	4	10	20	34	48	48

FIG. 25

called an *arithmetical pentagon*. By abuse of language, this name is also applied to the table associated with an infinite chess board bounded by two mutually perpendicular rays.

The basic property of an arithmetical pentagon is the same as the basic property of the arithmetical square (Table 3, p. 92). In both cases an element is the sum of the element directly above it and the element directly to the left of it. The difference between the two tables consists in the fact that the diagonals of the arithmetical pentagon above the bounding diagonal consist of zeros (in this respect the arithmetical pentagon resembles the arithmetical triangle, Table 5, p. 95).

Now we take a board bounded by two mutually perpendicular rays and introduce as extra boundaries two diagonals, one located q lines above the principal diagonal and the other located s lines below the principal diagonal. Then we write in each square the number of ways in which a rook located at the corner can reach the square in question. The resulting table is called an *arithmetical hexagon*. Figure 26 shows an arithmetical hexagon for which $q = 4, s = 3$.

1	1	1	1	1	0	0
1	2	3	4	5	5	0
1	3	6	10	15	20	20
1	4	10	20	35	55	75
0	4	14	34	69	124	199
0	0	14	48	117	241	340

FIG. 26

The arithmetical hexagon can also be thought of as arising in the following way: consider a chessboard bounded by a segment $q + s + 1$ squares long and by two rays perpendicular to the segment. A checker is placed q squares away from one corner and s squares away from the other corner. In each square reached by the checker we write the number of ways in which the checker can reach it. Rotation of this table through a 45° angle yields the arithmetical hexagon.

Geometric Proofs of Properties of Combinations

In Chap. II we proved certain properties of combinations. We now show how to prove these properties in a more intuitive way by using geometric arguments. First we give a geometric proof of the relation

$$C_n^0 + C_n^1 + C_n^2 + \cdots + C_n^n = 2^n. \tag{9}$$

To this end, we consider all paths leading from the point $A(0, 0)$ to the points $B_k(k, n - k)$, $0 \leqslant k \leqslant n$ (Fig. 27).

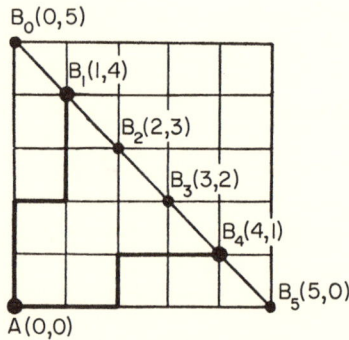

$B_0(0,5)$
$B_1(1,4)$
$B_2(2,3)$
$B_3(3,2)$
$B_4(4,1)$
$B_5(5,0)$
$A(0,0)$

Fig. 27

For each k, $0 \leqslant k \leqslant n$, the number of paths with end point B_k is C_n^k. Now we compute the number of all paths. Each path consists of n horizontal and vertical segments. If we mark each horizontal segment of a path with a zero and each vertical segment of that path with a one, we obtain a description of the path in terms of a sequence of n zeros and ones. Also, each n-sequence of zeros and ones determines just one of our paths. Since the number of n-sequences of zeros and ones is 2^n, relation (9) holds.

On p. 92 we gave geometric proofs of the relations

$$C_n^k = C_{n-1}^k + C_{n-1}^{k-1}, \quad \text{and} \quad C_n^k = C_n^{n-k}.$$

It is possible to give geometric proofs of more complex relations. Consider a vertical line with abscissa m, $0 \leqslant m \leqslant k$ (Fig. 28). Each path from $A(0, 0)$ to $B(k, n)$ has a point or a segment in common with this line. We separate the paths from A to B into classes and put into the sth class all paths for which the point $D_s(m, s)$ is the *last* point which they share with the line $x = m$.

Now we compute the number of paths joining A to B which belong to the sth class. A path in this class consists of a path from A to D_s, the segment joining $D_s(m, s)$ to $D'_s(m + 1, s)$ (D_s is the last point of the line $x = m$ which belongs to our path!), and a path from $D'_s(m + 1, s)$ to $B(k, n)$. By Formula (1), there are $P(m, s)$ paths from $A(0, 0)$ to $D_s(m, s)$. Similarly, there are $P(k - m - 1, n - s)$ paths from $D'_s(m + 1, s)$ to $B(k, n)$ (to get from D'_s to B we go $k - m - 1$ units to the right and $n - s$ units up). By the rule of product, the number of paths in the sth class is

$$P(m, s)P(k - m - 1, n - s).$$

The number of paths from A to B is $P(n, k)$. By the rule of sum, we obtain the equality

$$P(k, n) = P(m, 0)P(k - m - 1, n)$$
$$+ P(m, 1)P(k - m - 1, n - 1) + \cdots + P(m, n)P(k - m - 1, 0).$$

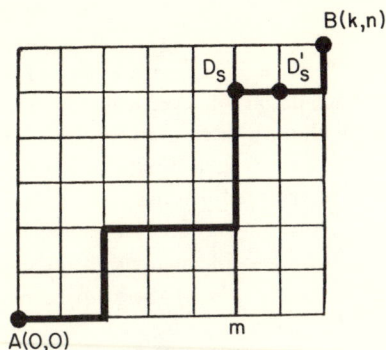

B(k,n)

D_s D'_s

A(0,0)   m

Fig. 28

This equality can be written as

$$C^k_{n+k} = C^m_m C^{k-m-1}_{n+k-m-1} + C^m_{m+1} C^{k-m-1}_{n+k-m-2} + \cdots + C^m_{m+n} C^{k-m-1}_{k-m-1} \tag{10}$$

(see Formula (24′), p. 38).

In particular, for $m = k - 1$, we obtain the relation

$$C^k_{n+k} = C^{k-1}_{k-1} + C^{k-1}_k + \cdots + C^{k-1}_{k+s-1} + \cdots + C^{k-1}_{k+n-1}. \tag{11}$$

We note that (10) and (11) can be obtained by repeated application of the relation $C_n^k = C_{n-1}^k + C_{n-1}^{k-1}$.

We leave it to the reader to give a geometric proof of relation (23) on p. 37 which we state in the following convenient form:

$$C_{n+k}^n = C_{n+k-s}^n C_s^0 + C_{n+k-s}^{n-1} C_s^1 + \cdots + C_{n+k-s}^{n-m} C_s^m + \cdots + C_{n+k-s}^{n-s} C_s^s,$$

$$\text{where} \quad 0 \leqslant s \leqslant k, \quad 0 \leqslant s \leqslant n. \quad (12)$$

Hint: Draw the line through the points $D(k - s, n)$ and $E(k, n - s)$ and separate the paths from A to B into classes of paths which go through the same point of that line.

By separating the paths from $A(0, 0)$ to $B(k, n)$ in different ways, it is possible to prove geometrically many more relations involving the numbers C_{n+k}^k.

Using multidimensional geometry, it is possible to give analogous proofs of relations involving the numbers $P(n_1, ..., n_k)$ (see (27) and (28) on p. 39). However, this exceeds the scope of our book.

It is also possible to give geometric proofs of the relations in the last section of Chap. III involving the numbers C_n^k. To this end, we consider the paths on a chessboard which do not cross a specified line parallel to the principal diagonal (but which may nevertheless have points in common with that line). Appropriate ways of separating these paths into classes lead to the required results.

The problem of queues in front of the ticket office also admits of a very simple geometric interpretation. If we associate with each half-dollar coin a horizontal (unit) segment and with each dollar bill a vertical segment, then the queues which file past the ticket office without a hitch are represented by paths which do not cross the principal diagonal. The transformations involved in our discussion of this problem can also be interpreted geometrically; placing a man with a half-dollar coin at the head of the queue and changing half-dollar coins to dollar bills and dollar bills to half-dollar coins in a part of the queue amounts to adding a horizontal segment at the beginning of the path corresponding to the queue and reflecting a part of the path in a diagonal parallel to and closest to the principal diagonal. We leave it to the reader to translate the arguments used in the solution of this problem into geometric language.

Random Walk

There is a close connection between the above problems dealing with motions of chess pieces and random walk problems which are of importance in physics. We consider the following problem (given in 1945 to participants in the VIII Moscow Mathematical Olympiad).

Consider the network of paths in Fig. 29. 2^N people depart from the point A. Half of the people go in the direction l and the other half go in the direction m. When either group reaches the first intersection half of its members go in the direction l and

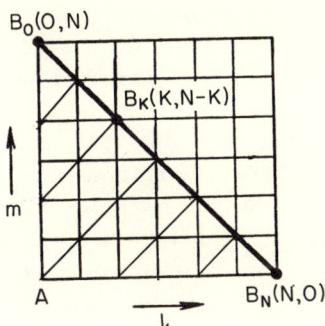

Fig. 29

the other half in the direction m. This process is repeated whenever a group reaches an intersection. Each person walks N segments. We wish to find the intersections reached by members of the original group and to determine the number of people who reach each of these intersections.

Since each person walks N segments, it follows that each person arrives at one of the points B_k with coordinates $(k, N - k)$, where k takes on the values $0, 1,..., N$. All of these points are located on the line joining the points $B_0(0, N)$ and $B_N(N, 0)$ (Fig. 29).

It remains to find out how many people arrive at the point $B_k(k, N - k)$. We code each path from $A(0, 0)$ to $B_k(k, N - k)$, $k = 0, 1,..., N$, by means of a sequence of zeros and ones. The result is the set of all N-sequences of zeros and ones. We know that there are 2^N such sequences. Since the number of people who departed from A was also 2^N, it follows that each path was traversed by exactly one person. But this means that the number of people who arrived at $B_k(k, N - k)$ is the same as the number of shortest paths joining A to $B_k(k, N - k)$. The number of such paths was computed in the beginning of this chapter and is equal to

$$P(k, N - k) = C_N^k = \frac{N!}{k! \, (N - k)!} \, .$$

To sum up: $C_N^k = N!/[k! \, (N - k)!]$ people arrived at $B_k(k, N - k)$. This number is the kth number of the Nth row in the arithmetical triangle.

Brownian Motion

The problem just solved is essentially equivalent to the following problem:
2^N *people depart from the origin 0 of a coordinatized line. Half of them go to the right and half go to the left. Such division takes places every hour. Find the number of people who arrive at various points after N hours.*

We assume that in one hour each person walks $\frac{1}{2}$ unit of length. Reasoning as before, we find that after N hours $C_N^k = N!/[k! \, (N - k)!]$ of the walkers will arrive at $B_k(k - N/2)$, $k = 0, 1,..., N$.

It is not likely that groups of people walk in the manner just described (by the way, in the folk version of our problem there is a pub at 0). However, in some physical problems this type of rambling arises in a natural way. Specifically, such rambling is the simplest model of what is known as Brownian motion of particles, which is the result of particles colliding with molecules.

Consider particles which are restricted to motion on a line. Since collisions with molecules are of a random kind, we can assume as a first approximation that in a unit of time, half of the particles move $\frac{1}{2}$ unit of length to the right and half move $\frac{1}{2}$ unit of length to the left (in reality the process is far more complicated and other displacements are possible). Therefore, if there are initially 2^N particles at 0, then they move roughly like the crowd in the preceding problem. Physicists apply the term *diffusion* to this type of dispersion of particles. In view of our solution of the random walk problem, we can say that after N units of time the number of particles at the point $B_k(k - N/2)$ is $C_N^k = N!/[k! \, (N - k)!]$.

As noted before, the numbers C_N^k are the elements of the Nth row of the arithmetical triangle. If the nature of the diffusion process is different, then we obtain instead the elements of the Nth row of the m-arithmetical triangle. Specifically, let the initial number of particles at the origin 0 of a coordinatized line be m^N. Instead of dividing the set of particles into halves, we divide them into m equal parts and place them on the line symmetrically with respect to 0, so that neighboring parts of the set are a unit of length apart. Then each of the m parts undergoes an analogous process of division (except that the center of symmetry is no longer 0 but the point at which the part in question is located). After N steps, the particles are at the points B_k with coordinates $k - (m - 1)N/2$, $k = 0, 1,..., (m - 1)N$, and the number of particles at B_k is $C_m(N, k)$.

For large values of N, it is rather difficult to compute the number of particles at each point. But, as often happens in mathematics, the more complex the law governing the distribution, the closer it gets to a simple limiting law. The larger the number of particles, and hence the greater the complexity of the precise law governing their distribution, the more accurate the description of the distribution furnished by the limiting law.

It is proved in the theory of probability that if N is large and a is small relative to N, then the segment $[x - a/2, x + a/2]$ contains approximately

$$\frac{12am^N}{\sqrt{2\pi\, N(m^2 - 1)}} \exp \left[-\frac{72x^2}{N^2(m^2 - 1)^2} \right]^*$$

particles. This statement is to be interpreted as follows: Consider the graph of a step function whose ordinate at the point $B_k(k - (m - 1)N/2)$ is $C_m(N, k)$. If we divide the abscissas of the points on the graph by the factor $N(m^2 - 1)/12$ and the ordinates by the factor $12am^N/[N(m^2 - 1)]$, then for large N we obtain the graph of a step function which is very close to the graph of the function

$$y = \frac{1}{\sqrt{2\pi}} e^{-x^2/2}.$$

The latter function was introduced into the theory of probability by the great German mathematician K. Gauss and is known as the *Gauss function*. This function plays an important role not only in problems of gas diffusion but also in the theory of heat conduction, in error theory, and so on.

In the Realm of the Tsarina of Shemakhan

We return to the crowd wandering on the directed line with origin at 0, except that now we assume that to the left of 0 there extends ... the realm of the tsarina of Shemakhan who played such a deplorable role in the affairs of the unlucky tsar Dodon and his sons.** The reader will undoubtedly remember that those who entered the tsarina's realm were never heard from again. We, too, assume that anyone who gets to the left of 0 stays there. Our problem is to determine the whereabouts of the people in the crowd N hours after the crowd left the point 0. Specifically, we wish to know how many people ended up in the realm of the tsarina of Shemakhan and how many people reached various other locations.

It turns out that this problem reduces to the problem of the queue in front of the box office. Indeed, consider a person leaving 0. This person moves $\frac{1}{2}$ unit of length per hour. If within the following N hours he does not end up in the realm of the tsarina of Shemakhan, then having made k moves to the right, he gets to the point $B_k(k - N/2)$. If we associate with a move to the right a person with a half-dollar coin and with a move to the left a person with a dollar bill, then we see that a walk which lands the walker in the realm of the tsarina of Shemakhan corresponds to a "bad" queue at the box office (that is a

---

* exp x stands for e^x.
** The reference is to a Russian folktale (Translators).

queue which is troubled by a lack of change), and that a walk which ends at $B_k(k - N/2)$ corresponds to a "good" queue at the box office in which k people have half-dollar coins and $N - k$ people have dollar bills. For a queue to be good it is necessary that $k \geqslant N - k$. The number of good queues is (see p. 59)

$$A(N - k, k) = C_N^{N-k} - C_N^{N-k-1} = \frac{N! \, (2k - N + 1)}{(N - k)! \, (k + 1)!} .$$

If follows that of the 2^N people leaving the point 0, $C_N^{N-k} - C_N^{N-k-1}$ people arrive at the point $B_k(k - N/2)$ for which $2k \geqslant N$.

To compute the number of people who end up in the realm of the tsarina of Shemakhan, we note that the number of lucky walkers is equal to the sum of all numbers of the form $C_N^{N-k} - C_N^{N-k-1}$ for integral k in the interval $N/2 \leqslant k \leqslant N$. Denote by $E(x)$ the integral part of x. Then k takes on the values $k = E[(N + 1)/2],..., N$, and the sum in question is

$$[C_N^{N-E[(N+1)/2]} - C_N^{N-E[(N+1)/2]-1}] + \cdots + [C_N^0 - C_N^{-1}] = C_N^{N-E[(N+1)/2]}.$$

Since the number of walkers is 2^N, it follows that the number of people who end up in the realm of the tsarina of Shemakhan is $2^N - C_N^{N-E[(N+1)/2]}$.

If the realm of Shemakhan stretched to the left of the point 0_1 with abscissa $-q/2$, then, in view of the solution of the problem on p. 60, $C_N^{N-k} - C_N^{N-k-q-1}$ people would arrive at all the points B_k, $N - q/2 \leqslant k \leqslant N$, and the remainder would end up in the realm of Shemakhan.

An Absorbing Wall

We already mentioned the fact that random walk problems are of great importance in physics because they yield the simplest models of diffusion of particles. In particular, the two cases of the problem about the tsarina of Shemakhan and her realm correspond to the physical problem of diffusion of particles on a line in the presence of an absorbing wall located at 0 or $q/2$ units away from 0.

At the time when the only applications of combinatorics and probability were to games of chance, the random walk problem with absorption was stated in terms of "games to the bitter end": Take two people playing, for example, the game of heads or tails. After each toss the loser pays the winner a dollar and the player who first becomes penniless quits the game. The mathematical problem was to compute the probabilities of various outcomes of the game if the players had initially p and q dollars, respectively. It is clear that this problem is related to the problem of diffusion of particles in a region bounded on two sides by absorbing walls.

Roaming on an Infinite Plane

So far we have considered motions of a rook where the rook was allowed to move forward or to the right only. We saw that this was equivalent to the consideration of what could be called roaming on an infinite line. Now we consider unrestricted motions of a rook on an infinite board. In other words, we propose to solve the following problem:

A rook is initially located at the square 0(0, 0) *of a chessboard supposed infinite in all directions. In how many ways can the rook reach the square* $A(p, q)$ *in N moves* (we assume that the rook moves a square at a time) ?

By symmetry, it suffices to consider the case $p \geqslant 0$, $q \geqslant 0$. Since each of the shortest paths from 0(0, 0) to $A(p, q)$ involves $p + q$ moves, we must have $N \geqslant p + q$. An N-move path differs from a shortest path by the existence of pairs of moves which cancel each other. Hence $N - p - q$ must be an even number. We put $N - p - q = 2k$.

Suppose the rook made s moves to the left. Then it must have made $p + s$ moves to the right. This leaves $N - p - 2s = q + 2(k - s)$ vertical moves. Of these, $k - s$ moves are "down" and $q + k - s$ moves are "up." Hence $0 \leqslant s \leqslant k$.

For every value of s satisfying this inequality, we obtain a number of paths consisting of s moves to the left, $p + s$ moves to the right, $k - s$ moves down, and $q + k - s$ moves up. These moves can be performed in any order. This implies that the number of paths determined by the various ways of combining the moves is $P(s, p + s, k - s, q + k - s)$. But then the totality T of N-move paths joining 0(0, 0) to $A(p, q)$ is

$$T = \sum_{s=0}^{k} P(s, p + s, k - s, q + k - s) = \sum_{s=0}^{k} \frac{(p + q + 2k)!}{s! \, (p + s)! \, (k - s)! \, (q + k - s)!} \cdot$$

To obtain a more convenient form for this expression, we note that

$$C_{p+q+2k}^{p+k} = \frac{(p + q + 2k)!}{(p + k)! \, (q + k)!}, \qquad C_{p+k}^{k-s} = \frac{(p + k)!}{(k - s)! \, (p + s)!},$$

$$C_{q+k}^{s} = \frac{(q + k)!}{s! \, (q + k - s)!},$$

and so

$$T = C_{p+q+2k}^{p+k} \sum_{s=0}^{k} C_{p+k}^{k-s} C_{q+k}^{s} \cdot$$

The sum on the right-hand side consists of products of the form $C_r^l C_n^m$, with fixed lower indices, and upper indices with index sum k. Applying Formula (23) on p. 37, we obtain for T the expression

$$T = C_{p+q+2k}^{p+k} C_{p+q+2k}^{k} ,$$

or, since $p + q + 2k = N$,

$$T = C_N^{p+k} C_N^k .$$

The General Rook Problem

Now we study a different class of combinatorial problems on a chessboard. A basic component of these problems is the computation of the number of ways of placing two pieces on a chessboard so that they can attack each other (or, the analogous computation of the number of ways of placing the two pieces so that they can *not* attack each other).

We discussed one such problem on p. 20, where we computed the number of ways of placing 8 rooks on an ordinary chessboard so that no two could attack one another. We now solve the more general problem of computing the number of ways of placing k rooks on an $m \times n$ chessboard (a chessboard with m rows and n columns) so that no two rooks can attack one another.

It is clear that if $k > m$ or $k > n$, then there will be two rooks in one row or in one column and the problem will have no solution. Suppose now that $k \leqslant m$ and $k \leqslant n$. We place our rooks in two stages. First we choose k of the m rows to be occupied by the k rooks. This can be done in C_m^k ways. Then we choose k of the n rows to be occupied by the k rooks. This can be done in C_n^k ways. Since the choice of columns is independent of the choice of rows, it follows, by the rule of product, that the k rows and k columns occupied by the rooks can be selected in $C_m^k C_n^k$ ways. Observe that for each selection of k rows and k columns we have k^2 squares in which the rows and columns intersect. Our k rooks are to be placed on k of these k^2 squares so that no two rooks can attack each other. Since the k^2 squares can be thought of as forming a $k \times k$ chessboard, we conclude (see p. 20) that this step can be carried out in $k!$ ways.

All in all, our problem admits

$$C_m^k C_n^k k! = \frac{n!\, m!}{k!\, (n-k)!\, (m-k)!} \tag{13}$$

solutions.

For example, 3 rooks can be placed in the required manner on an ordinary chessboard in

$$\frac{8!\, 8!}{3!\, 5!\, 5!} = 17{,}696$$

ways. Again, n rooks can be placed in the required manner on an $n \times n$ board in $n!$ ways (see p. 20).

If we remove the restriction that no two rooks can attack each other, then our problem reduces to the problem of selecting k out of mn squares. This can be done in

$$C_{mn}^k = \frac{(mn)!}{k!\,(mn - k)!}$$

ways. If the k rooks are different from each other, then the above answers must be multiplied by $k!$.

Symmetric Arrangements

We now complicate our rook problem by adding a symmetry requirement to the requirement that the rooks cannot attack one another.

The simplest requirement of this type is the requirement that the rooks are to be placed symmetrically about the origin. We denote by G_n the number of solutions of this problem when the number of rooks is n and the board is $n \times n$. We claim that

$$G_{2n} = 2nG_{2n-2}. \tag{14}$$

Consider a board with $2n$ rows and $2n$ columns. If a rook in the first column occupies the jth square, then there must be a rook in the last column occupying the $(2n - j + 1)$th square. Since $j \neq 2n - j + 1$, the rook can occupy any of the $2n$ squares in the first column without sharing a row with the rook in the last column. Now we delete the first and last columns as well as the rows occupied by the two rooks. The result is a board with $2n - 2$ rows and $2n - 2$ columns. It is clear that each symmetric arrangement on the new board determines $2n$ symmetric arrangements on the original board. But then $G_{2n} = 2nG_{2n-2}$, as asserted.

Repeated use of (14) shows that $G_{2n} = 2^n n!$.

Now consider the case of a board with $2n + 1$ rows and $2n + 1$ columns. In this case the central square of the board is symmetric with respect to the center of the board and must carry a rook. If we delete the central row and column, then we end up with $2n$ rooks arranged symmetrically on a $2n \times 2n$ board. This means that

$$G_{2n+1} = G_{2n} = 2^n n!. \tag{15}$$

Next we consider a somewhat more difficult problem involving arrangements which are invariant under a 90° rotation (Fig. 30 illustrates such an arrangement on an 8×8 board). Take the case of $4n$ rooks placed on a $4n \times 4n$ board. In this case the rook in the first column can occupy $4n - 2$ of the $4n$ squares in this column (if the rook occupied a corner square then, in view of the requirement of rotational symmetry, the remaining corner square in this column would also be occupied by a rook and the two rooks could attack one another). Rotations through 90°, 180°, and 270° associate with this rook three more rooks located, respectively, in the last row, the last column, and the first row. Deletion of these four border rows and columns yields a $(4n - 4) \times (4n - 4)$ board whose rooks satisfy the requirements of our problem. This means that for each admissible arrangement of the rooks on the smaller board there are $4n - 2$ admissible arrangements of the rooks on the original board. If R_n denotes the number of solutions of our problem for an $n \times n$ board, then we conclude that

$$R_{4n} = (4n - 2)R_{4n-4} .$$

Repeated use of this relation shows that

$$R_{4n} = 2^n(2n - 1)(2n - 3) \cdots 1. \tag{16}$$

Now consider the case of a $(4n + 1) \times (4n + 1)$ board. In this case our rotations carry the central square into itself and so this square must be occupied by a rook. Deletion of the central row and column leaves a $4n \times 4n$ board. It follows that

$$R_{4n+1} = R_{4n} . \tag{17}$$

In the remaining cases of a $(4n + 2) \times (4n + 2)$ and a $(4n + 3) \times (4n + 3)$ board there are no solutions. To see this, note that if a solution exists, then a rook not in the center of the board is one of four rooks which are permuted among themselves by 90° rotations of the board. It follows that the number of rooks is of the form $4n$ (when there is no rook at the center) or $4n + 1$ (when there is a rook at the center). This shows that $R_{4n+2} = R_{4n+3} = 0$, as asserted.

Finally we compute the number of arrangements of n rooks symmetric with respect to a diagonal. For definiteness, we choose the diagonal through the lower left corner of the board. We denote by Q_n the number of solutions for an $n \times n$ board. We claim that

$$Q_n = Q_{n-1} + (n - 1)Q_{n-2} . \tag{18}$$

For proof, note that a rook in the first column may or may not occupy the square in the lower left corner of the board. In the first case, deletion of the row and column determined by this square leaves us with a symmetric arrangement of $n - 1$ rooks on an $(n - 1) \times (n - 1)$ board, and the number of such arrangements is Q_{n-1}. In the second case, the rook may occupy any one of $n - 1$ squares in the first column and has a partner located symmetrically with respect to the diagonal.

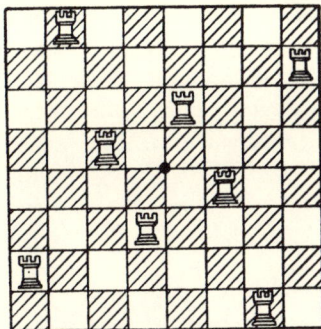

FIG. 30

Deletion of the rows and columns determined by the two corresponding rooks leaves us with a symmetric arrangement of $n - 2$ rooks on an $(n - 2) \times (n - 2)$ board. Since there are Q_{n-2} such arrangements and $n - 1$ possible locations of the rook in the first column, there are $(n - 1)Q_{n-2}$ solutions associated with this case. Relation (18) follows.

The number Q_n satisfies the relation

$$Q_n = 1 + C_n^2 + \frac{1}{1 \cdot 2} C_n^2 C_{n-2}^2 + \frac{1}{1 \cdot 2 \cdot 3} C_n^2 C_{n-2}^2 C_{n-4}^2 + \cdots. \qquad (19)$$

To prove (19), we separate the arrangements of the rooks into classes by putting into the sth class all arrangements in which s pairs of rooks are not on the diagonal.

Reasoning as in the previous problem, we can show that the number B_n of arrangements of n rooks on an $n \times n$ chessboard in which no rook can attack another and in which the rooks are symmetric with respect to both diagonals satisfies the relations

$$B_{2n} = 2B_{2n-2} + (2n - 2)B_{2n-4}, \qquad B_{2n+1} = B_{2n}.$$

Two Knights

In how many ways can a black and a white knight be placed on an $m \times n$ chessboard so that they cannot attack each other?

The complexity of the problem is due to the fact that the number of moves available to the knight varies with its location. For example, if $m \geqslant 5$ and $n \geqslant 5$, then a knight located at a corner square has a choice of two moves, a knight located at various edge squares has a choice of three or four moves, and a knight located at the center has a choice of eight moves. This variety of choices reflects the variety of ways in which a knight can move: for example, it can move one square forward and two squares up, or two squares back and one square down, and so on. All told, a knight can move in 8 ways which differ in the length and direction of their horizontal and vertical components. These 8 ways can be described as follows:

$$(2, 1), \quad (1, 2), \quad (-1, 2), \quad (-2, 1), \quad (-2, -1), \quad (-1, -2), \quad (1, -2), \quad (2, -1).$$

This suggests that we think of a knight as a combination of 8 pieces each of which moves in one way only. These "elementary" knights are relatively easy to work with. To compute the number of ways of placing a $(2, 1)$-knight on our board so that it represents a threat to a piece on the board, we note that we must ignore the last two columns and the last row. This means the column can be chosen in $n - 2$ ways, and the row can be chosen in $m - 1$ ways. Consequently, there are $(m - 1)(n - 2)$ ways of placing a $(2, 1)$-knight on an $m \times n$ board so that it can attack some piece on the board. By symmetry, this is also the number of ways of placing a white $(-2, -1)$-, $(-2, 1)$-, or a $(2, -1)$-knight so that it can attack a black knight. For a white $(1, 2)$-, $(-1, -2)$-, $(-1, 2)$-, or a $(1, -2)$-knight the corresponding number is $(m - 2)(n - 1)$. It follows that the number of ways of placing two knights of different color so that they can attack each other is

$$4[(m - 1)(n - 2) + (m - 2)(n - 1)] = 2[(2m - 3)(2n - 3) - 1].$$

The number of ways of placing two knights of the same color on the board so that they can protect each other is half of the above number (this is so because the knights are interchangeable). The number of ways of placing two knights of different color so that they can not attack each other is

$$m^2n^2 - 9mn + 12m + 12n - 16.$$

(Two knights can be placed on an $m \times n$ board in $mn(mn - 1)$ ways.)

Inventors of chess problems sometimes introduce "fairyland" pieces which do not move like their ordinary counterparts. Following this idea, we introduce a (p, q)-*knight*, $p \geqslant 0$, $q \geqslant 0$. A (p, q)-knight moves p squares horizontally (in either direction) and q squares vertically (in either direction); for example, an ordinary knight is the composite of a $(1, 2)$-knight and a $(2, 1)$-knight. Arguing as before we can show that, if $0 < p \leqslant n$, $0 < q \leqslant m$, then there are $4(n - p)(m - q)$ ways of placing two (p, q)-knights of different color on an $m \times n$ board so that they can attack each other. If $p = 0$ or $q = 0$, then the corresponding number is cut in half. The number of ways of placing two (p, q)-knights of the same color so that they protect each other is $2(n - p)(m - q)$.

Any chess piece can be viewed as a composite of (p, q)-knights for different values of p and q. For example, a chess king is composed of a $(0, 1)$-knight, a $(1, 0)$-knight, and a $(1, 1)$-knight. Hence the number of ways of placing two kings of different color on an $m \times n$ chessboard so that they can attack each other is

$$2[n(m - 1) + (n - 1)m + 2(n - 1)(m - 1)] = 8mn - 6m - 6n + 4.$$

It follows that there are $m^2n^2 - 9mn + 6m + 6n - 4$ ways of placing them so that they cannot attack each other.

A chess bishop is the composite of a $(1, 1)$-, a $(2, 2)$-,..., a (p, p)-knight, where p is the smaller one of the numbers m and n. We assume for definiteness that $m \leqslant n$. Then $p = m - 1$ and we see that the number of ways of placing two bishops of different color on the board so that they can attack one another is

$$4[(n - 1)(m - 1) + (n - 2)(m - 2) + \cdots + (n - m + 1) \cdot 1].$$

If we remove brackets and make use of the formulas for the sum of the natural numbers from 1 to $n - 1$ and for the sum of the squares of these numbers, then the above number of ways can be written as $2m(m - 1)(3n - m - 1)/3$. If $m \geqslant n$, then we must interchange the roles of m and n. In particular, if $m = n$, then the number of ways is $2m(m - 1)(2m - 1)/3$.

The number of arrangements of rooks can be computed more easily in the following manner: A white rook can be placed on each of the mn squares of the board. Then it threatens $m + n - 2$ squares. Each of these squares can be occupied by a black rook. Hence the number of ways of placing two rooks of different color on the board so that they can attack one another is $mn(m + n - 2)$.

Since a queen can be viewed as a composite of a rook and a bishop, it follows that the number of ways of placing two queens of different color on an $m \times n$ board, $m \leqslant n$, so that they can attack each other is

$$\tfrac{2}{3}m(m - 1)(3n - m - 1) + mn(m + n - 2).$$

If $m = n$, then this expression reduces to $\tfrac{2}{3}m(m - 1)(5m - 1)$. We leave it to the reader to compute the number of ways of placing these pieces so that they cannot attack each other.

VI

Recurrence Relations

On a number of occasions we solved combinatorial problems by reducing them to problems involving a smaller number of objects. We used this approach in deducing the formula for the number of arrangements with repetitions (see p. 3) and in solving most of the partition problems in Chap. IV. The method of reducing a problem to an analogous problem involving a smaller number of objects is called the *method of recurrence relations* (from the Latin *recurrere*—to return). Using a recurrence relation we can reduce a problem involving n objects to one involving $n - 1$ objects, then to one involving $n - 2$ objects, and so on. By successive reduction of the number of objects involved, we eventually end up with a problem which can be easily solved. In many cases it is possible to obtain from a recurrence relation an explicit formula for the solution of a combinatorial problem.

By way of an illustration we consider the problem of finding the number P_n of permutations of n elements.

In Chap. II (p. 19) we deduced the formula $P_n = n!$ from the formula for the number of arrangements without repetitions. Now we deduce the same formula by first finding a recurrence relation satisfied by P_n.

Let $a_1, \ldots, a_{n-1}, a_n$ be n objects. An arbitrary permutation of these objects can be obtained by adjoining a_n to a permuation of $a_1, \ldots, a_{n-1}$. It is clear that a_n can occupy various positions: it can be put to the left of a_1, between a_1 and a_2, and so on. In all, a_n can occupy n different positions. This means that each permutation of the $n - 1$ elements $a_1, \ldots, a_{n-1}$ gives rise to n permutations of the n elements. Hence the numbers P_n and P_{n-1} are connected by the recurrence relation $P_n = nP_{n-1}$.

Using this relation, we obtain the chain of equalities

$$P_n = nP_{n-1} = n(n-1)\,P_{n-2} = n(n-1)\cdots 2P_1\,.$$

Since $P_1 = 1$ (one element gives rise to just one permutation), we again obtain the formula $P_n = n!$.

We have encountered many recurrence relations in connection with partition problems, with problems involving pieces on a chessboard, and so on. Now we consider additional problems of this type, and at the end of the chapter we consider aspects of the general theory of recurrence relations.

Fibonacci Numbers

In his book "Liber Abaci" published in 1202, the Italian mathematician Fibonacci posed, among others, the following problem:

Each month the female of a pair of rabbits gives birth to a pair of rabbits (of different sexes). Two months later the female of the new pair gives birth to a pair of rabbits. Find the number of rabbits at the end of the year if there was one pair of rabbits in the beginning of the year.

We see that at the end of the first month there will be 2 pairs of rabbits. At the end of the second month just one of these 2 pairs will have offspring and so the number of pairs of rabbits will be 3. At the end of the third month the original pair of rabbits as well as the pair born at the end of the first month will have offspring and so the number of pairs of rabbits will be 5.

Let $F(n)$ denote the number of pairs of rabbits at the end of the nth month. At the end of the $(n + 1)$th month there will be $F(n)$ pairs of "old" rabbits and as many pairs of "new" rabbits as there were pairs of rabbits at the end of the $(n - 1)$th month, that is, $F(n - 1)$. In other words, we have the recurrence relation

$$F(n + 1) = F(n) + F(n - 1). \tag{1}$$

Since we know that $F(0) = 1$ and $F(1) = 2$, it follows that $F(2) = 3, F(3) = 5$, $F(4) = 8$, and so on.

In particular, $F(12) = 377$.

The numbers $F(n)$ are called *Fibonacci numbers*. These numbers have many remarkable properties. We are about to express the Fibonacci numbers in terms of the numbers C_m^k. To this end, we establish a connection between the Fibonacci numbers and the following combinatorial problem:

Find the number of n-sequences of zeros and ones with no two ones in a row.

To establish the required connection, we associate with a sequence of this type the pair of rabbits for which the sequence serves as a "genealogical tree."

By this we mean that the ones in the sequence correspond to the months of birth of the various pairs of "forbears" of our pair of rabbits (including the pair itself), and the zeros in the sequence correspond to the remaining months. For example, the sequence 010010100010 determines the following genealogical tree: the pair of rabbits was born at the end of the 11th month, its parents were born at the end of the 7th month, its "grandparents" were born at the end of the 5th month, and its "great grandparents" were born at the end of the 2nd month. The genealogical tree of the original pair of rabbits is 000000000000.

It is clear that none of our sequences can contain two ones in a row (a new pair of rabbits cannot have offspring in a month). Also, different sequences correspond to different pairs of rabbits and, conversely, different pairs of rabbits have different genealogical trees. The latter fact follows if we bear in mind that a litter invariably consists of a single pair of rabbits.

The correspondence just established shows that the number of n-sequences satisfying the prescribed conditions is $F(n)$.

Now we show that

$$F(n) = C_{n+1}^0 + C_n^1 + C_{n-1}^2 + \cdots + C_{n-p+1}^p \,, \tag{2}$$

where $p = E[(n + 1)/2]$. Here, as usual, $E(\alpha)$ denotes the integral part of α.

$F(n)$ denotes the number of n-sequences of zeros and ones with no two ones in a row. The number of such sequences with just k ones and $n - k$ zeros is C_{n-k+1}^k (see p. 45). Since $k \leqslant n - k + 1$, it follows that k varies from 0 to $E[(n + 1)/2]$. Now relation (2) is implied by the rule of sum.

Relation (2) can be deduced in a different manner. We put

$$G(n) = C_{n+1}^0 + C_n^1 + C_{n-1}^2 + \cdots + C_{n-p+1}^p ,$$

where $p = E[(n + 1)/2]$. The equality $C_n^k = C_{n-1}^k + C_{n-1}^{k-1}$ implies readily that

$$G(n) = G(n - 1) + G(n - 2). \tag{3}$$

Also, $G(1) = 2 = F(1)$ and $G(2) = 3 = F(2)$. Since the sequences $G(n)$ and $F(n)$ satisfy the recurrence relation $X(n) = X(n - 1) + X(n - 2)$, it follows that

$$G(3) = G(2) + G(1) = F(2) + F(1) = F(3),$$

and, in general, $G(n) = F(n)$.

Another Method of Proof

In the previous section we established directly a connection between the Fibonacci problem about the rabbits and a combinatorial problem. Another way of establishing this connection is to show directly that $T(n)$, the number of solutions of the combinatorial problem for a given n, satisfies the recurrence relation

$$T(n + 1) = T(n) + T(n - 1), \tag{4}$$

which is the analog of the recurrence relation (1) satisfied by the Fibonacci numbers.

Consider an $(n + 1)$-sequence of zeros and ones with no two ones in a row. This sequence can end in a 0 or a 1. If the sequence ends in a 0, then deletion of this 0 yields an n-sequence satisfying our condition. Conversely, if we add a 0 at the end of an n-sequence of zeros and ones with no two ones in a row, then we obtain an $(n + 1)$-sequence satisfying the same condition and ending in a 0. This shows that the number of "good" $(n + 1)$-sequences ending in a 0 is $T(n)$.

Now consider a sequence ending in a 1. Since there are no two ones in a row, this 1 must be preceded by a 0. In other words, our $(n + 1)$-sequence ends in 01. Deletion of these two digits yields a "good" $(n - 1)$-sequence. Hence the number of good $(n + 1)$-sequences ending in a 1 is $T(n - 1)$. Since our sequences must end in either a 0 or a 1, it follows by the rule of sum that $T(n + 1) = T(n) + T(n - 1)$.

We see that $T(n)$ and $F(n)$ satisfy the same recurrence relation. This does not prove that the sequences $F(n)$ and $T(n)$ coincide. After all, factorials and sub-factorials (see p. 52) satisfy the same recurrence relation

$$X(n + 1) = n[X(n) + X(n - 1)]. \tag{5}$$

However, for factorials $0! = 1$, $1! = 1$, and for subfactorials $D(0) = 1$, $D(1) = 0$. That is why all subsequent entries in one of these sequences differ from the corresponding entries in the other sequence.

To show that the sequences $T(n)$ and $F(n)$ coincide, it remains to show that $T(1) = F(1)$ and $T(2) = F(2)$. To see this, note that there are just two good 1-sequences, namely 0 and 1, and just three good 2-sequences, namely 00, 01, and 10. But then $T(1) = 2 = F(1)$ and $T(2) = 3 = F(2)$. This completes the proof of the coincidence of the sequences $T(n)$ and $F(n)$.

Successive Partitions

Many combinatorial problems can be solved by the method employed in the previous section. If $f(n)$ denotes the number of solutions of a combinatorial problem, then it may be possible to find $f(n)$ by showing that $f(n)$ satisfies the same recurrence relation as the available solution $g(n)$ of another combinatorial problem, and that both sequences have in common sufficiently many initial terms. (The exact meaning of the term "sufficiently many" will be clarified in the sequel.)

We apply the technique just outlined to obtain the solution of the following problem: Given n objects arranged in some definite order. We partition this ordered set into two nonempty subsets so that the elements of one subset are to the left of the elements of the other subset. Then each subset with more than one element is again subdivided in the same manner into two nonempty subsets. This process continues until we end up with one-element subsets. *We are required to find the number of (distinct) partition procedures* (two procedures are defined as different if they yield different results at one or more steps).

Let B_n denote the number of ways of partitioning $n + 1$ ordered objects. The first step in the partition process can be accomplished in n ways (the left subset may contain one object, two objects,..., n objects). We separate accordingly our partition procedures into n classes and put in the sth class all those partition procedures which start by assigning s elements to the left subset.

We compute the number of procedures in the sth class. Any procedure in this class separates the initial ordered set into a left subset with s elements and a right subset with $n - s + 1$ elements. The number of partition procedures applicable to the initial left subset is B_{s-1} and the corresponding number for the initial right subset is B_{n-s}. By the rule of product, the number of procedures in the sth class is $B_{s-1}B_{n-s}$. By the rule of sum we have

$$B_n = B_0B_{n-1} + B_1B_{n-2} + \cdots + B_{n-1}B_0. \tag{6}$$

We have obtained a recurrence relation for B_n. This recurrence relation occurred in connection with the solution of the problem of the queue in front of the box office (see p. 63). There we showed that this recurrence relation is satisfied by the numbers

$$T_n = \frac{1}{n+1} C_{2n}^n.$$

To prove the equality

$$B_n = T_n = \frac{1}{n+1} C_{2n}^n \tag{7}$$

for all $n \geqslant 0$, that is, the equality of the sequences B_0, B_1,... and T_0, T_1,..., it it remains to prove that $B_0 = T_0$. But this follows readily from the fact that $T_0 = C_0^0 = 1$ and $B_0 = 1$ (the one partitioning procedure applicable to an initial set consisting of a single element is to leave it alone). To sum up:

The number of procedures for the successive subdivision of an ordered set of $n + 1$ elements is

$$T_n = \frac{1}{n+1} C_{2n}^n$$

Multiplication and Division of Numbers

Consider n numbers a_1,..., a_n arranged in a definite order. In view of the associativity of multiplication, the product of these numbers can be computed in various ways (without upsetting their initial order). For example, the product of three numbers can be computed in two ways $((ab) c = a(bc))$, the product of four numbers can be computed in five ways, and so on. We are required *to find the number of ways of computing the product of n numbers arranged in some definite order.*

It is clear that each multiplication scheme corresponds to a partition procedure of the n numbers into n one-number parts. For example, the multiplication scheme $(ab)(cd)$ for four numbers corresponds to the partition procedure $a \mid b \mid c \mid d$, and the multiplication scheme $((ab)c) d$ corresponds to the partition procedure $a \mid b \mid c \mid d$. It follows that the number of ways of computing the product of n numbers (arranged in definite order) is equal to the number of partition procedures of a set of n elements, that is, $T_{n-1} = (1/n) C_{2n-2}^{n-1}$.

If we take into consideration the commutativity of multiplication, then the number of ways of computing our product is increased by a factor of n! This is so because there are n! ways of permuting our n numbers, and each permutation gives rises to T_{n-1} multiplication schemes. It follows that our n numbers can be multiplied in $n! T_{n-1} = (n-1)! \, C^{n-1}_{2n-2}$ ways.

This result can be arrived at directly without the use of the formula for the number of partition procedures. The direct derivation which we are about to present gives a new method for obtaining the number of partition procedures and, therefore, also a new method for obtaining the number of good queues at the box office (in the special case when the number of people with dollar bills is equal to the number of people with half-dollar coins).

Suppose we have already found the number $\Phi(n)$ of ways of multiplying n numbers. We introduce an additional factor a_{n+1} and compute the number of ways of adjoining a_{n+1} to a definite product of the numbers $a_1, ..., a_n$.

Two obvious ways of adjoining a_{n+1} are to premultiply or postmultiply the product of $a_1, ..., a_n$ by a_{n+1}. As for other ways of adjoining a_{n+1}, we note that multiplication of n numbers involves $n-1$ successive multiplications of two numbers each. The number a_{n+1} can enter each of these two-number multiplications in 4 ways (on either side of either factor). It follows that a_{n+1} can be adjoined to our product in $2 + 4(n-1) = 4n - 2$ ways.

This implies the recurrence relation

$$\Phi(n+1) = (4n-2)\,\Phi(n)$$

Using this relation and the fact that $\Phi(1) = 1$, we see that

$$\Phi(n) = 2 \cdot 6 \cdots (4n-6) = 2^{n-1} \cdot 1 \cdot 3 \cdots (2n-3).$$

Since

$$\Phi(n) = 2^{n-1} \cdot 1 \cdot 3 \cdots (2n-3) = \frac{(2n-2)!}{(n-1)!} = (n-1)! \, C^{n-1}_{2n-2},$$

we see that our present argument yields the same answer to our problem as the earlier argument.

Now we turn to division. To give meaning to the expression

$$\cfrac{a_1}{\cfrac{a_2}{\cfrac{a_3}{\cfrac{\vdots}{a_n}}}} \tag{8}$$

we must indicate the order in which the various divisions are to be carried out. To see in how many ways we can assign meaning to (8), we observe that each assignment of the order in which the divisions are to be carried out corresponds to a procedure for partitioning an ordered set of n elements into one-element subsets. Since we know that the number of such procedures is $(1/n)\,C_{2n-2}^{n-1}$, it follows that the number of ways of assigning meaning to the expression in (8) is likewise $(1/n)\,C_{2n-2}^{n-1}$.

Polygon Problems

Some issues in quantum chemistry give rise to the following problems:

A regular 2n-gon is inscribed in a circle. In how many ways is it possible to join its vertices in pairs so that the resulting segments do not intersect one another?

For $n = 0$, we define the number of ways to be one. For $n = 1$, the number of ways is one.* For $n = 2$, the number of ways is two (see Fig. 31). To find the

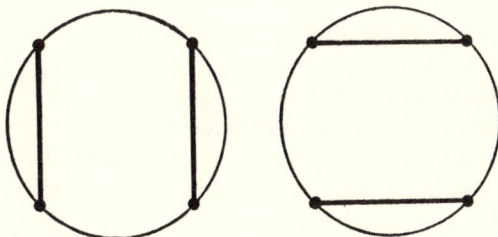

FIG. 31

number of ways $F(n)$ for an arbitrary n, we deduce a recurrence relation for $F(n)$. A particular vertex A of our polygon can be joined to a vertex B if and only if the number of vertices on either side of AB is even (see Fig. 32). The various ways of joining the vertices can be separated into classes in accordance with the number of vertices, say, to the left of the segment AB.

If the number of vertices to the left of AB is $2s$, then the number of vertices to the right of AB is $2(n - s - 1)$. Hence our $2n$-gon is subdivided into a $2s$-gon and a $2(n - s - 1)$-gon. Now the number of ways of joining the vertices of a $2s$-gon in pairs without segments intersecting one another is $F(s)$. The corresponding number for a $2(n - s - 1)$-gon is $F(n - s - 1)$. By the rule of product, the number of procedures in the sth class is $F(s)\,F(n - s - 1)$. This implies that

$$F(n) = F(0)\,F(n - 1) + F(1)\,F(n - 2) + \cdots + F(n - 1)\,F(0).$$

The recurrence relation just established for the numbers $F(n)$ is the same as the

---

* We regard a diameter as a regular 2-gon.

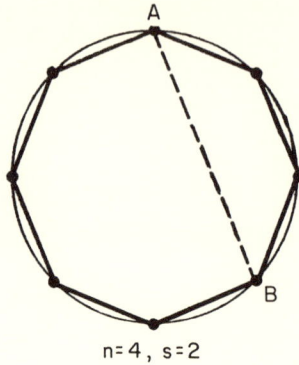

n= 4 , s = 2

FIG. 32

recurrence relation satisfied by the numbers $T_n = [1/(n + 1)] C_{2n}^n$. Since $F(0) = T_0 = 1$, it follows that $F(n) = T_n$ for all $n \geqslant 0$. In other words, there are $T_n = [1/(n + 1)] C_{2n}^n$ ways of joining the vertices of a $2n$-gon so that no two segments intersect one another.

The same answer holds for the following problem:

In how many ways is it possible to subdivide a convex $(n + 2)$-gon into triangles by means of diagonals which do not intersect in the interior of the polygon?

Let $\Phi(n)$ denote the number of ways of subdividing the $(n + 2)$-gon for $n > 0$, and let $\Phi(0) = 1$. Designate one side of the polygon as a special side. Each subdivision includes a triangle one of whose sides is the special side of the polygon. Designate this triangle as the special triangle. Now put in the same class all subdivision procedures which share a special triangle (Fig. 33). If deletion of the special triangle associated with a class of subdivision procedures leaves us with an $(s + 2)$-gon and an $(n - s + 1)$-gon, then, by the rule of

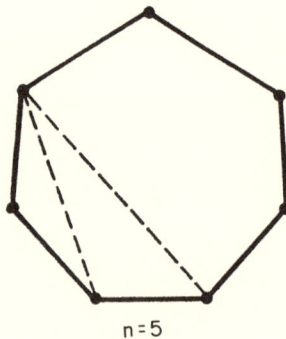

n = 5

FIG. 33

product, the number of procedures in this class is $\Phi(s)\,\Phi(n-s-1)$. By the rule of sum, the total number of subdivision procedures of our $(n+2)$-gon is

$$\Phi(n) = \Phi(0)\,\Phi(n-1) + \Phi(1)\,\Phi(n-2) + \cdots + \Phi(n-1)\,\Phi(0).$$

Since $\Phi(0) = 1$, we see that

$$\Phi(n) = T_n = \frac{1}{n+1}\,C_{2n}^n\,.$$

The Head Steward Faces a Difficulty

In some combinatorial problems we must set up not one recurrence relation but a system of recurrence relations connecting a number of sequences. These relations express the $(n+1)$th term of each sequence in terms of the lower terms of all the sequences.

King Arthur's head steward was faced with the task of seating 6 pairs of hostile knights at the Round Table. In how many ways could he seat the knights without seating two foes side by side?

Any acceptable seating arrangement of the knights gives rise to 11 more acceptable seating arrangements obtained by moving each knight the same number of seats in the same direction. We do not regard two acceptable seating arrangements as different if they differ only by a cyclic rearrangement.

Let the number of knights to be seated be $2n$. Let A_n be the number of seating arrangements with no hostile neighbors. Let B_n be the number of seating arrangements with just one pair of hostile neighbors. Finally, let C_n denote the number of seating arrangements with just two pairs of hostile neighbors.

We deduce a formula which expresses A_{n+1} in terms of A_n, B_n, and C_n. Consider a seating arrangement of $n + 1$ pairs of hostile knights with no hostile neighbors. We suppose the pairs of hostile knights numbered. We ask the hostile pair numbered $n + 1$ to leave the table. Now there are three possibilities: either there are no hostile neighbors among the remaining $2n$ knights, or there is exactly one pair of hostile neighbors, or there are exactly two pairs of hostile neighbors (each of the two departing knights could have sat between two foes).*

We explain in how many ways we can again seat the $(n + 1)$th pair of knights so as to end up with a seating arrangement with no hostile neighbors.

The case when there are two pairs of hostile neighbors among the $2n$ remaining knights is easiest to deal with. Then each of the two returning knights must separate one pair of hostile neighbors. This can be done in two ways. Since the number of seating arrangements of $2n$ knights with exactly two pairs of hostile neighbors is C_n, the number of seating arrangements in this case is $2C_n$.

Now consider the case when there is exactly one pair of hostile neighbors among the $2n$ remaining knights. Then one of the returning knights must sit between the two hostile neighbors. There are now $2n + 1$ knights at the table and there are $2n + 1$ seats between them. Since the second returning knight must not sit to either side of his foe, he can choose one of $2n - 1$ seats. Since either one of the knights can be seated first, there are $2(2n - 1)$ ways of seating the two knights at the table. In view of the fact that the number of seating arrangements of $2n$ knights with exactly one pair of hostile neighbors is B_n, the number of seating arrangements in this case is $2(2n - 1) B_n$.

Finally, consider the case when there are no hostile neighbors among the $2n$ remaining knights. Then the first returning knight can take any of the $2n$ available seats and the second returning knight can take one of $2n - 1$ seats (he must not sit next to his foe). We see that the two returning knights can be seated at the table in $2n(2n - 1)$ ways, and so the number of seating arrangements in this case is $2n(2n - 1) A_n$.

As already noted, the three possibilities investigated above exhaust all the possibilities. This means that A_{n+1} satisfies the recurrence relation

$$A_{n+1} = 2n(2n - 1) A_n + 2(2n - 1) B_n + 2C_n. \qquad (9)$$

This relation alone does not determine A_n for all values of n. Analogs of (9) are needed for B_{n+1} and C_{n+1}. We deduce these analogs next.

Consider a seating arrangement of the $2n + 2$ knights, $n > 1$, with exactly one pair of hostile neighbors. The number of such seating arrangements is B_{n+1}. To avoid an argument at the table, we ask the two hostile neighbors to leave the table. Now there are two possibilities: either there are no hostile neighbors

---

* Here and in the sequel, we suppose $n > 1$. For $n = 1$, the argument which follows is irrelevant.

among the remaining $2n$ knights or there is exactly one pair of hostile neighbors (they were separated by the two hostile neighbors who were asked to leave the table). In the latter case, the two returning foes must again occupy the two seats they vacated when asked to leave the table (otherwise we would end up with two pairs of hostile neighbors at the table), and this they can do in two ways. Since there are B_n ways of seating $2n$ knights so that there is exactly one pair of hostile neighbors, we end up with $2B_n$ seating arrangements. In the former case, the returning pair can be seated in any of the $2n$ available seats. In view of the possibility of members of the pair interchanging their seats, the two foes can be seated at the table in $4n$ ways. Combining this with the A_n seating arrangements of $2n$ knights with no hostile neighbors we obtain $4nA_n$ seating arrangements. Also, in either of the two cases under discussion, the number of the distinguished pair of foes can be chosen in $n + 1$ ways. It follows that B_{n+1} satisfies the recurrence relation

$$B_{n+1} = 4n(n + 1)\, A_n + 2(n + 1)\, B_n \,. \tag{10}$$

Finally we consider the case when there are two pairs of hostile neighbors among the $2n + 2$ knights at the table. The numbers of the two hostile pairs can be chosen in $C_{n+1}^2 = n(n + 1)/2$ ways. If we replace each hostile pair of knights with one new knight and regard the two new knights as foes, then we end up with $2n$ knights and no hostile neighbors (this case arises when the two new knights are not seated side by side) or exactly one pair of hostile neighbors.

There are A_n seating arrangements of $2n$ knights with no hostile neighbors. We can go from one of these seating arrangements to the original seating pattern in four ways due to the possibility of interchanging the knights in each of the two hostile pairs. This means that the A_n seating arrangements give rise to $4C_{n+1}^2 A_n$ seating arrangements which conform to the original pattern.

There are $(1/n)\, B_n$ seating arrangements with exactly one *designated* pair of hostile neighbors. Reasoning as before, we see that these $(1/n)\, B_n$ seating arrangements give rise to $4C_{n+1}^2 \cdot (1/n)\, B_n = 2(n + 1)\, B_n$ seating arrangements which conform to the original pattern. With both cases taken into consideration, we see that for $n \geqslant 1$

$$C_{n+1} = 2n(n + 1)\, A_n + 2(n + 1)\, B_n \,. \tag{11}$$

All in all, we obtain the following system of recurrence relations valid for $n \geqslant 2$:

$$A_{n+1} = 2(2n - 1)(nA_n + B_n) + 2C_n \,, \tag{9}$$

$$B_{n+1} = 2(n + 1)(2nA_n + B_n), \tag{10}$$

$$C_{n+1} = 2(n + 1)(nA_n + B_n). \tag{11}$$

A simple computation shows that $A_2 = 2$, $B_2 = 0$, $C_2 = 4$. Now relations (9)–(11) give the values $A_3 = 32$, $B_3 = 48$, $C_3 = 24$. Continuing, we find that the 6 pairs of hostile knights confronting the head steward could be seated in $A_6 = 12,771,840$ ways.

The problem just solved is similar to the following problem frequently referred to as "the guest problem."

In how many ways is it possible to seat n married couples at a round table so that men and women alternate and so that no man is seated next to his wife?

This problem is solved in a manner similar to that used to solve the problem of the chief steward. We begin by seating the women. If we number the seats then the women occupy all the even-numbered seats or all the odd-numbered seats. In either case the women can be seated in $n!$ ways. This means that altogether the women can be seated in $2 \cdot (n!)$ ways. Then we consider the number of seating arrangements with no couple sitting together, with just one couple sitting together, and with just two couples sitting together. We leave it to the reader to deduce the appropriate system of recurrence relations.

Lucky Bus Ticket Numbers

Some people regard a bus ticket as lucky if it shows a six-digit number with the property that the sum of the digits in the even positions is equal to the sum of the digits in the odd positions. For example, the ticket numbered 631,752 is lucky because $6 + 1 + 5 = 3 + 7 + 2 = 12$. We compute the number of lucky tickets showing every possible six-digit number (from 000,000 to 999,999).

First we consider the problem of computing the number of three-digit numbers with digit sum N (here we include numbers like 075 and 000). This problem is similar to the problem solved on p. 77. Stated in terms of that problem, our present problem is to compute the number of ordered partitions of the number N into 3 summands drawn from the numbers 0, 1,..., 9. The number $F(3, 9; N)$ of solutions of this problem satisfies the recurrence relation

$$F(3, 9; N) = F(2, 9; N) + F(2, 9; N - 1)$$
$$+ F(2, 9; N - 2) + F(2, 9; N - 3) + F(2, 9; N - 4)$$
$$+ F(2, 9; N - 5) + F(2, 9; N - 6) + F(2, 9; N - 7)$$
$$+ F(2, 9; N - 8) + F(2, 9; N - 9).$$

Similarly,

$$F(2, 9; N) = F(1, 9; N) + F(1, 9; N - 1) + \cdots + F(1, 9; N - 9).$$

It is clear that $F(1, 9; N) = 1$, for $0 \leqslant N \leqslant 9$, and that $F(1, 9; N) = 0$ otherwise, Using these relations we can easily complete the following table:

$\diagdown N$ $k\diagdown$	0	1	2	3	4	5	6	7	8	9	10	11	12	13	14	15	16	17	18	19	20	21	22	23	24	25	26	27
1	1	1	1	1	1	1	1	1	1	1	0	0	0	0	0	0	0	0	0	0	0	0	0	0	0	0	0	0
2	1	2	3	4	5	6	7	8	9	10	9	8	7	6	5	4	2	3	1	0	0	0	0	0	0	0	0	0
3	1	3	6	10	15	21	28	36	45	55	63	69	73	75	75	73	69	63	55	45	36	28	21	15	10	6	3	1

To find the number of lucky tickets we must square each entry in the third row of our table and add the results. In fact, for each lucky ticket, the sum of the digits in the even positions is equal to the sum of the digits in the odd positions. Suppose this sum is N. The entry in the Nth position in the third row of our table tells us how many three-digit numbers have digit sum N. Hence this entry also tells us in how many ways we can choose the digits in either the even or the odd positions of those lucky numbers for which the sum of the digits in either the even or the odd positions is N. Since these two sets of choices are independent of each other, the number of lucky numbers for which the sum of the digits in the even positions is N is equal to $[F(3, 9; N)]^2$. By the rule of sum, the total of lucky numbers is

$$2[1^2 + 3^2 + 6^2 + 10^2 + 15^2 + 21^2 + 28^2 + 36^2 + 45^2$$
$$+ 55^2 + 63^2 + 69^2 + 73^2 + 75^2].$$

This number is equal to 55,252.

Recurrent Tables

Many of the magnitudes in combinatorics depend on more than one number. For example, C_n^k depends on n and k. If a magnitude $F(n, k)$ depends on two natural numbers n and k, then its values can be arranged in a table by placing the number $F(n, k)$ at the intersection of the nth row and the kth column. In Chap. V, we encountered a number of such magnitudes as well as their tables. The arithmetical square, the arithmetical triangle, and the generalized arithmetical triangles are instances of just such tables.

In all of the examples in Chap. V, the elements in each table were connected by certain relations. These relations made it possible to compute the elements in the nth row from the elements in the preceding row and, possibly, from a few initial elements in the nth row. Therefore, given the elements of the first row and the initial elements of the other rows, it was possible to compute successively the entries in all the remaining rows. Such tables resemble recurrent sequences and therefore we refer to them in the sequel as *recurrent tables*.

For the arithmetical square, we had the recurrence relation

$$F(n, k) = F(n - 1, k) + F(n, k - 1), \tag{12}$$

and the boundary conditions $F(n, 0) = 1$, $F(0, k) = 0$, for $k > 0$ (we recall that in the case of the arithmetical square the count of rows and columns started with zero and not with one).

The recurrence relations which govern the arithmetical pentagon and hexagon are also of the form (12). This is hardly surprising if we recall that these tables tell us in how many ways a certain square can be reached by a rook moving on a chessboard bounded by two perpendicular rays and by one or two lines parallel to the principal diagonal. A rook can reach the square (n, k) from the square $(n - 1, k)$ or from the square $(k - 1, n)$. This means that regardless of the restrictions imposed on the rook, the number of ways in which it can reach a certain square is governed by relation (12). The restrictions assign the value 0 to certain entries in the table. In the case of the arithmetical pentagon the entries in question are those above a certain line parallel to the principal diagonal, and in the case of the arithmetical hexagon the entries in question are those outside a region bounded by two lines parallel to the principal diagonal.

The recurrence relations governing the arithmetical triangle and the m-arithmetical triangle are different from relation (12). The recurrence relation for the m-arithmetical triangle is

$$F(n, k) = F(n - 1, k - m + 1) + F(n - 1, k - m + 2) + \cdots + F(n - 1, k), \tag{13}$$

with $F(0, 0) = 1$ and $F(0, k) = 0$, for $k > 0$.

Another Solution of the Problem of the Chief Steward

As an additional example of the use of recurrence relations, we give another solution of the problem of the chief steward (see p. 127). The reader will recall that the problem concerned the number of ways of seating $2n$ knights at a round table so that no two foes sat side by side (the $2n$ knights made up n pairs of foes).

We denote by $F(m, n)$ the number of seating arrangements with exactly m pairs of hostile neighbors. We deduce a recurrence relation which expresses $F(m, n + 1)$ in terms of $F(k, n)$, $k = m - 1, m, m + 1, m + 2$.

We assume that, to begin with, n pairs of knights are seated at the table and that subsequently an $(n + 1)$th pair of foes arrives and is seated at the table. We compute the number of ways in which we can obtain a seating arrangement with m pairs of hostile neighbors. The following cases arise:

(a) There are initially $m - 1$ pairs of hostile neighbors at the table. This can happen in $F(m - 1, n)$ ways. If we are to end up with m pairs of hostile neighbors, then the new pair must stay together and not separate a pair of hostile neighbors already seated at the table. This means that of the $2n$ spaces between the $2n$ knights at the table $m - 1$ spaces are barred to the newcomers. This leaves $2n - m + 1$ spaces for the newcomers. Each of these spaces can be occupied by either one of the two newcomers. It follows that the number of seating arrangements arising in this case is

$$2(2n - m + 1)\,F(m - 1, n). \tag{14}$$

(b) There are m pairs of hostile neighbors at the table. The newcomers can sit apart without separating a pair of hostile neighbors or they can sit together between two hostile neighbors. It is easy to see that the first possibility can be realized in $(2n - m)(2n - m - 1)$ ways, and the second possibility can be realized in $2m$ ways. This gives a total of $(2n - m)^2 - 2n + 3m$ possibilities. Since the number of seating arrangements of $2n$ knights with exactly m pairs of hostile neighbors is $F(m, n)$, this case gives rise to

$$[(2n - m)^2 - 2n + 3m]\,F(m, n) \tag{15}$$

seating arrangements.

(c) There are $m + 1$ pairs of hostile neighbors at the table. This can happen in $F(m + 1, n)$ ways. One of the newcomers must separate a pair of hostile neighbors and the other must not separate a pair of hostile neighbors or sit next to the first newcomer. It follows that the number of seats available to the first newcomer is $m + 1$ and the number of seats available to the second newcomer is $2n - m - 1$. The number of seating possibilities for the pair of newcomers is $2(m + 1)(2n - m - 1)$ (the factor 2 appears because either one of the two newcomers can separate a pair of hostile neighbors). It follows that this case gives rise to

$$2(m + 1)(2n - m - 1)\,F(m + 1, n) \tag{16}$$

seating arrangements.

(d) There are $m + 2$ pairs of hostile neighbors at the table. This can happen in $F(m + 2, n)$ ways. If we are to end up with a seating arrangement with m pairs of hostile neighbors, then each of the two newcomers must separate a pair of foes. This means that the first newcomer can be seated in $m + 2$ ways and the second newcomer can be seated in $m + 1$ ways. It follows that in this case we end up with

$$(m + 1)(m + 2)\,F(m + 2, n) \tag{17}$$

seating arrangements.

It is easy to see that we have considered all possible ways of seating $2n + 2$ knights at a round table so as to end up with m pairs of hostile neighbors. But then $F(m, n)$ must satisfy the recurrence relation

$$
\begin{aligned}
F(m, n + 1) = {} & 2(2n - m + 1) F(m - 1, n) \\
& + [(2n - m)^2 - 2n + 3m] F(m, n) \\
& + 2(m + 1)(2n - m - 1) F(m + 1, n) \\
& + (m + 1)(m + 2) F(m + 2, n).
\end{aligned} \tag{18}
$$

Direct computation shows that

$$
F(0, 2) = 2, \qquad F(1, 2) = 0, \qquad F(2, 2) = 4
$$

(we do not distinguish between seating arrangements which differ by a rotation). Using Formula (18), we find that $F(0, 6) = 12{,}771{,}840$.

Solution of Recurrence Relations

We say that a recurrence relation is of *order* k if it allows us to express $f(n + k)$ in terms of $f(n), f(n + 1), \dots, f(n + k - 1)$. For example,

$$
f(n + 2) = f(n) f(n + 1) - 3f^2(n + 1) + 1
$$

is a recurrence relation of order two, and

$$
f(n + 3) = 6f(n) f(n + 2) + f(n + 1)
$$

is a recurrence relation of order three.

A recurrence relation of order k is satisfied by infinitely many sequences. The reason for this is that the first k elements can be assigned in an arbitrary manner, for they are not restricted by any relations whatsoever. On the other hand, if the first k elements are prescribed, then the remaining elements are uniquely determined; the recurrence relation enables us to express the element $f(k + 1)$ in terms of $f(1), \dots, f(k)$, the element $f(k + 2)$ in terms of $f(2), \dots, f(k + 1)$, and so on.

By means of a recurrence relation and initial terms we can compute successively all the terms of the sequence satisfying this relation. However, in many cases we are interested in a definite term of this sequence and not in all its predecessors. In such cases it is convenient to have an explicit formula for the nth term of the sequence. This brings us to the question of solutions of recurrence relations.

We call a sequence a *solution* of a given recurrence relation if the sequence satisfies the recurrence relation. The sense of this nebulous definition is made clear by an example. Consider the sequence $f(n) = 2^n$, $n = 1, 2, \dots$, and the recurrence relation

$$
f(n + 2) = 3f(n + 1) - 2f(n).
$$

We have $f(n + 2) = 2^{n+2}, f(n + 1) = 2^{n+1}, f(n) = 2^n$. We say that the sequence $f(n) = 2^n$, $n = 1, 2, \dots$, is a solution of the recurrence relation because $2^{n+2} = 3 \cdot 2^{n+1} - 2 \cdot 2^n$, for all $n = 1, 2, \dots$.

A solution of a recurrence relation is called *general* if it depends on k arbitrary constants and if every solution of the recurrence relation can be obtained by assigning suitable values to the arbitrary constants. For example, consider the recurrence relation

$$f(n + 2) = 5f(n + 1) - 6f(n). \tag{19}$$

We claim that

$$f(n) = C_1 2^n + C_2 3^n \tag{20}$$

is the general solution of (19). It is easy to verify the fact that, whatevever the values of C_1 and C_2, the sequence (20) satisfies the recurrence relation (19). It remains to show that every solution of (19) is uniquely determined by the choice of $f(1)$ and $f(2)$. Therefore we must show that for every choice of numbers a and b it is possible to find numbers C_1 and C_2 such that

$$2C_1 + 3C_2 = a$$
$$2^2 C_1 + 3^2 C_2 = b.$$

This system of equations is the same as the system of equations

$$\begin{cases} 2C_1 + 3C_2 = a, \\ 4C_1 + 9C_2 = b. \end{cases} \tag{21}$$

Since the system (21) has a (unique) solution C_1, C_2 for every choice of the numbers a and b, we conclude that $f(n)$ in (20) is indeed the general solution of (19).

Linear Recurrence Relations with Constant Coefficients

There are no general rules for solving all recurrence relations. However, there is a class of recurrence relations which can be solved by a uniform method. The class in question is the class of recurrence relations of the form

$$f(n + k) = a_1 f(n + k - 1) + a_2 f(n + k - 2) + \cdots + a_k f(n), \tag{22}$$

where $a_1, a_2, ..., a_k$ are constants. Such relations are called *linear recurrence relations with constant coefficients*.

First we consider the problem of solving such relations for $k = 2$, that is we consider the problem of solving recurrence relations of the form

$$f(n + 2) = a_1 f(n + 1) + a_1 f(n). \tag{23}$$

The solution of such relations is based on the following two assertions:

[1] If two sequences $f_1(n)$ and $f_2(n)$ are solutions of the recurrence relation (23), then for every choice of numbers A and B the sequence $f(n) = Af_1(n) + Bf_2(n)$ is also a solution of this relation.

[2] If r_1 is a solution of the quadratic equation

$$r^2 = a_1 r + a_2$$

then the sequence

$$f(n) = r_1^{n-1}, \qquad n = 1, 2, ...,$$

is a solution of the recurrence relation

$$f(n + 2) = a_1 f(n + 1) + a_2 f(n).$$

To prove [1], note that

$$f_1(n + 2) = a_1 f_1(n + 1) + a_2 f_1(n)$$
$$f_2(n + 2) = a_1 f_2(n + 1) + a_2 f_2(n).$$

If we multiply the first of these equalities by A and the second by B and add the resulting equalities, then we obtain the equality

$$A f_1(n + 2) + B f_2(n + 2) = a_1[A f_1(n + 1) + B f_2(n + 1)] + a_2[A f_1(n) + B f_2(n)]),$$

which states that $A f_1(n) + B f_2(n)$ is indeed a solution of (23).

To prove [2], note that if $f(n) = r_1^{n-1}$, then $f(n + 1) = r_1^n$ and $f(n + 2) = r_1^{n+1}$. Multiplying the equality $r_1^2 = a_1 r_1 + a_2$ by r_1^{n-1}, we obtain the equality

$$r_1^{n+1} = a_1 r_1^n + a_2 r_1^{n-1},$$

that is, the equality

$$f(n + 2) = a_1 f(n + 1) + a_2 f(n).$$

We note that if the sequence $\{r_1^{n-1}\}$ is a solution of (23), then any sequence of the form

$$f(n) = r_1^{n+m}, \qquad n = 1, 2, \dots$$

is also a solution of (23). For proof use assertion [1] with $A = r_1^{m+1}$, $B = 0$.

Together, assertions [1] and [2] imply the following rule for the solution of a linear recurrence relation of order two with constant coefficients:

To solve the recurrence relation

$$f(n + 2) = a_1 f(n + 1) + a_2 f(n) \tag{23}$$

form the quadratic equation

$$r^2 = a_1 r + a_2. \tag{24}$$

(This equation is called the *characteristic equation* of the recurrence relation (23).) If Equation (24) has two distinct roots r_1 and r_2, then the general solution of (23) is given by

$$f(n) = C_1 r_1^{n-1} + C_2 r_2^{n-1}.$$

To justify this rule, note that in view of [2] the sequences $f_1(n) = r_1^{n-1}$ and $f_2(n) = r_2^{n-1}$ are solutions of our recurrence relation. But then, in view of [1], the same is true of the sequence $C_1 r_1^{n-1} + C_2 r_2^{n-1}$. It remains to show that every solution of (23) can be put in this form. Any solution $f(n)$ of a recurrence relation of order one is uniquely determined by the values $f(1)$ and $f(2)$. In other words, it suffices to show that the system of equations

$$\begin{cases} C_1 + C_2 = a, \\ C_1 r_1 + C_2 r_2 = b \end{cases}$$

has a solution for every choice of a and b. The reader can easily verify that the numbers

$$C_1 = \frac{b - ar_2}{r_1 - r_2}, \qquad C_2 = \frac{ar_1 - b}{r_1 - r_2}$$

are indeed a solution of this system of equations.

The case when the roots of the characteristic equation (24) coincide will be treated in the next section.

We illustrate the use of our rule by solving the recurrence relation for the Fibonacci numbers:

$$f(n) = f(n-1) + f(n-2). \tag{25}$$

The characteristic equation for this relation is

$$r^2 = r + 1.$$

The roots of this equation are

$$r_1 = \frac{1 + \sqrt{5}}{2}, \qquad r_2 = \frac{1 - \sqrt{5}}{2}.$$

Therefore the general solution of the Fibonacci relation is

$$f(n) = C_1 \left(\frac{1 + \sqrt{5}}{2} \right)^n + C_2 \left(\frac{1 - \sqrt{5}}{2} \right)^n. \tag{26}$$

(Note that in the general solution we used the exponent n rather than $n-1$; this is justified by the remark following the proof of [2] above.)

We defined the Fibonacci numbers as the solution of the recurrence relation (25) satisfying the initial conditions $f(0) = 1$, $f(1) = 2$, that is, as the sequence 1, 2, 3, 5, 8, 13,... . It is frequently convenient to put in the beginning of this sequence the numbers 0 and 1, that is, to consider the sequence 0, 1, 1, 2, 3, 5, 8, 13,... . It is clear that the new sequence satisfies relation (25) and the initial conditions $f(0) = 0$, $f(1) = 1$. Putting $n = 0$ and $n = 1$ in (26), we obtain for C_1 and C_2 the system of equations

$$\begin{cases} C_1 + C_2 = 0, \\ \dfrac{\sqrt{5}}{2}(C_1 - C_2) = 1. \end{cases}$$

The solution of this system is given by the numbers $C_1 = 1/\sqrt{5}$, $C_2 = -1/\sqrt{5}$. Therefore the corresponding solution of relation (25) is

$$f(n) = \frac{1}{\sqrt{5}} \left[\left(\frac{1 + \sqrt{5}}{2} \right)^n - \left(\frac{1 - \sqrt{5}}{2} \right)^n \right]. \tag{27}$$

It is surely surprising (at least at first sight) that this expression takes on integral values for all natural numbers n.

The Case of Equal Roots of the Characteristic Equation

It remains to consider the case when the roots r_1, r_2 of the characteristic equation are equal. In that case the expression $C_1 r_1^{n-1} + C_2 r_2^{n-1}$ is no longer the general solution, for it reduces to

$$f(n) = (C_1 + C_2) r_1^{n-1} = C r_1^{n-1}.$$

The above expression contains a single arbitrary constant C and, in general, it is not possible to choose C so as to satisfy two initial conditions $f(1) = a$, $f(2) = b$.

It is clear that we must find a "suitable" solution different from the solution $f_1(n) = r_1^{n-1}$. In turns out that such a solution is $f_2(n) = n r_1^{n-1}$. In fact, if the quadratic equation $r^2 = a_1 r + a_2$ has two equal roots $r_1 = r_2$, then $a_1 = 2 r_1$, $a_2 = -r_1^2$. Now our equation takes the form

$$r^2 = 2 r_1 r - r_1^2.$$

But then the recurrence relation takes the form

$$f(n + 2) = 2 r_1 f(n + 1) - r_1^2 f(n). \tag{28}$$

With $f_2(n) = n r_1^{n-1}$ we have

$$f_2(n + 2) = (n + 2) r_1^{n+1}, \qquad f_2(n + 1) = (n + 1) r_1^n.$$

Substituting these values in (28), we obtain the obvious identity

$$(n + 2) r_1^{n+1} = 2(n + 1) r_1^{n+1} - n r_1^{n+1}.$$

This shows that $f_2(n) = n r_1^{n-1}$ is a solution of (28). It is a "suitable" second solution of (28) in the sense that

$$f(n) = C_1 r_1^{n-1} + C_2 n r_1^{n-1} = r_1^{n-1}(C_1 + C_2 n)$$

is the general solution of (28). We leave the verification of this fact to the reader.

Linear recurrence relations with constant coefficients whose order exceeds two are solved in much the same way as linear relations of order two. Consider a relation of order k of the form

$$f(n + k) = a_1 f(n + k - 1) + \cdots + a_k f(n). \tag{29}$$

Its characteristic equation is

$$r^k = a_1 r^{k-1} + \cdots + a_k.$$

If the roots $r_1, \ldots, r_k$ of this equation are all distinct, then the general solution of (29) is given by

$$f(n) = C_1 r_1^{n-1} + C_2 r_2^{n-1} + \cdots + C_k r_k^{n-1}.$$

On the other hand, if $r_1 = r_2 = \cdots = r_s$, say, then this root is said to have multiplicity s and we associate with it the s solutions

$$f_1(n) = r_1^{n-1}, \qquad f_2(n) = n r_1^{n-1}, \qquad f_3(n) = n^2 r_1^{n-1}, \qquad \ldots, \qquad f_s(n) = n^{s-1} r_1^{n-1}$$

of the recurrence relation (29). In the general solution of (29) the part corresponding to this root is

$$r_1^{n-1}[C_1 + C_2 n + C_3 n^2 + \cdots + C_s n^{s-1}].$$

To illustrate the general theory, we apply it to the recurrence relation

$$f(n + 4) = 5f(n + 3) - 6f(n + 2) - 4f(n + 1) + 8f(n).$$

Its characteristic equation is

$$r^4 - 5r^3 + 6r^2 + 4r - 8 = 0.$$

The roots of this equation are

$$r_1 = 2, \quad r_2 = 2, \quad r_3 = 2, \quad r_4 = 1.$$

This implies that the general solution of our relation is given by

$$f(n) = 2^{n-1}[C_1 + C_2 n + C_3 n^2] + C_4(-1)^{n-1}.$$

Application of the Theory of Recurrence Relations to Information Theory

Earlier (see p. 75) we considered the problem of computing the number of different messages which can be transmitted in time T if one knows how long it takes to transmit the individual signals used to make up the messages. In this connection, we obtained the recurrence relation

$$f(T) = f(T - t_1) + f(T - t_2) + \cdots + f(T - t_n), \tag{30}$$

with $f(0) = 1$ and $f(T) = 0$, for $T < 0$.

Let T, t_1 ,..., t_n be integers, and let the roots of the characteristic equation of (30) be λ_1 ,..., λ_k . With the roots different, the general solution of (30) can be written in the form

$$f(T) = C_1 \lambda_1^T + \cdots + C_k \lambda_k^T.$$

If λ_1 has the largest absolute value of all the roots, then for large values of T, we have

$$f(T) \sim C_1 \lambda_1^T.$$

This equality enables us to estimate the number of messages which can be transmitted in time T by means of a given system of signals.

Third Solution of the Problem of the Chief Steward

The two solutions of the problem of the chief steward involved recurrence relations. We now derive a formula which yields a solution of these recurrence relations; one which enables us to compute directly the number of acceptable ways of seating the feuding knights. Our derivation makes use of the principle of inclusion and exclusion. Let α_k denote the event when the hostile knights in the kth pair sit side by side. We compute $N(\alpha_1, ..., \alpha_k)$, that is, the number of seating arrangements in which there are k pairs of hostile neighbors. The first pair can be seated at the table in $4n$ ways (the first knight can be seated in $2n$ ways, the second can be seated next to him, on his right, say, and then the two knights can exchange seats). There are $2n - 2$ seats left for the remaining knights who are to be seated so that the feuding knights in the second, third, ..., kth pair are seated side by side. We think of each of these $k - 1$ pairs of knights as a single "object." These $k - 1$ pairs of knights and the $2n - 2k$ remaining knights can be permuted in $(2n - k - 1)!$ ways. If we choose one of these permutations and seat the knights in the free seats at the table in the order determined by the permutation, then we end up with a seating arrangement in which the feuding knights of each of the selected $k - 1$ pairs sit side by side. This remains the case even if some of the hostile neighbors decide to exchange seats. Such exchanges can be carried out in 2^{k-1} ways. It follows that the number of seating arrangements in question is $4n \cdot 2^{k-1}(2n - k - 1)!$. Equivalently,

$$N(\alpha_1 \cdots \alpha_k) = 2^{k+1}n(2n - k - 1)!.$$

Ultimately, we are interested in the number $N(\alpha_1' \cdots \alpha_n')$ of seating arrangements with no hostile neighbors. Since k pairs can be selected in C_n^k ways, the the principle of inclusion and exclusion yields the relation

$$A_n = N(\alpha_1' \cdots \alpha_n')$$

$$= (2n)! - C_n^1 2^2 n(2n - 2)! + C_n^2 2^3 n(2n - 3)! -$$

$$\cdots + (-1)^k C_n^k 2^{k+1} n(2n - k - 1)! + \cdots + (-1)^n 2^{n+1} n!$$

VII

Combinatorics and Series

The method of recurrence relations enables us to solve many combinatorial problems. However, in many cases the necessary recurrence relations are difficult to set up and even more difficult to solve. It is frequently possible to get around these difficulties by using generating functions. Since generating functions are connected with infinite series, we will find it necessary to familiarize ourselves with such series.

Division of Polynomials

The reader is doubtless familiar with division of polynomials. Given two polynomials $f(x)$ and $\varphi(x)$ there always exist polynomials $q(x)$ (*quotient*) and $r(x)$ (*remainder*) such that $f(x) = \varphi(x)\,q(x) + r(x)$ and such that either $r(x) = 0$ or the degree of $r(x)$ is less than the degree of $\varphi(x)$. The polynomial $f(x)$ is called the *dividend* and $\varphi(x)$ is called the *divisor*. If we insist on a zero remainder, then we must admit as quotients infinite power series. Consider, for example, division of 1 by $1 - x$. We have

$$
\begin{array}{c|l}
1 & 1 - x \\
\underline{\mp 1 \pm x} & \overline{1 + x + x^2 + \cdots} \\
\quad x & \\
\quad \underline{\mp x \pm x^2} & \\
\qquad x^2 & \\
\qquad \underline{\mp x^2 \pm x^3} & \\
\qquad\quad x^3 \cdots &
\end{array}
$$

It is clear that this division never ends (much as in the process of obtaining the decimal expansion of the number $\frac{1}{3}$). Using induction, it is easy to prove that all the coefficients of the quotient are equal to 1. It follows that the quotient is the infinite series

$$1 + x + x^2 + \cdots + x^n + \cdots.$$

In general, if $f(x)$ and $\varphi(x)$ are two polynomials,

$$f(x) = a_0 + \cdots + a_n x^n, \qquad \varphi(x) = b_0 + \cdots + b_m x^m,$$

and the constant term b_0 of the polynomial $\varphi(x)$ is different from zero, then division of $f(x)$ by $\varphi(x)$ yields an infinite series

$$c_0 + c_1 x + \cdots + c_k x^k + \cdots. \tag{1}$$

For example, taking $f(x) = 6x^3 - 2x^2 + x + 3$ and $\varphi(x) = x^2 - x + 1$, we obtain

$$
\begin{array}{ll}
\begin{array}{l}
3 + x \ - 2x^2 + 6x^3 \\
\mp 3 \pm 3x \mp 3x^2 \\
\hline
\qquad 4x - 5x^2 + 6x^3 \\
\qquad \pm 4x \pm 4x^2 \mp 4x^3 \\
\hline
\qquad\qquad - x^2 + 2x^3 \\
\qquad\qquad \pm x^2 \mp x^3 \pm x^4 \\
\hline
\qquad\qquad\qquad x^3 + x^4 \\
\qquad\qquad\qquad \mp x^3 \pm x^4 \mp x^5 \\
\hline
\qquad\qquad\qquad\qquad 2x^4 - x^5 \\
\qquad\qquad\qquad\qquad \cdots
\end{array}
&
\left|\ \dfrac{1 - x + x^2}{3 + 4x - x^2 + x^3 + 2x^4 + \cdots}\right.
\end{array}
$$

This situation arises whenever $b_0 \neq 0$ and $r(x) \neq 0$. Only when $\varphi(x)$ divides $f(x)$ (that is, only when $r(x) = 0$) does the series (1) terminate, and the quotient is a polynomial.

Algebraic Fractions and Power Series

Division of a polynomial $f(x)$ by a polynomial $\varphi(x)$ yielded a power series. There arises the question of the connection between this power series and the fraction $f(x)/\varphi(x)$, that is, the question of the meaning to be assigned to the expression

$$\frac{f(x)}{\varphi(x)} = c_0 + c_1 x + \cdots + c_n x^n + \cdots. \tag{2}$$

For example, consider the expansion

$$\frac{1}{1-x} \simeq 1 + x + x^2 + \cdots + x^n + \cdots. \tag{3}$$

We do not write an equality sign, for we have not yet assigned any meaning to a sum with infinitely many summands. In an effort to clear up this issue we substitute different numerical values for x on both sides of relation (3). We start with the value $x = 1/10$. Then the left-hand side of (3) has the value 10/9 and the right-hand side reduces to the infinite series of numbers

$$1 + 0.1 + 0.01 + \cdots + 0.000 \cdots 01 + \cdots.$$

Since we cannot add infinitely many summands, we take first one, then two, then three, and so on, summands. The resulting sums are

$$1, \quad 1.1, \quad 1.11, \quad ..., \quad \underbrace{1.111 \cdots 1}_{n \text{ ones}}, \quad ....$$

It is clear that as n increases these sums come ever closer to $10/9 = 1.11...$, that is, the value of $1/(1-x)$ for $x = 1/10$.

A similar situation arises when we replace x in (3) with the number 1/2. Then the left-hand side of (3) takes on the value 2, and the right-hand side reduces to the infinite series of numbers $1 + 1/2 + 1/4 + 1/8 + \cdots + 1/2^n + \cdots$. Addition of one, two, three, four,..., summands in this series yields the numbers $1, 1\frac{1}{2}, 1\frac{3}{4}, 1\frac{7}{8}, ..., 2 - 1/2^n$. It is clear that as n increases these numbers tend to the number 2.

A different state of affairs arises when we put $x = 4$. Then the left-hand side of (3) takes on the value $-1/3$, and the right-hand side reduces to the infinite series of numbers $1 + 4 + 4^2 + \cdots + 4^n + \cdots$. Successive addition of the terms of this series yields the numbers 1, 5, 21, 85,.... It is clear that these numbers increase indefinitely and so do not approach the number $-1/3$.

We are obviously confronted with two different cases. To distinguish between these cases we introduce the notion of *convergence and divergence of a series of numbers*. Consider the infinite series of numbers

$$a_1 + a_2 + \cdots + a_n + \cdots. \tag{4}$$

We say that this series *converges to the number b* if the difference

$$b - (a_1 + a_2 + \cdots + a_n)$$

tends to zero with increasing n. In other words, for every $\epsilon > 0$, we can find an index N so large that the sum $a_1 + \cdots + a_n$ with N or more summands deviates from b by less than ϵ:

$$|b - (a_1 + \cdots + a_n)| < \epsilon, \quad \text{if} \quad n \geqslant N.$$

If this is the case, then we say that the number b is the sum of the infinite series $a_1 + \cdots + a_n + \cdots$ and we write

$$b = a_1 + \cdots + a_n + \cdots.$$

If the series (4) does not converge to any number b, then we say that this series *diverges*.

Our computations show that

$$\frac{10}{9} = 1 + 0.1 + 0.01 + \cdots + 0.00 \cdots 01 + \cdots,$$

$$2 = 1 + \frac{1}{2} + \frac{1}{4} + \cdots + \frac{1}{2^n} + \cdots,$$

and that the series $1 + 4 + 16 + \cdots + 4^n + \cdots$ diverges.

A more thorough investigation shows that the series $1 + x + \cdots + x^n + \cdots$ converges to $1/(1 - x)$ for $|x| < 1$ and diverges for $|x| \geqslant 1$.

To prove this assertion if suffices to note that

$$1 + x + \cdots + x^n = \frac{1 - x^{n+1}}{1 - x},$$

and that as $n \to \infty$ the term x^{n+1} tends to zero or to infinity according as $|x| < 1$ or $|x| > 1$. For $x = \pm 1$, we obtain the divergent series $1 + 1 + \cdots + 1 + \cdots$ and $1 - 1 + \cdots + 1 - 1 + \cdots$.

Thus for $|x| < 1$,

$$\frac{1}{1 - x} = 1 + x + \cdots + x^n + \cdots. \tag{5}$$

We note that Eq. (5) is the high school formula for the sum of a decreasing geometric progression.

By now we have assigned meaning to the equality

$$\frac{1}{1 - x} = 1 + x + \cdots + x^n + \cdots.$$

This equality states that for $|x| < 1$ the series on the right-hand side converges to $1/(1 - x)$. One frequently says that for $|x| < 1$ the function $1/(1 - x)$ can be expanded in a power series $1 + x + \cdots + x^n + \cdots$.

Quite generally, if division of a polynomial $f(x)$ by a polynomial $\varphi(x)$ yields a power series

$$c_0 + c_1 x + \cdots + c_n x^n + \cdots, \tag{6}$$

then for sufficiently small $|x|$ the series (6) converges to $f(x)/\varphi(x)$.

The size of the domain of convergence depends on the roots of the denominator, that is, on the numbers for which the denominator takes on the value zero. Specifically, if the numbers in question are $x_1, \ldots, x_k$ and r is the least of the numbers $|x_1|, \ldots, |x_k|$, then the series converges for all x with $|x| < r$. For example, the function $1 - x$ takes on the value zero for $x = 1$, and so the power series expansion of $1/(1 - x)$ is valid only for $|x| < 1$. Again, the function $x^2 - 7x + 10$ vanishes for $x_1 = 2$, $x_2 = 5$, and so the power series expansion of $(x - 1)/(x^2 - 7x + 10)$ converges for $|x| < 2$.

We recall the stipulation that the constant term in the denominator $\varphi(x)$ of $f(x)/\varphi(x)$ must be different from zero. Since $\varphi(0) = b_0$, this amounts to saying that zero is not a root of the denominator of the algebraic fraction $f(x)/\varphi(x)$.

In other words, there is always an r such that for $|x| < r$ we have the equality

$$\frac{f(x)}{\varphi(x)} = c_0 + c_1 x + \cdots + c_n x^n + \cdots. \tag{7}$$

Algebraic fractions are not the only functions which can be expanded in power series. For example, in analysis one proves that for all x we have

$$\sin x = x - \frac{x^3}{3!} + \frac{x^5}{5!} - \cdots, \tag{8}$$

$$\cos x = 1 - \frac{x^2}{2!} + \frac{x^4}{4!} - \cdots, \tag{9}$$

$$e^x = 1 + x + \frac{x^2}{2!} + \frac{x^3}{3!} + \cdots. \tag{10}$$

Of greatest interest to us is the expansion (10). Putting $x = 1$, we obtain the expansion

$$e = 1 + 1 + \frac{1}{2!} + \frac{1}{3!} + \cdots. \tag{11}$$

By adding sufficiently many terms in the series (11) we can find the value of e with arbitrary accuracy. A rather accurate value of e is

$$2.7182818289045\ldots .$$

We note one more important result:

A function $f(x)$ cannot have two different power series expansions. In other words, if

$$f(x) = a_0 + a_1 x + \cdots + a_n x^n + \cdots$$

and

$$f(x) = b_0 + b_1 x + \cdots + b_n x^n + \cdots,$$

then

$$a_0 = b_0, \qquad a_1 = b_1, \ldots, \quad a_n = b_n, \ldots.$$

Operations on Power Series

Next we consider operations on power series. Let $f(x)$ and $\varphi(x)$ admit expansions in power series:

$$f(x) = a_0 + a_1 x + \cdots + a_n x^n + \cdots, \tag{12}$$

and

$$\varphi(x) = b_0 + b_1 x + \cdots + b_n x^n + \cdots. \tag{13}$$

Then

$$f(x) + \varphi(x) = (a_0 + a_1 x + \cdots + a_n x^n + \cdots)$$
$$+ (b_0 + b_1 x + \cdots + b_n x^n + \cdots).$$

It turns out that we may rearrange and group terms with the same power of x appearing on the right-hand side of the last equality (this is far from obvious; after all, we are dealing with infinite sums and rearrangement of terms may affect the value of an infinite sum). Then we obtain the equality

$$f(x) + \varphi(x) = (a_0 + b_0) + (a_1 + b_1) x + \cdots + (a_n + b_n) x^n + \cdots. \tag{14}$$

The power series in (14) is called the *sum* of the power series in (12) and (13).

Now we investigate the power series expansion of a product of two functions $f(x)$ and $\varphi(x)$. We have

$$f(x)\,\varphi(x) = (a_0 + a_1 x + \cdots + a_n x^n + \cdots)(b_0 + b_1 x + \cdots + b_n x^n + \cdots). \tag{15}$$

It turns out that, just as in the case of polynomials, the series on the right-hand side of (15) may be multiplied termwise (we omit the proof of this assertion). Now we find the series resulting from termwise multiplication. The constant term of this series is $a_0 b_0$. There are two terms containing x; one is the result of multiplying a_0 by b_1x, and the other is the result of multiplying b_0 by a_1x. Addition of these two terms yields

$$a_0b_1x + a_1b_0x = (a_0b_1 + a_1b_0)\,x.$$

Addition of the terms containing x^2 yields

$$a_0b_2x^2 + a_1b_1x^2 + a_2b_0x^2 = (a_0b_2 + a_1b_1 + a_2b_0)\,x^2.$$

Quite generally, the coefficient of x^n is

$$a_0b_n + a_1b_{n-1} + \cdots + a_kb_{n-k} + \cdots + a_nb_0\,.$$

It follows that

$$f(x)\,\varphi(x) = a_0b_0 + (a_0b_1 + a_1b_0)\,x + \cdots + (a_0b_n + \cdots + a_nb_0)\,x^n + \cdots. \tag{16}$$

The series on the right-hand side of Eq. (16) is called the *product* of the series (12) and (13).

In particular, the square of the series (12) is the series

$$f^2(x) = a_0^2 + 2a_0a_1x + (a_1^2 + 2a_0a_2)x^2 + 2(a_0a_3 + a_1a_2)x^3 + \cdots. \tag{17}$$

Next we consider division of power series. Suppose that the constant term in the series (13) is different from zero. We show that in that case there exists a power series

$$c_0 + c_1x + \cdots + c_nx^n + \cdots, \tag{18}$$

such that

$$(b_0 + b_1x + \cdots + b_nx^n + \cdots)(c_0 + c_1x + \cdots + c_nx^n + \cdots)$$
$$= a_0 + a_1x + \cdots + a_nx^n + \cdots. \tag{19}$$

For proof, we multiply the series on the left-hand side of Eq. (19). The result is the series

$$b_0c_0 + (b_0c_1 + b_1c_0)x + \cdots + (b_0c_n + \cdots + b_nc_0)x^n + \cdots.$$

For this series to coincide with the series (12), it is necessary and sufficient that the following equalities hold:

$$b_0c_0 = a_0\,,$$
$$b_0c_1 + b_1c_0 = a_1\,,$$
$$\cdots \cdots \cdots \cdots \cdots \cdots$$
$$b_0c_n + \cdots + b_nc_0 = a_n\,,$$
$$\cdots \cdots \cdots \cdots \cdots \cdots$$

This is an infinite system of equations for the coefficients c_0, c_1, ..., c_n, ... From the first of these equations, we obtain $c_0 = a_0/b_0$. Substitution of this value of c_0 in the second equation yields the equation

$$b_0 c_1 = a_1 - \frac{b_1 a_0}{b_0},$$

from which we find $c_1 = (a_1 b_0 - b_1 a_0)/b_0^2$. Quite generally, having computed the values of c_0, ..., c_{n-1}, we obtain the following equation for c_n :

$$b_0 c_n = a_n - b_1 c_{n-1} - \cdots - b_n c_0.$$

In view of the assumption $b_0 \neq 0$, this equation can be solved for c_n.

We have proved the existence of the series (18) satisfying relation (19). The series (18) is called the *quotient* of the series (12) and (13), in this order. It can be shown that this series coincides with the power series expansion of the function $f(x)/\varphi(x)$. It follows that power series can be added, multiplied, and divided (the latter if the constant term of the divisor is not zero). Also, the results of these operations correspond to the results of the corresponding operations applied to the functions defined by the various series.

We note that we can now give a new interpretation to the expansion

$$\frac{a_0 + \cdots + a_n x^n}{b_0 + \cdots + b_m x^m} = c_0 + c_1 + \cdots + c_k x^k + \cdots. \tag{20}$$

We can say that the power series $c_0 + c_1 x + \cdots + c_k x^k + \cdots$ is the result of dividing the finite power series $a_0 + \cdots + a_n x^n$ by the finite power series $b_0 + \cdots + b_m x^m$. In other words, Eq. (19) means that

$$(b_0 + \cdots + b_m x^m)(c_0 + c_1 x + \cdots + c_k x^k + \cdots) = a_0 + \cdots + a_n x^n, \tag{21}$$

where the product on the left-hand side of (21) is defined by means of a formula analogous to (16).

Application of Power Series to Proofs of Identities

Power series can be used to prove many identities. The idea here is to expand a function in two different ways in a power series. Since a function has exactly one power series expansion, the corresponding coefficients of the two expansions must be equal. This yields the required identities.

For example, consider the familiar expansion

$$\frac{1}{1-x} = 1 + x + x^2 + \cdots + x^n + \cdots.$$

Squaring both sides, we obtain the equality

$$\frac{1}{(1-x)^2} = 1 + 2x + 3x^2 + \cdots + (n+1)x^n + \cdots. \tag{22}$$

Replacing x by $-x$ in (22), we see that

$$\frac{1}{(1+x)^2} = 1 - 2x + 3x^2 - \cdots + (-1)^n(n+1)x^n + \cdots. \tag{22'}$$

Multiplying the expansions (22) and (22'), we obtain

$$\frac{1}{(1-x)^2} \frac{1}{(1+x)^2} = 1 + [1(-2) + 2 \cdot 1]x + [1 \cdot 3 + 2(-2) + 3 \cdot 1]x^2 + \cdots$$
$$+ [1(-1)^n(n+1) + 2(-1)^{n-2}n + \cdots$$
$$+ (-1)^n(n+1) \cdot 1]x^n + \cdots. \tag{23}$$

It is clear that the coefficient of each odd power of x is zero (these coefficients are sums in which each summand appears twice: once with a plus sign and once with a minus sign). The coefficient of x^{2n} is

$$1(2n+1) - 2 \cdot 2n + 3(2n-1) - \cdots + (2n+1).$$

Now, the function $1/(1-x)^2 (1+x)^2$ can be expanded in a power series differently. We have

$$\frac{1}{(1-x)^2(1+x)^2} = \frac{1}{(1-x^2)^2}.$$

If we replace x by x^2 in (22), then we obtain the following expansion of the function $1/(1-x^2)^2$:

$$\frac{1}{(1-x^2)^2} = 1 + 2x^2 + 3x^4 + \cdots + (n+1)x^{2n} + \cdots. \tag{24}$$

In view of the uniqueness of the power series expansion of a function, the coefficient of x^{2n} in the expansion (23) must be equal to the coefficient of x^{2n} in the expansion (24). This implies the identity

$$1(2n+1) - 2 \cdot 2n + 3(2n-1) - \cdots + (2n+1) \cdot 1 = n+1.$$

Generating Functions

We are now in a position to introduce the main topic of this chapter, namely, the concept of a generating function. Let a_0, a_1,..., a_n,... be a given sequence of numbers. We form the power series

$$a_0 + a_1 x + \cdots + a_n x^n + \cdots .$$

In this series converges in some domain to a function $f(x)$, then this function is called a *generating function* for the sequence a_0, a_1, ..., a_n, ... For example, the formula

$$\frac{1}{1-x} = 1 + x + \cdots + x^n + \cdots$$

implies that the function $1/(1-x)$ is a generating function for the sequence 1, 1, 1, ..., 1, ... Again, Formula (22) shows that the function $1/(1-x)^2$ is a generating function for the sequence 1, 2, 3, 4, ..., n, ...

We are interested in generating functions for sequences a_0, a_1, ..., a_n, ... which are connected in some way with recurrence relations. Using generating functions it is possible to obtain many properties of such sequences. Another topic which we explore in the sequel is the connection between generating functions and solutions of recurrence relations.

The Binomial Expansion

We now obtain a generating function for the finite sequence of numbers C_n^o, C_n^1, ..., C_n^n.

We know from elementary algebra that

$$(a + x)^2 = a^2 + 2ax + x^2$$

and that

$$(a + x)^3 = a^3 + 3a^2 x + 3ax^2 + x^3.$$

These equalities are special cases of a more general formula for the expansion of $(a + x)^n$. We write $(a + x)^n$ as

$$(a + x)^n = \underbrace{(a + x)(a + x) \cdots (a + x)}_{n \text{ times}}. \tag{25}$$

We remove the brackets on the right-hand side of (25) but write out the factors in each summand in the order in which they occur. To get a clue as to the form of the resulting sum we observe that

$$(a + x)^2 = (a + x)(a + x) = aa + ax + xa + xx, \tag{26}$$

and that

$$(a + x)^3 = (a + x)(a + x)(a + x)$$
$$= aaa + aax + axa + axx + xaa + xax + xxa + xxx. \quad (27)$$

It is clear that the sum in (26) consists of all 2-letter arrangements with repetitions of the letters x and a, and that the sum in (27) consists of all 3-letter arrangements with repetitions of the letters x and a. The same type of statement holds in the general case: *after removing brackets in (25) (and prior to any reductions) we obtain a sum consisting of all n-letter arrangements with repetitions of the letters x and a.* Now we count the number of summands which contain k letters x, and thus $n - k$ letters a. Each of these summands is a permutation with repetitions of k letters x and $n - k$ letters a. By Formula (5) in Chap. II, the number of such permutations is

$$P(k, n - k) = C_n^k = \frac{n!}{k!\,(n - k)!}.$$

It follows that after reduction the coefficient of x^k is

$$C_n^k = \frac{n!}{k!\,(n - k)!}.$$

This proves that

$$(a + x)^n = C_n^0 a^n + C_n^1 a^{n-1} x + \cdots + C_n^k a^{n-k} x^k + \cdots + C_n^n x^n. \quad (28)$$

Equation (28) is commonly referred to as *the binomial expansion.* Putting $a = 1$ in this equality, we obtain the relation

$$(1 + x)^n = C_n^0 + C_n^1 x + \cdots + C_n^k x^k + \cdots + C_n^n x^n. \quad (29)$$

This relation shows that the function $(1 + x)^n$ is a generating function for the numbers C_n^k, $k = 0, 1, ..., n$.

This generating function enables us to prove with relative ease many properties of the numbers C_n^k obtained earlier by means of rather intricate arguments.

We prove first that

$$C_{n+1}^k = C_n^k + C_n^{k-1}. \quad (30)$$

For this it suffices to multiply both sides of (29) by $1 + x$. The result is the equality

$$(1 + x)^{n+1} = (C_n^0 + C_n^1 x + \cdots + C_n^k x^k + \cdots + C_n^n x^n)(1 + x).$$

Now we compute the coefficients of x^k in the polynomials on both sides of our equality. By the binomial expansion [see (28) with $n + 1$ in place of n] the coefficient of x^k on the left-hand side is C_{n+1}^k. If we expand the product on the right-hand side, then x^k turns up twice: once as a result of multiplication of $C_n^k x^k$ by 1, and once as a result of multiplication of $C_n^{k-1} x^{k-1}$ by x. Hence the coefficient of x^k on the right-hand side is $C_n^k + C_n^{k-1}$. Since our two polynomials are equal, it follows that $C_{n+1}^k = C_n^k + C_n^{k-1}$.

In proving this relation on p. 35 we made use of combinatorial arguments. Another relation proved (on p. 35) by means of relatively complicated combinatorial arguments is the relation

$$2^n = C_n^0 + C_n^1 + \cdots + C_n^k + \cdots + C_n^n. \tag{31}$$

With Formula (29) at our disposal, all we need do to prove this relation is to put $x = 1$ in (29).

If we put $x = -1$ in (29), then we obtain the relation

$$0 = C_n^0 - C_n^1 + C_n^2 - C_n^3 + \cdots + (-1)^k C_n^k + \cdots + (-1)^n C_n^n.$$

This relation states that *the sum of the C_n^k with k even is equal to the sum of the C_n^k with k odd*:

$$C_n^0 + C_n^2 + C_n^4 + \cdots + C_n^{2m} + \cdots = C_n^1 + C_n^3 + \cdots + C_n^{2m+1} + \cdots. \tag{32}$$

Both sums in (32) are finite. The sum on the left-hand side terminates when $2m > n$, and the sum on the right-hand side terminates when $2m + 1 > n$.

If we put $x = i$, $n = 4m$ in (29), then we obtain a rather interesting result: A simple computation shows that $(1 + i)^4 = -4$, and so, $(1 + i)^{4m} = (-4)^m$. It follows that

$$(-4)^m = C_{4m}^0 + C_{4m}^1 i + C_{4m}^2 i^2 + C_{4m}^3 i^3 + C_{4m}^4 i^4 + \cdots + C_{4m}^{4m} i^{4m}$$

$$= C_{4m}^0 + C_{4m}^1 i - C_{4m}^2 - C_{4m}^3 i + C_{4m}^4 + \cdots + C_{4m}^{4m}.$$

Separating real and imaginary parts, we obtain the relations

$$C_{4m}^1 - C_{4m}^3 + C_{4m}^5 - \cdots - C_{4m}^{4m-1} = 0, \tag{33}$$

$$C_{4m}^0 - C_{4m}^2 + C_{4m}^4 + \cdots + C_{4m}^{4m} = (-4)^m. \tag{34}$$

We leave it to the reader to see what relations are obtained for $n = 4m + 1$, $4m + 2$, $4m + 3$.

The same generating function yields an easy proof of the relation

$$C_{n+m}^s = C_n^0 C_m^s + C_n^1 C_m^{s-1} + \cdots + C_n^k C_m^{s-k} + \cdots + C_n^n C_m^{s-n} \tag{35}$$

(here for $s - k < 0$ we put $C_m^{s-k} = 0$, so that, actually, k varies from 0 to the smaller of the numbers s, n). For proof, we multiply the left- and right-hand sides of the expansions

$$(1 + x)^n = C_n^0 + C_n^1 x + \cdots + C_n^k x^k + \cdots + C_n^n x^n,$$

and

$$(1 + x)^m = C_m^0 + C_m^1 x + \cdots + C_m^s x^s + \cdots + C_m^m x^m,$$

and consider the resulting equality

$$(1 + x)^{n+m} = [C_n^0 + C_n^1 x + \cdots + C_n^k x^k + \cdots + C_n^n x^n]$$

$$\times [C_m^0 + C_m^1 x + \cdots + C_m^s x^s + \cdots + C_m^m x^m].$$

To obtain (35) we expand both sides of this equality and equate the coefficients of x^s on both sides of the resulting equality. A special case of (35) is

$$C_{2n}^n = (C_n^0)^2 + (C_n^1)^2 + \cdots + (C_n^n)^2 \tag{35'}$$

(we recall that $C_n^k = C_n^{n-k}$).

The Multinomial Expansion

The binomial expansion can be used to obtain expansions of rather complex expressions. Consider, for example, the following expansion of $(x + y + z)^4$:

$$(x + y + z)^4 = [(x + y) + z]^4$$

$$= (x + y)^4 + C_4^1 (x + y)^3 z$$

$$+ C_4^2 (x + y)^2 z^2 + C_4^3 (x + y) z^3 + C_4^4 z^4.$$

Now we again apply the binomial expansion to $(x + y)^4$, $(x + y)^3$, $(x + y)^2$, and obtain, ultimately, the expansion

$$(x + y + z)^4 = x^4 + y^4 + z^4 + 4x^3 y + 4x^3 z + 4xy^3$$

$$+ 4y^3 z + 4xz^3 + 4yz^3 + 6x^2 y^2 + 6x^2 z^2$$

$$+ 6y^2 z^2 + 12x^2 yz + 12xy^2 z + 12xyz^2. \tag{36}$$

This stepwise procedure is rather awkward. It does not give us the means of computing directly the coefficient of, say, $x^2 y^4 z^3$ in the expansion of $(x + y + z)^9$.

It is therefore desirable to obtain an analog of the binomial expansion for the expression

$$(x_1 + x_2 + \cdots + x_m)^n. \tag{37}$$

It is not difficult to guess the form of such a formula. When we proved the binomial expansion we saw that in the expansion of $(a + x)^n$ the coefficient of $x^k a^{n-k}$ was $P(k, n - k)$. It is reasonable to guess that in the expansion of $(x_1 + x_2 + \cdots + x_m)^n$ the coefficient of $x_1^{k_1} x_2^{k_2} \cdots x_m^{k_m}$ is $P(k_1, k_2, \ldots, k_m)$. We now show this to be the case.

We write $(x_1 + x_2 + \cdots + x_m)^n$ as a product of n factors and remove brackets. When doing this we write the factors in each term in the order in which they appear. It is clear that the terms of our expansion yield all the n-arrangements with repetitions of the letters $x_1, \ldots, x_m$. Similar terms are terms in which corresponding letters enter the same number of times. It follows that the coefficient of $x_1^{k_1} x_2^{k_2} \cdots x_m^{k_m}$ in our expansion coincides with the number of n-arrangements with repetitions of k_1 replicas of the letter x_1, k_2 replicas of the letter x_2, ..., k_m replicas of the letter x_m. Clearly, this number is the number $P(k_1, k_2, \ldots, k_m)$ of n-permutations with repetitions of k_1 replicas of x_1, k_2 replicas of x_2, ..., k_m replicas of x_m; here $k_1 + k_2 + \cdots + k_m = n$. This proves the *multinomial expansion* formula

$$(x_1 + x_2 + \cdots + x_m)^n = \sum P(k_1, k_2, \ldots, k_m) x_1^{k_1} x_2^{k_2} \cdots x_m^{k_m}, \tag{38}$$

where we sum over all ordered partitions $k_1 + k_2 + \cdots + k_m$ of n into m nonnegative summands. We recall that

$$P(k_1, k_2, \ldots, k_m) = \frac{(k_1 + k_2 + \cdots + k_m)!}{k_1! \, k_2! \cdots k_m!}. \tag{39}$$

It is clear that if the numbers $s_1, s_2, \ldots, s_m$ are a permutation of the numbers $k_1, \ldots, k_m$, then $P(s_1, \ldots, s_m) = P(k_1, \ldots, k_m)$. That is why in the example (36) the coefficients, say, of $x^2 yz$ and xyz^2 are the same. This remark simplifies the computation of the terms of the expansion (37): we find the coefficients corresponding to the partitions $n = k_1 + k_2 + \cdots + k_m$, with $k_1 \geqslant k_2 \geqslant \cdots \geqslant k_m$, and then permute the exponents in all possible ways.

For example, let us compute the expansion of $(x + y + z)^5$. If we disregard the order of the summands, then the number 5 admits the following five partitions into 3 summands each:

$$5 = 5 + 0 + 0, \quad 5 = 4 + 1 + 0, \quad 5 = 3 + 2 + 0,$$
$$5 = 3 + 1 + 1, \quad 5 = 2 + 2 + 1.$$

Now, $P(5, 0, 0) = 1$, $P(4, 1, 0) = 5$, $P(3, 2, 0) = 10$, $P(3, 1, 1) = 20$,
$P(2, 2, 1) = 30$. Therefore,

$$
\begin{aligned}
(x + y + z)^5 = {}& x^5 + y^5 + z^5 + 5x^4y + 5xy^4 + 5x^4z + 5xz^4 \\
& + 5y^4z + 5yz^4 + 10x^3y^2 + 10x^2y^3 + 10x^3z^2 \\
& + 10x^2z^3 + 10y^3z^2 + 10y^2z^3 + 20x^3yz + 20xy^3z \\
& + 20xyz^3 + 30x^2y^2z + 30x^2yz^2 + 30xy^2z^2.
\end{aligned}
$$

Formula (38) enables us to prove certain properties of the number $P(k_1, k_2, ..., k_m)$. For example, if we put in this formula $x_1 = x_2 = \cdots = x_m = 1$, then we see that

$$
m^n = \sum P(k_1, ..., k_m), \tag{40}
$$

where the sum is taken over all ordered partitions of the number n into m nonnegative summands, $n = k_1 + k_2 + \cdots + k_m$.

To obtain another property of these numbers, we multiply both sides of (38) by $x_1 + \cdots + x_m$, expand the left-hand side by the multinomial expansion (38), remove brackets on the right-hand side, and equate corresponding coefficients. The result is the recurrence relation

$$
\begin{aligned}
P(k_1, k_2, ..., k_m) = {}& P(k_1 - 1, k_2, ..., k_m) + P(k_1, k_2 - 1, ..., k_m) + \cdots \\
& + P(k_1, k_2, ..., k_m - 1).
\end{aligned} \tag{41}
$$

One more property of our numbers is obtained by multiplying corresponding sides of the expansions

$$
(x_1 + x_2 + \cdots + x_m)^n = \sum P(k_1, k_2, ..., k_m) x_1^{k_1} x_2^{k_2} \cdots x_m^{k_m}
$$

and

$$
(x_1 + x_2 + \cdots + x_m)^s = \sum P(l_1, l_2, ..., l_m) x_1^{l_1} x_2^{l_2} \cdots x_m^{l_m}
$$

and equating the coefficients of $x_1^{r_1} x_2^{r_2} \cdots x_m^{r_m}$ on both sides of the resulting equality. In this way we obtain the identity

$$
P(r_1, r_2, ..., r_m) = \sum_{k_p + l_p = r_p} P(k_1, k_2, ..., k_m) P(l_1, l_2, ..., l_m). \tag{42}
$$

Here we sum over all arrangements of nonnegative integers $k_1, k_2, ..., k_m$; $l_1, l_2, ..., l_m$ such that $k_1 + k_2 + \cdots + k_m = n$, $l_1 + l_2 + \cdots + l_m = s$ and $k_1 + l_1 = r_1$, $k_2 + l_2 = r_2, ..., k_m + l_m = r_m$. We leave the details of this derivation to the reader.

Formulas (40)–(42) could, of course, be derived without the use of the generating function (38). But then it would be necessary to use n-dimensional analogs of the arguments used to derive Formula (10) on p. 105. Using a generating function we obtain these identities in a practically mechanical way by carrying out some relatively simple algebraic manipulations.

The Binomial Series Expansion

Newton is usually credited with the discovery of the binomial expansion. This is historically incorrect, for the expansion formula for $(a + x)^n$ for positive integral n was well known to Nasir Eddin, Omar Khayyam, and others in the Near East, and, long before Newton, to Blaise Pascal in Western Europe. What Newton did was to generalize the expansion of $(x + a)^n$ to the case of nonintegral exponents. Specifically, Newton was able to prove that *if a is a positive number and* $|x| < a$, *then for any real number α we have the equality*

$$(x + a)^\alpha = a^\alpha + \alpha a^{\alpha-1}x + \frac{\alpha(\alpha - 1)}{1 \cdot 2} a^{\alpha-2}x^2 + \cdots$$

$$+ \frac{\alpha(\alpha - 1) \cdots (\alpha - k + 1)}{1 \cdot 2 \cdots k} a^{\alpha-k}x^k + \cdots. \qquad (43)$$

In general, the expansion (43) yields an infinite series. If α is a natural number n, then the coefficients of x^n, x^{n+1},..., in (43) contain the factor $\alpha - n = 0$, and (43) reduces to the binomial expansion (28).

We prove Formula (43) for $\alpha = -n$, a negative integer, and do not attempt to prove it for arbitrary α.

We wish to show that

$$(x + a)^{-n} = a^{-n} - na^{-n-1}x + \frac{n(n + 1)}{1 \cdot 2} a^{-n-2}x^2$$

$$- \frac{n(n + 1)(n + 2)}{1 \cdot 2 \cdot 3} a^{-n-3}x^3 + \cdots$$

$$+ (-1)^k \frac{n(n + 1) \cdots (n + k - 1)}{1 \cdot 2 \cdots k} a^{-n-k}x^k + \cdots. \qquad (44)$$

If we bear in mind the fact that

$$C^k_{n+k-1} = \frac{n(n + 1) \cdots (n + k - 1)}{1 \cdot 2 \cdots k},$$

then we can replace (44) with

$$\left(1 + \frac{x}{a}\right)^{-n} = 1 - C_n^1\left(\frac{x}{a}\right) + C_{n+1}^2\left(\frac{x}{a}\right)^2$$

$$- C_{n+2}^3\left(\frac{x}{a}\right)^3 + \cdots + (-1)^k C_{n+k-1}^k\left(\frac{x}{a}\right)^k + \cdots. \qquad (44')$$

Instead of proving (44′) we find it convenient to replace x/a in (44′) with $-t$ and to prove the resulting equality

$$(1 - t)^{-n} = 1 + C_n^1 t + C_{n+1}^2 t^2 + \cdots + C_{n+k-1}^k t^k + \cdots. \qquad (45)$$

Our proof is by induction on n. For $n = 1$, we have $C_{n+k-1}^k = C_k^k = 1$, and so (45) reduces to

$$\frac{1}{1-t} = 1 + t + t^2 + \cdots + t^k + \cdots, \qquad (46)$$

which is the well-known formula for the sum of an infinite decreasing geometric progression (we recall that $|t| = |x/a| < 1$).

It remains to show that the validity of (45) for any n implies its validity for $n + 1$, that is, the validity of

$$(1 - t)^{-n-1} = 1 + C_{n+1}^1 t + C_{n+2}^2 t^2 + \cdots + C_{n+k}^k t^k + \cdots. \qquad (47)$$

Multiplication of (47) by $1 - t$ yields

$$(1 - t)^{-n} = [1 + C_{n+1}^1 t + C_{n+2}^2 t^2 + \cdots + C_{n+k-1}^{k-1} t^{k-1} + C_{n+k}^k t^k + \cdots](1 - t).$$

It is easy to see that the coefficient of t^k on the right-hand side is $C_{n+k}^k - C_{n+k-1}^{k-1}$. Since

$$C_{n+k}^k - C_{n+k-1}^{k-1} = C_{n+k-1}^k,$$

(Formula (11) on p. 35), our equality reduces to (45). Just as multiplication of (47) by $1 - t$ yields (45), so conversely, multiplication of (45) by $1/(1 - t)$ must yield (47). Having proved the validity of (45) for $n = 1$, and the fact that whenever (45) is valid for some natural number n it is also valid for $n + 1$, we may conclude that (45) is valid for all natural numbers n.

To obtain (47) directly from (45), we can multiply corresponding sides of (45) and (46). The result is the equality

$$(1 - t)^{-n-1} = (1 + C_n^1 t + C_{n+1}^2 t^2 + \cdots + C_{n+k-1}^k t^k + \cdots)$$

$$\times (1 + t + t^2 + \cdots + t^k + \cdots).$$

If we remove brackets on the right-hand side and make use of the identity

$$C_{n-1}^0 + C_n^1 + C_{n+1}^2 + \cdots + C_{n+k-1}^k = C_{n+k}^k$$

(see Formula (15) p. 36), then we obtain (47).

We have proved (45) for $|t| < 1$. Substitution of the value $t = -1$ in (45) leads to the "remarkable" formula

$$\frac{1}{2^n} = 1 - C_n^1 + C_{n+1}^2 - C_{n+2}^3 + \cdots + (-1)^k C_{n+k-1}^k + \cdots, \qquad (48)$$

which is nonsense; if the series of integers on the right-hand side tended to any number at all, this number would have to be an integer and not the fraction $1/2^n$.

In the 18th century, when the theory of infinite series was not yet fully understood, such mistakes were made even by first rate mathematicians. It took decades of intensive investigations to clarify the concept of the sum of an infinite series, of its existence and nonexistence. It should be noted that at the end of the 19th century the concept of the sum of an infinite series was significantly generalized; in fact, there are definitions of summability in which Formula (48) holds. However, all these questions are beyond the scope of this book.

Comparison of the expansions

$$(1 + t)^{-n} = 1 - C_n^1 t + C_{n+1}^2 t^2 - \cdots + (-1)^k C_{n+k-1}^k t^k + \cdots \qquad (49)$$

and

$$(1 + t)^n = 1 + C_n^1 t + C_n^2 t^2 + \cdots + C_n^k t^k + \cdots + t^n \qquad (50)$$

indicates the appropriateness of the definition

$$C_{-n}^k = (-1)^k C_{n+k-1}^k$$

introduced on p. 96 (Formula (2)). Again, the absence of negative powers of t in (49) and (50) suggests that it is reasonable to put for negative k, $C_n^k = 0$. Similar considerations support the definition $C_n^k = 0$, for $0 \leqslant n < k$.

Computation of Square Roots

While we proved the binomial expansion (Formula (43)) for integral exponents only, this formula, as already mentioned, is also valid for fractional (as well as irrational) values of the exponent. We consider two expansions, for $n = 1/2$ and $n = -1/2$, and take their validity for granted.

For $n = 1/2$, the binomial expansion formula yields the equality

$$(1 + x)^{1/2} = 1 + \frac{1}{2}x + \frac{\frac{1}{2}(\frac{1}{2} - 1)}{1 \cdot 2}x^2 + \frac{\frac{1}{2}(\frac{1}{2} - 1)(\frac{1}{2} - 2)}{1 \cdot 2 \cdot 3}x^3 + \cdots$$
$$+ \frac{\frac{1}{2}(\frac{1}{2} - 1) \cdots (\frac{1}{2} - k + 1)}{1 \cdot 2 \cdots k}x^k + \cdots, \tag{51}$$

which can be rewritten in the form

$$(1 + x)^{1/2} = 1 + \frac{1}{2}x - \frac{1}{2 \cdot 4}x^2 + \frac{1 \cdot 3}{2 \cdot 4 \cdot 6}x^3 - \cdots$$
$$+ (-1)^{k-1}\frac{1 \cdot 3 \cdots (2k - 3)}{2 \cdot 4 \cdots 2k}x^k + \cdots.$$

Similarly, for $n = -1/2$ we obtain

$$(1 + x)^{-1/2} = 1 - \frac{1}{2}x + \frac{1 \cdot 3}{2 \cdot 4}x^2 - \cdots$$
$$+ (-1)^k\frac{1 \cdot 3 \cdots (2k - 1)}{2 \cdot 4 \cdots 2k}x^k + \cdots. \tag{52}$$

By making use of the identity

$$\frac{1 \cdot 3 \cdots (2k - 1)}{2 \cdot 4 \cdots 2k} = \frac{(2k)!}{2^{2k}(k!)^2} = \frac{1}{2^{2k}}C_{2k}^k$$

we can write our two expansions as

$$(1 + x)^{-1/2} = 1 - \frac{1}{2^2}C_2^1 x + \frac{1}{2^4}C_4^2 x^2 - \cdots$$
$$+ \frac{(-1)^k}{2^{2k}}C_{2k}^k x^k + \cdots \tag{53}$$

and

$$(1 + x)^{1/2} = 1 + \frac{1}{2}x - \frac{1}{2 \cdot 2^3}C_2^1 x^2$$
$$+ \frac{1}{3 \cdot 2^5}C_4^2 x^3 - \cdots + \frac{(-1)^{k-1}}{k \cdot 2^{2k-1}}C_{2k-2}^{k-1} x^k + \cdots. \tag{54}$$

These expansions are valid for $|x| < 1$. They can be used to obtain square roots of numbers with arbitrary accuracy. For example,

$$\sqrt{30} = \sqrt{25 + 5} = 5\sqrt{1 + 0.2} = 5(1 + 0.2)^{1/2}$$
$$= 5\left[1 + \frac{1}{2} \cdot 0.2 - \frac{1}{2 \cdot 4}0.2^2\right.$$
$$\left. + \frac{1 \cdot 3}{2 \cdot 4 \cdot 6} \cdot 0.2^3 - \cdots\right] = 5.4775 \cdots.$$

What interests us more than the application of these formulas to the extraction of square roots is their application to the derivation of relations connecting the binomial coefficients. To obtain one such relation we square (53). The rule for multiplication of power series implies that the coefficient of x^k on the right-hand side of the resulting equality is

$$\frac{(-1)^k}{2^{2k}} [C_{2k}^k + C_2^1 C_{2k-2}^{k-1} + C_4^2 C_{2k-4}^{k-2} + \cdots + C_{2k}^k].$$

On the left-hand side we obtain

$$[(1+x)^{-1/2}]^2 = \frac{1}{1+x}.$$

We know that

$$\frac{1}{1+x} = 1 - x + x^2 - \cdots + (-1)^k x^k + \cdots.$$

Equating the coefficients of x^k in the two expansions, we obtain the identity

$$C_{2k}^k + C_2^1 C_{2k-2}^{k-1} + C_4^2 C_{2k-4}^{k-2} + \cdots + C_{2k}^k = 2^{2k}. \tag{55}$$

Similarly, (54) yields the identity

$$\frac{C_{2k-4}^{k-2}}{1 \cdot (k-1)} + \frac{C_2^1 C_{2k-6}^{k-3}}{2(k-2)} + \frac{C_4^2 C_{2k-8}^{k-4}}{3(k-3)} + \cdots + \frac{C_{2k-4}^{k-2}}{(k-1) \cdot 1} = \frac{C_{2k-2}^{k-1}}{k} \tag{56}$$

which holds for $k \geqslant 2$.

Multiplication of corresponding sides of the expansions (53) and (54) yields

$$1 = \left[1 + \frac{1}{2} x - \frac{1}{2 \cdot 2^3} C_2^1 x^2 + \frac{1}{3 \cdot 2^5} C_4^2 x^3 - \cdots \right.$$
$$+ \frac{(-1)^{k-1}}{k \cdot 2^{2k-1}} C_{2k-2}^{k-1} x^k + \cdots \left]\right[1 - \frac{1}{2^2} C_2^1 x$$
$$+ \frac{1}{2^4} C_4^2 x^2 + \cdots + \frac{(-1)^4}{2^{2k}} C_{2k}^k x^k + \cdots \right]. \tag{57}$$

If we remove brackets on the right-hand side of this equality, then we obtain a power series all of whose coefficients (other than the constant term) are zero. This implies the identity

$$C_{2k-2}^{k-1} + \frac{1}{2} C_2^1 C_{2k-4}^{k-2} + \frac{1}{3} C_4^2 C_{2k-6}^{k-3} + \cdots + \frac{1}{k} C_{2k-2}^{k-1} = \frac{1}{2} C_{2k}^k, \tag{58}$$

which holds for $k \geqslant 1$.

Finally we note that

$$(1 + x)^{1/2} (1 + x)^{-1} = (1 + x)^{-1/2}.$$

Replacing the factors by appropriate power series, we have

$$\left(1 + \frac{1}{2}x - \frac{1}{2 \cdot 2^3} C_2^1 x^2 + \cdots + \frac{(-1)^{k-1}}{k \cdot 2^{2k-1}} C_{2k-2}^{k-1} x^k + \cdots\right)$$

$$\times (1 - x + x^2 - x^3 + \cdots + (-1)^k x^k + \cdots)$$

$$= 1 - \frac{1}{2^2} C_2^1 x + \frac{1}{2^4} C_4^2 x^2 - \cdots + \frac{(-1)^k}{2^{2k}} C_{2k}^k x^k + \cdots.$$

Removing brackets and equating the coefficients of x^k on both sides of this equality, we obtain the identity

$$1 - \frac{1}{2 \cdot 2^2} C_2^1 - \frac{1}{3 \cdot 2^4} C_4^2 - \cdots - \frac{1}{k \cdot 2^{2k-2}} C_{2k-2}^{k-1} = \frac{1}{2^{2k-1}} C_{2k}^k. \quad (59)$$

Generating Functions and Recurrence Relations

We have already mentioned the existence of a close connection between generating functions and recurrence relations. Our familiarity with division of polynomials permits us to consider this issue in some detail.

Let

$$f(x) = a_0 + a_1 x + \cdots + a_n x^n$$

and

$$\varphi(x) = b_0 + b_1 x + \cdots + b_m x^m$$

be two polynomials and let $b_0 \neq 0$. We also assume that $n < m$, that is, that the algebraic fraction $f(x)/\varphi(x)$ is proper (otherwise we could always write $f(x)/\varphi(x)$ as the sum of a polynomial and a proper algebraic fraction).

We know that if

$$\frac{f(x)}{\varphi(x)} = c_0 + c_1 x + \cdots + c_k x^k + \cdots, \quad (60)$$

then

$$a_0 + a_1 x + \cdots + a_n x^n = (b_0 + b_1 x + \cdots + b_m x^m)(c_0 + c_1 x + \cdots + c_k x^k + \cdots).$$

We remove brackets on the right-hand side of this equality and equate the coefficients of equal powers of x on both sides. As a result we obtain two sets of relations. The first m relations are

$$b_0 c_0 = a_0,$$
$$b_0 c_1 + b_1 c_0 = a_1,$$
$$b_0 c_2 + b_1 c_1 + b_2 c_0 = a_2, \tag{61}$$
$$\cdot \cdot \cdot \cdot \cdot \cdot \cdot \cdot \cdot \cdot \cdot \cdot \cdot$$
$$b_0 c_{m-1} + b_1 c_{m-2} + \cdots + b_{m-1} c_0 = a_{m-1}$$

(if $n < m - 1$, then we put $a_{n+1} = \cdots + a_{m-1} = 0$). The remaining relations are of the form

$$b_0 c_{m+k} + b_1 c_{m+k-1} + \cdots + b_m c_k = 0, \qquad k = 0, 1,\dots \tag{62}$$

($f(x)$ contains no terms with x^m, x^{m+1}, and so on). Thus the coefficients $c_0, c_1, \dots, c_k, \dots$ of the series (60) satisfy the recurrence relation (62). The coefficients of this recurrence relation depend only on the denominator of the algebraic fraction. The coefficients of the numerator of the fraction enter relations (61) from which we compute the first m terms $c_0, c_1, \dots, c_{m-1}$ of the recurrent sequence.

Conversely, given the recurrence relation (62) and the terms $c_0, c_1, \dots, c_{m-1}$ we can use Formulas (61) to compute $a_0, \dots, a_{m-1}$. Then the algebraic fraction

$$\frac{f(x)}{\varphi(x)} = \frac{a_0 + a_1 x + \cdots + a_{m-1} x^{m-1}}{b_0 + b_1 x + \cdots + b_m x^m} \tag{63}$$

is a generating function for the sequence of numbers $c_0, c_1, \dots, c_k, \dots$.

At first sight it would seem that little is gained from replacing the recurrence relation by a generating function. After all, division of the numerator by the denominator brings us right back to the recurrence relation (62). But what counts is the fact that we can apply to the fraction (63) certain algebraic transformations which simplify the problem of finding the numbers c_k.

Decomposition into Elementary Fractions (Partial Fraction Decomposition)

We show how to solve recurrence relations by applying algebraic transformations to the corresponding generating functions.

Assume that we have managed to write the denominator of the fraction (63) as a product of linear factors

$$\varphi(x) = b_m (x - \alpha_1)^r \cdots (x - \alpha_k)^s.$$

(This requires the solution of the characteristic equation $b_0 + \cdots + b_m x^m = 0$ of relation (62).) Then it is clear that the fraction (63) is the sum of elementary algebraic fractions of the form

$$\frac{A_{11}}{(x - \alpha_1)^r}, \quad \frac{A_{12}}{(x - \alpha_1)^{r-1}}, \quad \ldots, \quad \frac{A_{1r}}{x - \alpha_1},$$

$$\cdots \cdots \cdots \cdots \cdots \cdots \cdots \cdots \cdots \cdots$$

$$\frac{A_{k1}}{(x - \alpha_k)^s}, \quad \frac{A_{k2}}{(x - \alpha_k)^{s-1}}, \quad \ldots, \quad \frac{A_{ks}}{x - \alpha_k};$$

in other words,

$$\frac{a_0 + \cdots + a_{m-1} x^{m-1}}{b_m (x - \alpha_1)^r \cdots (x - \alpha_k)^s} = \frac{A_{11}}{(x - \alpha_1)^r} + \cdots + \frac{A_{1r}}{x - \alpha_1} + \cdots$$

$$+ \frac{A_{k1}}{(x - \alpha_k)^s} + \cdots + \frac{A_{ks}}{x - \alpha_k}. \qquad (64)$$

To find the coefficients $A_{11}, \ldots, A_{ks}$ we must multiply both sides of (64) by the denominator $(x - \alpha_1)^r \cdots (x - \alpha_k)^s$, remove brackets, and equate coefficients of the same power of x. The required coefficients can be computed from the resulting system of equations.

Sometimes we can get by without solving a system of equations. For example, suppose we wish to obtain the decomposition into elementary fractions of

$$\frac{x^3 - 2x^2 + 6x + 1}{x^4 - 5x^2 + 4}.$$

Since

$$x^4 - 5x^2 + 4 = (x^2 - 1)(x^2 - 4) = (x - 1)(x + 1)(x - 2)(x + 2),$$

the required decomposition must be of the form

$$\frac{x^3 - 2x^2 + 6x + 1}{(x - 1)(x + 1)(x - 2)(x + 2)} = \frac{A}{x - 1} + \frac{B}{x + 1} + \frac{C}{x - 2} + \frac{D}{x + 2}.$$

Multiplication by the common denominator yields the relation

$$x^3 - 2x^2 + 6x + 1 = A(x + 1)(x - 2)(x + 2) + B(x - 1)(x - 2)(x + 2)$$
$$+ C(x - 1)(x + 1)(x + 2) + D(x - 1)(x + 1)(x - 2).$$

This relation is to be an identity in x. For $x = 1$, all the terms of the right-hand side beginning with the second term vanish, and we have $-6A = 6$, so that

$A = -1$. Similarly, putting $x = -1$, $x = 2$, $x = -2$, we find that $B = -4/3$, $C = 13/12$, $D = 9/4$. It follows that

$$\frac{x^3 - 2x^2 + 6x + 1}{x^4 - 5x^2 + 4} = -\frac{1}{x - 1} - \frac{4}{3(x + 1)} + \frac{13}{12(x - 2)} + \frac{9}{4(x + 2)}.$$

(65)

Fractions of the form $A/(x - \alpha)^r$ are expanded by means of the binomial series formula. For example,

$$\frac{13}{12(x - 2)} = -\frac{13}{24}\left(1 - \frac{x}{2}\right)^{-1} = -\frac{13}{24}\left[1 + \frac{x}{2} + \frac{x^2}{2^2} + \cdots + \frac{x^n}{2^n} + \cdots\right].$$

Applying this expansion to the various fractions in (65), we obtain

$$\frac{x^3 - 2x^2 + 6x + 1}{x^4 - 5x^2 + 4} = (1 + x + x^2 + \cdots + x^n + \cdots)$$

$$-\frac{4}{3}(1 - x + x^2 - \cdots + (-1)^n x^n + \cdots)$$

$$-\frac{13}{24}\left(1 + \frac{x}{2} + \frac{x^2}{2^2} + \cdots + \frac{x^n}{2^n} + \cdots\right)$$

$$+\frac{9}{8}\left(1 - \frac{x}{2} + \frac{x^2}{2^2} - \cdots + \frac{(-1)^n x^n}{2^n} + \cdots\right).$$

Grouping terms involving the same power of x, we see that the coefficient of x^n is

$$c_n = 1 - \frac{4}{3}(-1)^n - \frac{13}{24 \cdot 2^n} + \frac{9(-1)^n}{8 \cdot 2^n}.$$

To sum up: The problem of expanding an algebraic fraction into a power series is equivalent to the problem of solving a certain linear recurrence relation with prescribed initial conditions. Hence one way of solving a linear recurrence relation with constant coefficients is to decompose the appropriate algebraic fraction into elementary fractions and then expand the elementary fractions into power series by the binomial series formula.

Specifically, given the recurrence relation (62) and the initial values $c_0, \ldots, c_{m-1}$, we use (61) to compute $a_0, \ldots, a_{m-1}$. These numbers determine the numerator of the fraction

$$\frac{f(x)}{\varphi(x)} = c_0 + c_1 x + \cdots + c_k x^k + \cdots$$

whose denominator is $b_0 + \cdots + b_m x^m$.

Next we decompose the fraction $f(x)/\varphi(x)$ into elementary fractions which we expand in power series by the binomial series formula. The element c_k of the solution of our recurrence relation is the coefficient of x^k in the power series expansion of $f(x)/\varphi(x)$.

For example, consider the recurrence relation

$$c_{k+2} - 5c_{k+1} + 6c_k = 0, \tag{66}$$

with initial conditions $c_0 = 1$, $c_1 = -2$. Here $b_0 = 1$, $b_1 = -5$, $b_2 = 6$. Formula (61) implies that

$$a_0 = b_0 c_0 = 1, \qquad a_1 = b_0 c_1 + b_1 c_0 = -7.$$

It follows that the numerator of the fraction

$$\frac{f(x)}{\varphi(x)} = c_0 + c_1 x + \cdots + c_k x^k + \cdots$$

is $1 - 7x$. From our recurrence relation (66) it follows that the denominator of our fraction is $1 - 5x + 6x^2$. This means that to obtain the solution of our problem we must expand the fraction

$$\frac{1 - 7x}{6x^2 - 5x + 1}$$

in a power series. Now, $6x^2 - 5x + 1 = (2x - 1)(3x - 1)$. Hence,

$$\frac{1 - 7x}{6x^2 - 5x + 1} = \frac{A}{2x - 1} + \frac{B}{3x - 1}.$$

But then

$$1 - 7x = A(3x - 1) + B(2x - 1).$$

Putting $x = 1/3$, we obtain $B = 4$. Putting $x = 1/2$, we obtain $A = -5$, so that

$$\frac{1 - 7x}{6x^2 - 5x + 1} = \frac{-5}{2x - 1} + \frac{4}{3x - 1}$$

$$= 5(1 - 2x)^{-1} - 4(1 - 3x)^{-1}$$

$$= 5(1 + 2x + \cdots + 2^n x^n + \cdots) - 4(1 + 3x + \cdots + 3^n x^n + \cdots).$$

This implies that

$$c_n = 5 \cdot 2^n - 4 \cdot 3^n.$$

On a Certain Nonlinear Recurrence Relation

The problem of successive partitions of a sequence (see p. 123) involved the recurrence relation

$$T_n = T_0 T_{n-1} + T_1 T_{n-2} + \cdots + T_{n-1} T_0. \tag{67}$$

with $T_0 = 1$. To solve this relation we reduced our problem to one variant of the problem of queues (see p. 63). This approach is not very satisfactory since it is indirect and since the solution of the problem of queues is not particularly elegant.

We now present a direct solution of the recurrence relation (67) which consists in computing the generating function

$$f(x) = T_0 + T_1 x + T_2 x^2 + \cdots + T_n x^n + \cdots. \tag{68}$$

We put

$$F(x) \equiv xf(x) = T_0 x + T_1 x^2 + \cdots + T_n x^{n+1} + \cdots \tag{69}$$

and square $F(x)$. The result is

$$F^2(x) = T_0^2 x^2 + (T_0 T_1 + T_1 T_0)x^3 + \cdots + (T_0 T_{n-1} + \cdots + T_{n-1} T_0)x^{n+1} + \cdots.$$

In view of the recurrence relation (67),

$$T_0 T_{n-1} + \cdots + T_{n-1} T_0 = T_n.$$

Hence

$$F^2(x) = T_1 x^2 + T_2 x^3 + \cdots + T_n x^{n+1}.$$

This series is equal to $F(x) - T_0 x$. Since $T_0 = 1$, we obtain

$$F^2(x) = F(x) - x. \tag{70}$$

Solving this quadratic equation for $F(x)$, we find that

$$F(x) = \frac{1 - \sqrt{1 - 4x}}{2}.$$

Our choice of the minus sign before the square root is dictated by the fact that choice of the plus sign implies $F(0) = 1$, whereas (69) implies that $F(0) = 0$. By Formula (54),

$$\sqrt{1 - 4x} = (1 - 4x)^{1/2} = 1 - 2x - \frac{2}{2} C_2^1 x^2$$

$$- \frac{2}{3} C_4^2 x^3 - \cdots - \frac{2}{n+1} C_{2n}^n x^{n+1} - \cdots.$$

Hence

$$F(x) = \frac{1}{2}\left[1 - \left(1 - 2x - \cdots - \frac{2}{n+1}\, C_{2n}^n x^{n+1} - \cdots\right)\right]$$

$$= x + C_2^1 x^2 + \cdots + \frac{1}{n+1}\, C_{2n}^n x^{n+1} + \cdots. \tag{71}$$

Equating coefficients in (69) and (71), we see that $T_n = [1/(n+1)]\, C_{2n}^n$, in full agreement with the solution obtained earlier by means of combinatorial arguments (see p. 123).

Generating Functions and Partitions of Numbers

In Chap. IV we solved various combinatorial problems involving partitions of numbers. Such problems can be easily solved by means of *generating functions*. To explain the underlying idea, we consider the series

$$a_0 + a_1 x + \cdots + a_n x^n + \cdots,$$

where a_n denotes the number of (unordered) partitions of n. It is frequently possible to form an algebraic expression $f(x)$ such that the result of removing brackets in $f(x)$ is a sum in which x^n enters exactly a_n times. Then

$$f(x) = a_0 + a_1 x + \cdots + a_n x^n + \cdots,$$

that is, $f(x)$ is a generating function for the sequence $a_0, a_1, \ldots, a_n, \ldots$.

For example, consider the problem of computing *the number of partitions of a positive integer N into summands $n_1, \ldots, n_k$ subject to the restriction that each of these numbers enters a partition at most once and the order of the summands in a partition is disregarded.*

To solve this problem, we form the product

$$(1 + x^{n_1})(1 + x^{n_2}) \cdots (1 + x^{n_k}). \tag{72}$$

If we remove brackets in this product, then we obtain a sum of the form $1 + x^{m_1} + \cdots + x^{m_s}$, where the numbers $m_1, \ldots, m_s$ are sums of the different combinations of numbers selected from among the numbers $n_1, \ldots, n_k$. It follows that the number of appearances of x^N in this sum is equal to the number of partitions of N of the required type.

For example, to compute the number of ways of paying 78 cents using each of the coins 1, 2, 3, 5, 10, 15, 20, 50 at most once, we form the product

$$(1 + x)(1 + x^2)(1 + x^3)(1 + x^5)(1 + x^{10})(1 + x^{15})(1 + x^{20})(1 + x^{50}) \tag{73}$$

remove brackets, and compute the coefficient of x^{78}.

Next we use generating functions to solve the following problem:

In how many ways can we pay 29 cents using 3– and 5-cent coins?

In this problem we are required to find the number of partitions of 29 into summands 3 and 5 without regard to order. Another way of stating this problem is to say that we wish to find the number of nonnegative solutions of the equation $3m + 5n = 29$.

We form the product

$$f(x) = (1 + x^3 + x^6 + \cdots + x^{3m} + \cdots)(1 + x^5 + x^{10} + \cdots + x^{5n} + \cdots).$$
$$(74)$$

Here the exponents of x in the first factor are the successive nonnegative multiples of 3, and the exponents of x in the second factor are the successive nonnegative multiples of 5. But then, after removal of brackets, the coefficient of x^N is equal to the number of nonnegative solutions of the equation $3m + 5n = N$. In particular, the coefficient of x^{29} furnishes the solution of our problem.

To avoid removing brackets, we can use the formula for an infinite geometric progression and write (74) as

$$f(x) = \frac{1}{1 - x^3} \frac{1}{1 - x^5} = \frac{1}{1 - x^3 - x^5 + x^8}.$$

Then we divide the numerator by the denominator in the usual way except that we arrange the two polynomials in increasing, rather than decreasing, powers of x. The first few steps of the division process are:

$$
\begin{array}{l}
1 \\ \hline
x^3 + x^5 - x^8 \\ \hline
\quad x^5 + x^6 - x^{11} \\ \hline
\quad\quad x^6 + x^8 + x^{10} - x^{11} - x^{13} \\ \hline
\quad\quad\quad x^8 + x^9 + x^{10} - x^{13} - x^{14}
\end{array}
\qquad
\begin{array}{l}
\left| \dfrac{1 - x^3 - x^5 + x^8}{1 + x^3 + x^5 + x^6 + x^8 + \cdots} \right.
\end{array}
$$

Continued division ultimately yields the required coefficient of x^{29}.

This problem is a special case of the following problem:

Find the number of unordered partitions of N into summands a, b,..., m.

The relevant generating function is

$$f(x) = (1 + x^a + x^{2a} + \cdots + x^{ta} + \cdots)$$
$$\times (1 + x^b + x^{2b} + \cdots + x^{sb} + \cdots)$$
$$\times (1 + x^m + x^{2m} + \cdots + x^{qm} + \cdots)$$
$$= \frac{1}{(1 - x^a)(1 - x^b) \cdots (1 - x^m)}.$$
$$(75)$$

For example, in the problem of changing a dime into coins of smaller denominations (see p. 80) the required generating function is

$$f(x) = \frac{1}{(1 - x)(1 - x^2)(1 - x^3)(1 - x^5)},$$

that is,

$$f(x) = \frac{1}{1 - x - x^2 + x^7 - x^9 - x^{10} + x^{11}}.$$

The first few steps in the division process are

$$
\begin{array}{l|l}
1 & 1 - x - x^2 + x^7 - x^9 - x^{10} + x^{11} \\
\hline
x + x^2 - x^7 + x^9 + x^{10} - x^{11} & 1 + x + 2x^2 + 3x^3 + 5x^4 + \cdots \\
\hline
2x^2 + 3x^3 - x^7 - x^8 + x^9 + 2x^{10} - x^{12} \\
\hline
3x^3 + 2x^4 - x^7 - x^8 - x^9 + 2x^{10} + 2x^{11} + x^{12} - 2x^{13} \\
\hline
5x^4 + 3x^5 - x^7 - x^8 - x^9 - x^{10} + 2x^{11} - 4x^{12} + x^{13} - 3x^{14}
\end{array}
$$

The coefficient of x^{10} yields the answer to our problem.

Getting the answer by means of the usual division process can be quite a chore. A different procedure consists in writing the algebraic fraction as a power series with undetermined coefficients

$$\frac{1}{1 - x - x^2 + x^7 - x^9 - x^{10} + x^{11}} = A_0 + A_1 x + A_2 x^2 + \cdots + A_n x^n + \cdots$$

and multiplying both sides of this equality by the denominator on the left-hand side. Then the coefficient of x^n is

$$A_n - A_{n-1} - A_{n-2} + A_{n-7} - A_{n-9} - A_{n-10} + A_{n-11}.$$

For $n \geqslant 1$, the coefficient of x^n on the left-hand side is zero. Hence for $n \geqslant 1$, the coefficients A_n must satisfy the recurrence relation

$$A_n = A_{n-1} + A_{n-2} - A_{n-7} + A_{n-9} + A_{n-10} - A_{n-11}.$$

The initial conditions are $A_n = 0$, for $n < 0$, and $A_0 = 1$. It is now easy to compute successively the various A_n.

The problems just considered involved unordered partitions of a number. As an illustration of the use of generating functions to count ordered partitions we consider the problem of the university applicant (see p. 76). Here we are required to find the number of ordered partitions of 17 into four summands which take on the values $3, 4, 5$. The required generating function is $(x^3 + x^4 + x^5)^4$. In fact, when we remove brackets in the expression $f(x) = (x^3 + x^4 + x^5)^4$, then we obtain a sum of all possible terms of the form $x^\alpha x^\beta x^\gamma x^\delta$, where each entry in the quadruple α, β, γ, δ is one of the numbers 3, 4, 5. In other words, the coefficient of x^N gives the number of *ordered* partitions of N into four summands whose values are 3, 4, or 5.

When expanding the expression $(x^3 + x^4 + x^5)^4 = x^{12}(1 + x + x^2)^4$ we can make use of the multinomial expansion formula. A simpler way is to note that $1 + x + x^2 = (1 - x^3)/(1 - x)$, so that

$$f(x) = \frac{x^{12}(1 - x^3)^4}{(1 - x)^4} = x^{12}(1 - x^3)^4(1 - x)^{-4}.$$

By the binomial expansion, we have

$$(1 - x^3)^4 = 1 - 4x^3 + 6x^6 - 4x^9 + x^{12},$$

and, by the binomial series expansion, we have

$$(1 - x)^{-4} = 1 + 4x + 10x^2 + 20x^3 + \cdots + \frac{4 \cdot 5 \cdots (n + 3)}{1 \cdot 2 \cdots n} x^n + \cdots,$$

so that

$$f(x) = x^{12}(1 - 4x^3 + 6x^6 - 4x^9 + x^{12})$$
$$\times (1 + 4x + 10x^2 + 20x^3 + 35x^4 + 56x^5 + \cdots).$$

From this we find that the coefficient of x^{17} has the value 16. This means that the number of required partitions of 17 is 16.

In general, to find the number of ordered partitions of n into k summands which take on the values $n_1, \ldots, n_s$ we use the generating function

$$f(x) = (x^{n_1} + x^{n_2} + \cdots + x^{n_s})^k. \tag{76}$$

The problem is simplified if the numbers $n_1, \ldots, n_s$ form an arithmetic progression, for then $x^{n_1}, \ldots, x^{n_s}$ form a geometric progression and the expression for $f(x)$ is relatively easy to deal with.

Consider, for example, the problem of computing the number of ways in which it is possible to obtain a score of 25 when throwing 7 dice. In this case the generating function is

$$f(x) = (x + x^2 + \cdots + x^6)^7. \tag{77}$$

Using the formula for the sum of a geometric progression, we see that

$$f(x) = \frac{x^7(1 - x^6)^7}{(1 - x)^7} = x^7(1 - x^6)^7(1 - x)^{-7}.$$

Now we apply the binomial expansion to $(1 - x^6)^7$ and the binomial series expansion to $(1 - x)^{-7}$ and obtain the equality

$$f(x) = x^7(1 - 7x^6 + 21x^{12} - 35x^{18} + 35x^{24} - 21x^{30} + 7x^{36} - x^{42})$$
$$\times \ (1 + 7x + 28x^2 + 84x^3 + 210x^4 + 462x^5 + \cdots),$$

from which it is easy to find the coefficient of x^{25}, the answer to our problem.

Many other problems discussed in Chap. IV can be solved by means of generating functions using techniques similar to those employed above.

Summary of Results on Combinatorial Distributions

(1)   The number of distributions of n different objects (no two alike) in r different boxes (no two alike) with empty boxes allowed is r^n.

(2)   The number of distributions of n different objects in r different boxes with no box empty is equal to the product of $n!$ by the coefficient of x^n in the power series expansion of $(e^x - 1)^r$. This number can be written as

$$r^n - \frac{r}{1}(r - 1)^n + \frac{r(r - 1)}{1 \cdot 2}(r - 2)^n - \cdots.$$

(3)   If in (2) we suppose the boxes all alike, then the number of distributions is $r!$ times smaller (than in (2)).

(4)   The number of distributions of n like objects in r different boxes with no box empty is C_{n-1}^{r-1}.

(5)   The number of distributions of n like objects in r different boxes with empty boxes allowed is C_{n+r-1}^{r-1}.

(6)   The number of distributions of n like objects in r different boxes with each box containing at least q objects is $C_{n-1-r(q-1)}^{r-1}$.

(7)   The number of distributions of n like objects in r different boxes where the number of objects in a box is a least q and at most $q + s - 1$ is equal to the coefficient of x^{n-rq} in the power series expansion of $[(1 - x^s)/(1 - x)]^r$.

(8)   Let Π_n^r denote the number of distributions of n like objects in r like boxes with no box empty. Then we have the recurrence relation

$$\Pi_n^r = \Pi_{n-1}^{r-1} + \Pi_{n-r-1}^{r-1} + \Pi_{n-2r-1}^{r-1} + \cdots$$

Also,

$$\Pi_n^r = \Pi_{n-1}^{r-1} + \Pi_{n-r}^r.$$

For $n - r < r$, we have $\Pi_n^r = \Pi_{n-1}^{r-1}$.

So far we have ignored the order of the objects in a box. For such "ordered distributions" the following assertions hold:

(9)   The number of ordered distributions of n different objects in r different boxes with empty boxes allowed is

$$r(r + 1) \cdots (r + n - 1).$$

(10)   The number of ordered distributions of n different objects in r different boxes with no box empty is

$$n! \, C_{n-1}^{r-1} = \frac{n! \, (n - 1)!}{(n - r)! \, (r - 1)!} \, .$$

If the boxes are all alike, then the number of such distributions is $(n!/r!) \, C_{n-1}^{r-1}$.

(11)   The number of ordered distributions of n or fewer different objects in r different boxes with empty boxes allowed is

$$n! \left[\frac{1}{n!} + \frac{r}{1! \, (n - 1)!} + \frac{r(r + 1)}{2! \, (n - 2)!} + \cdots \right].$$

This number is the product of $n!$ by the coefficient of x^n in the power series expansion of $e^x(1 - x)^{-r}$.

(12)   If in (11) we disallow empty boxes, then the number of ordered distributions is equal to the product of $n!$ by the coefficient of x^{n-r} in the power series expansion of $e^x(1 - x)^{-r}$.

VIII

Problems in Combinatorics

1. Five roads link towns A and B. Three roads link towns B and C. How many roads link towns A and C?

2. Each of 2 fencing clubs has 100 members and each selects 1 member to represent it at a fencing meet. In how many ways can the 2 men be chosen?

3. There are 5 kinds of envelopes and 4 kinds of stamps of the same denomination. In how many ways can we choose an envelope and a stamp for sending a letter?

4. In how many ways can we choose a consonant and a vowel in the word "paste"?

5. Answer the same question in connection with the word "bike."

6. We cast a 6-sided die and spin an 8-sided top. In how many ways can the two fall?

7. Five roads lead to the top of a mountain. In how many ways can a tourist go up and down the mountain? What is the answer if the tourist must not retrace his steps?

8. There are 20 sheep and 24 pigs on a farm. In how many ways can we choose a sheep and a pig? Having chosen a sheep and a pig, in how many ways can we choose another sheep and another pig?

9. In how many ways can we choose a black square and a white square on a chessboard? In how many ways can we choose a pair of squares?

10. In how many ways can we choose a black square and a white square on a chessboard if the two squares must not belong to the same row or column?

11. In how many ways can we choose 3 words, one each from 12 three-letter words, 9 four-letter words, and 10 five-letter words?

12. Given 6 pairs of gloves of different sizes, in how many ways can we choose 2 gloves of different sizes?

13. In how many ways can we choose 3 textbooks, one each from 3 copies of an algebra textbook, 7 copies of a geometry textbook, and 7 copies of a trigonometry textbook?

14. A bookstore has 6 copies of I. S. Turgenev's novel "Rudin," 3 copies of his novel "Nobleman's Nest," and 4 copies of his novel "Fathers and Sons." In addition, the bookstore has 5 copies of a combined edition of "Rudin" and "Nobleman's Nest," and 7 copies of a combined edition of "Nobleman's Nest" and "Fathers and Sons." In how many ways can we purchase a selection one each of the 3 novels?

15. What is the answer to the preceding problem if we assume that the bookstore also has 3 copies of a combined edition of "Rudin" and "Fathers and Sons"?

16. A basket contains 12 apples and 10 oranges. Jack chooses an apple or an orange. Then Jill chooses an apple and an orange. In which case does Jill have a greater number of choices?

17. Given 3 tops with 6, 8, and 10 numbered sides respectively. In how many ways can the 3 tops fall? What if at least 2 of the tops fall on the side marked with the number 1?

18. In how many ways can we choose 3 out of 5 different colors?

19. How many 3-stripe flags can be made of materials of 5 different colors if the stripes are to be all different? What if one of the stripes is to be red?

20. How many dictionaries are needed for direct translation from each of 5 languages into another?

21. How many additional dictionaries are needed if the number of languages is 10?

22. In how many ways is it possible to choose 4 cards of different suits from a deck of (52) cards? What is the answer to the preceding question if no 2 of the selected cards can have the same value (for example, 2 kings or 2 tens)?

23. In how many ways is it possible to choose from a deck of cards 4 cards of different suits, so that the diamonds-selection has the same value as the hearts-selection, and the spades-selection has the same value as the clubs-selection?

24. A newborn child can be given 1, 2, or 3 names. In how many ways can a child be named if we can choose from 300 names?

25. We regard two seating arrangements at a round table as being the same if each person has the same neighbors in both seating arrangements. What is the number of different seating arrangements of 4 people? Of 7 people? In how many cases will a certain 2 of the 7 people turn out to be neighbors? In how many cases will a certain 1 of the 7 people have a certain 2 neighbors?

26. In how many ways can 5 girls and 3 boys be divided into 2 teams of 4 if each team is to include at least 1 boy?

27. In how many ways can 6 urgent letters be dispatched if we can make use of 3 couriers?

28. One person has 7 mathematics books and the other has 9 mathematics books. In how many ways can the two people exchange a mathematics book?

29. What is the answer to the preceding problem if the two people exchange 2 books?

30. A meeting is to be addressed by 5 speakers, A, B, C, D, E. In how many ways can the speakers be ordered if B must not precede A?

31. What is the answer to the preceding problem if B is to speak immediately after A?

32. In how many ways can 5 men and 5 women be seated at a round table if neighbors are to be of different sexes?

33. What is the answer to the preceding problem if the table is replaced by a merry-go-round and we identify arrangements which differ by a rotation?

34. Ten cards are selected out of a deck of 52 cards. In how many cases do the 10 cards contain at least 1 ace? Exactly 1 ace? At least 2 aces? Exactly 2 aces?

35. Each of m lightposts at a railroad station can be made to beam a red, yellow, or green light. How many different signals can be beamed by the array of lightposts?

36. No two inhabitants of an imaginary country have the same number of teeth. What is the maximal number of inhabitants of this country (a person can have at most 32 teeth)?

37. A railroad compartment has 10 seats, 5 facing the locomotive and 5 facing away from the locomotive. Of 10 passengers, 4 prefer to face the locomotive, 3 prefer to face away from the locomotive, and 3 have no preference. In how many ways can the passengers be seated?

38. There are 9 candidates running for 4 different offices. What is the number of possible outcomes of the election?

39. A delegation of 5 is to be elected from among 52 persons at a conference. In how many ways can this be done?

40. An automobile license number contains 1, 2, or 3 letters followed by a 4-digit number. Compute the maximal number of different licenses (the Latin alphabet consists of 26 letters).

41. Mother has 2 apples and 3 pears. On each of 5 days she gives her child 1 fruit. In how many ways can this be done?

42. What is the answer to the preceding problem if the number of apples is m and the number of pears is n?

43. Solve the analog of Problem 41 if there are 2 apples, 3 pears, and 4 oranges.

44. Father distributes 5 different oranges among 8 sons and each son gets at most 1 orange. In how many ways can this be done?

45. What is the answer to the preceding problem if we do not restrict the number of oranges received by each son?

46. How many different "words" can we obtain by permuting the letters in "mathematics"? "parabola"? "ingredient"?

47. A 30-member sports club delegates a 4-man team to take part in a 1000-meter race. In how many ways can the team be chosen? In how many ways can a 4-man team be chosen to take part in a relay race $100 + 200 + 400 + 800$?

48. In how many ways can we place the white pieces (2 knights, 2 bishops, 2 rooks, 1 queen, and 1 king) in the first row of the chessboard?

49. There are n telephone subscribers. Six subscribers happen to be talking in pairs. In how many ways can this happen?

50. The post office sells 10 types of postcards. In how many ways can one buy 12 postcards? 8 postcards? 8 different postcards?

51. In how many ways can we choose 6 people including at least 2 women out of a group of 7 men and 4 women?

52. Find the number of 4-digit multiples of 4 composed of the digits 1, 2, 3, 4, 5.

53. A train with n passengers aboard makes m stops. In how many ways can the passengers distribute themselves among these m stops? What is the answer to the preceding question if we are concerned only with the *number* of passengers who get off at each stop?

54. Compute the number of permutations of n symbols in which 2 symbols a and b do not appear together. Also, the number of permutations in which 3 symbols a, b, c do not appear together. Also, the number of permutations in which no 2 of the 3 symbols a, b, c appear together.

55. Ten people take part in an athletic meet. Each of 3 judges assigns 10 different ratings to the 10 athletes. For an athlete to be pronounced winner he must be given the highest rating by at least 2 judges. Compute the fraction of cases in which a winner is named.

56. Each of 4 students taking an examination is assigned one of the grades A, B, or C. What is the number of possible scores?

57. How many different 7-bead necklaces can be made with 7 different beads?

58. How many different 7-bead necklaces can be made with 2 large beads and 5 small beads?

59. There are 2000 inhabitants in a village. Show that at least 2 of them have the same initials.

60. Seven boys and 10 girls are at a dance. A certain dance number involves all the boys. In how many ways can the boys select partners? What is the number of outcomes if we consider only the girls left without partners? Answer the same questions if a certain 2 of the girls are sure to be asked to dance.

61. A company consists of 3 officers, 6 sergeants, and 60 soldiers. In how many ways can we form a detachment consisting of 1 officer, 2 sergeants, and

20 soldiers? What is the answer if the detachment is to include the commander and the senior sergeant?

62. Twelve girls and 15 boys attend a school party. In how many ways can we select 4 pairs to dance?

63. In how many ways can we choose at least one each of 3 chickens, 4 ducks, and 2 geese?

64. In how many ways can $m + n + p$ different objects be divided into 3 batches containing m, n, and p objects respectively?

65. There are $m + n$ different books on a shelf. m of the books have black bindings and n have red bindings. In how many ways can we arrange the books so that the books with black bindings occupy the first m positions? In how many ways can we arrange the books so that the books with black bindings are together?

66. In how many ways can we select a detachment of 1, 2, 3,..., 15 people from a group of 15 people?

67. Let p_1 ,..., p_n be distinct primes. What is the number of divisors of the number

$$q = p_1^{\alpha_1} , \cdots, p_n^{\alpha_n}$$

where α_1 ,..., α_n are natural numbers (here we include the divisors 1 and q). What is the sum of these divisors?

68. In how many ways can we divide 12 half-dollar coins into 5 numbered nonempty batches?

69. In how many ways can we place 20 books on 5 shelves of a bookcase if each shelf can accomodate all the 20 books?

70. In how many ways can we place 5 rings on the fingers of one hand excluding the thumb?

71. Thirty people vote on 5 issues. Each person votes in favor of 1 issue. What is the number of voting outcomes if we are merely concerned with the number of votes in favor of each issue?

72. A bookbinder is to bind 12 different books in red, green, and brown cloth. In how many ways can he do this if each color of cloth is to be used for at least one book?

73. In how many ways can we make up 6 words out of the 26 letters of the Latin alphabet if each letter is to be used exactly once?

74. In how many ways can we select 12 out of 17 people if a certain 2 people must not both be selected?

75. Five emeralds, 6 rubies, and 7 sapphires are made into a bracelet. In how many was can this be done? (The stones in each category are supposed alike.)

76. In how many ways can one choose 3 of the 18 stones in the preceding problem for a ring?

77. Three students share a room in a dormitory. They have 4 cups, 5 saucers, and 3 teaspoons, all different. In how many ways can they set the table for tea (each setting consists of a cup, a saucer, and a teaspoon)?

78.   A man has 5 female and 7 male acquaintances, and his wife has 7 female and 5 male acquaintances. In how many ways can they invite 6 male and 6 female acquaintances if husband and wife are to invite 6 acquaintances each?

79.   Four people can sit on each side of a boat. In how many ways can one select a crew for the boat if out of 31 candidates 10 prefer portseats, 12 prefer starboard seats, and 9 have no preference?

80.   An urn contains 10 counters numbered 1,..., 10. A person removes 3 of the counters from the urn. In how many cases will the sum of the numbers on these counters be 9? At least 9?

81.   In how many ways can we withdraw 6 out of 52 cards in a deck so that each suit is represented?

82.   A choir consists of 10 persons. On 3 successive days, 6 persons are to be selected for singing assignments. In how many ways can this be done if the 3 selections are to be different?

83.   A man has 6 friends. Every day for 20 days in a row he invites 3 of his friends. In how many ways can he do this if he does not want to have identical company on any two days?

84.   Three men and 2 women choose employment. The town has 3 factories with openings (in the casting shops) for men, 2 knitting mills with openings for women, and 2 factories with openings for workers of either sex. In how many ways can these people choose employment?

85.   How many 5-letter words (with repetitions) can be composed of the 26 letters of the alphabet if no two neighboring letters may be the same?

86.   The winners in a mathematical contest are awarded 3 copies of one book, 2 copies of another book, and 1 copy of a third book. In how many ways can the prizes be distributed if none of the 20 participants may receive 2 books? What is the answer if a participant can be awarded 1, 2, or 3 different books (but no 2 copies of the same book)?

87.   Consider all the dominos from $(0, 0)$ to (n, n). Show that the number of dominos with face-sum $n - r$ is equal to the number of dominos with face-sum $n + r$, and that this number is $E[(n - r + 2)/2]$. How many dominos are there altogether?

88.   In how many ways can 7 men and 7 women be seated at a round table if no 2 women may be seated side by side?

89.   In how many ways can one select a team of 6 out of 16 horses if the team must include 3 of the 6 horses marked A, B, C, A', B', C', but none of the pairs A, A'; B, B'; C, C'?

90.   What is the number of 7-letter words composed of 9 consonants and 7 vowels if each word includes 4 different consonants and 3 different vowels? How many of these words contain no 2 consecutive consonants?

91. Each member of a division of a research institute knows at least one foreign language. Six know Russian, 6 know German, and 7 know French. Four know Russian and German, 3 know German and French, and 2 know French and Russian. One person knows all 3 languages. How many people are employed in the division of the institute? How many of them know only Russian? How many know only French?

92. Ninety-two people went for an outing. Forty-seven took along salami sandwiches, 38 took along cheese sandwiches, 42 took along ham sandwiches, 28 took along cheese sandwiches and salami sandwiches, 31 took along salami sandwiches and ham sandwiches, 26 took along cheese sandwhiches and ham sandwiches. Twenty five people took along sandwiches of all three kinds. Some preferred pies to sandwiches. How many people took along pies?

93. Ten couples go on a boat trip. Each boat carries 4 people. In how many ways can the people occupy the boats if each boat is to carry 2 men and 2 women?

94. In how many of the cases considered in the preceding problem will a certain man and his wife occupy the same boat?

95. In how many of the cases considered in Problem 93 will each of a certain 2 men be in the same boat with his wife?

96. How many 4-digit numbers (with repetitions) can be composed of the digits 0, 1, 2, 3, 4, 5, 6? (No number can start with the digit 0.)

97. Find the number of 6-digit numbers such that the sum of the 3-digit number formed by the first 3 digits and the 3-digit number formed by the last 3 digits is less than 1000.

98. In how many ways can we place 12 white and 12 black checkers on the black squares of a chessboard?

99. In how many permutations of the word "Jupiter" do the vowels appear in alphabetical order?

100. In how many permutations of the word "Alabama" do 4 letters "*a*" not appear together?

101. In how many ways can we permute the word "opossum" so that the letter "*p*" directly follows the letter "o"?

102. In how many ways can we permute the word "Mississippi" so that no 2 letters "*i*" appear together?

103. In how many ways can we permute the word "karakule" so that no 2 vowels are together?

104. In how many ways can we permute the letters in the word "bivouac" without changing the order of the vowels?

105. In how many ways can we permute the letters in the word "parallelism" without changing the order of the vowels?

106. In how many ways can we permute the letters in the word "poster" so that there are 2 consonants between the 2 vowels?

107. In how many ways can we permute the word "paroxysm" so that the letters in the second, fourth, and sixth positions are consonants?

108. In how many ways can we choose in the word "paroxysm" 2 consonants and 1 vowel? What is the answer to the preceding question if one of the selected consonants is to be "*s*"?

109. In how many ways can we permute the word "Oporto" so that the letter "o" does not appear 3 times in a row?

110. What is the answer to Problem 109 if the letter "o" must not appear twice in a row?

111. Consider the phrase: "An eye for an eye, a tooth for a tooth." In how many ways can we select from this phrase an assortment of letters if we pay no attention to order?

112. In how many ways can we choose 3 letters from the phrase in Problem 111?

113. In how many ways can we choose 3 letters from the phrase in Problem 111 if order is taken into consideration?

114. In how many ways can we permute the word "bitumen" so that the vowels and the consonants appear in alphabetical order?

115. In how many ways can we permute the word "triannual" so that vowels and consonants alternate? What is the answer if the word in question is "samovar"?

116. In how many ways can we permute the word "Abakan" if the consonants are to appear in alphabetical order? What is the answer to the preceding question if we add the restriction that two letters "*a*" must not appear together?

117. In how many was can we permute the letters in the word "zigzag" so that the same letters don't appear together? Answer the same question for the word "Tartar."

118. In how many ways can we select 4 letters out of the 6 letters in the word "Tartar" if we disregard order? How many 4-digit numbers can we compose of the digits in the number 132,132?

119. How many nonnegative integers smaller than 1,000,000 include all four of the digits 1, 2, 3, 4? How many of these numbers consist of the digits 1, 2, 3, 4 alone?

120. Compute the sum of all 4-digit numbers obtained by permuting the digits 1, 2, 3, 4.

121. Solve Problem 120 for the digits 1, 2, 2, 5.

122. Solve Problem 120 for the digits 1, 3, 3, 3.

123. Solve Problem 120 for the digits 1, 1, 4, 4.

124. Solve Problem 120 for all 5-digit numbers obtained by permuting the digits 0, 1, 2, 3, 4. No number can begin with the digit 0.

125. How many numbers smaller than 1,000,000 can we make using the digits 8 and 9?

126. Solve Problem 125 for the digits 9, 8, 7.

127. Solve Problem 125 for the digits 9, 8, 0. No number can begin with the digit 0.

128. Compute the sum of all 3-digit numbers composed of the digits 1, 2, 3, 4.

129. Compute the sum of all 5-digit numbers composed of the digits 1, 2, 3, 4, 5 if each digit appears in a number exactly once. Compute the corresponding sum for the digits 1, 2, 3, 4, 5, 6, 7, 8, 9.

130. How many odd 4-digit numbers can be composed of the digits in the number 3694 if each digit can appear at most once in a particular number? Compute the number of even 4-digit numbers composed in this way.

131. Compute the number of 6-digit numbers with 3 even digits and 3 odd digits.

132. Solve Problem 131 if one allows 6-digit numbers starting with the digit 0.

133. Compute the number of 6-digit numbers which do not begin with the digit 0 and have even digit sum. Solve the same problem for the numbers from 1 to 999,999.

134. How many 10-digit numbers (not beginning with the digit 0) have digit sum 3? What is the answer if instead of 10-digit numbers we consider all the numbers from 1 to 9,999,999,999?

135. Compute the number of 9-digit numbers all of whose digits are different.

136. How many numbers from 0 to 999 are not divisible by either 5 or 7?

137. How many numbers from 0 to 999 are not divisible by 2, 3, 5, or 7?

138. Compute the number of numbers from 0 to 999 which include the digit 9; which include the digit 9 twice; which include the digit 0; which include the digit 0 twice; which include the digits 0 and 9; which include the digits 8 and 9. How many of the numbers from 0 to 999,999 do not have the same digit twice in a row?

139. How many 4-digit numbers can be composed of the digits in the 123,153?

140. How many 5-digit numbers can be composed of the digits in the number 12,335,233?

141. How many 6-digit numbers can be composed of the digits in the number 1,233,145,254 if the same digit must not appear twice in a row?

142. How many 5-digit numbers can be composed of the digits in the number 12,312,343 if the digit 3 must not appear 3 times in a row?

143. In how many ways can we permute the digits in the number 12,341,234 if the same digit must not appear twice in a row?

144. Solve Problem 143 for the number 12,345,254.

145. In how many ways can we permute the digits in the number 1,234,114,546 if the same digit must not appear 3 times in a row?

146. Solve Problem 145 if the same digit must not appear twice in a row.

147. In how many ways can we choose 2 natural numbers from the numbers 1 to 20 so that their sum is odd?

148. In how many ways can we choose 3 natural numbers from the numbers 1 to 30 so that their sum is even?

149. There are 2 highways from London to Brighton which connect 10 country roads (Fig. 34). How many routes from London to Brighton are free of self-intersections?

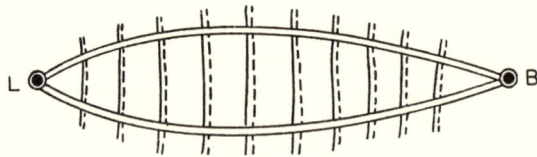

Fig. 34

150. The same setting as in Problem 149. Two travelers set out from London along different highways. In how many cases will neither of the 2 travelers traverse the same portion of highway in the same direction?

151. There are 3 highways from London to Cambridge which connect 4 country roads (Fig. 35). How many of the routes from London to Cambridge do not require a traveler either to traverse a portion of highway in the direction from Cambridge to London or to traverse a portion of highway twice?

152. There are unlimited numbers of 10-, 15-, and 20-cent coins. In how many ways can we select 20 coins?

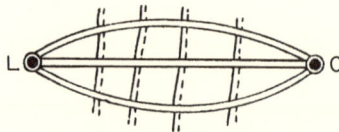

Fig. 35

153. Consider a game in which a person hides 5 coins of the following denominations: 1, 2, 3, 5, 10, 15, 20, 50 cents, and 1 dollar, and another person guesses the denominations of the hidden coins. What is the maximal number of false guesses?

154. Compute the number of 5-digit numbers. In how many of these numbers are all the digits even ? In how many of these numbers are all the digits odd ? How many of these numbers contain no digits smaller than 6 ? How many of these numbers contain no digits larger than 3 ? How many of these numbers contain all of the digits 1, 2, 3, 4, 5 ? How many of these numbers contain all of the digits 0, 2, 4, 6, 8 ?

155. The numbers on the faces of each of 2 dice are 0, 1, 3, 7, 15, 31. Compute the number of different sums obtained by throwing the 2 dice.

156. The numbers on the faces of each of 3 dice are 1, 4, 13, 40, 121, 364. Compute the number of different sums obtained by throwing the three dice.

157. The numbers on the faces of each of 6 dice are 1, 2, 3, 4, 5, 6. Regard the dice as different. Compute the number of throws in which the dice turn up the same number, 2 different numbers, 3 different numbers, 4 different numbers, 5 different numbers, 6 different numbers.

158. What are the possible outcomes in throwing n ordinary dice ? (Here we disregard order of the dice.)

159. In how many ways can we write the number 1,000,000 as a product of 3 ordered factors ?

160. Solve Problem 159 if order of the factors is disregarded.

161. In how many ways can one distribute 9 coins of different denominations in 2 pockets ?

162. In how many ways can we distribute $3n$ different objects among 3 people so that each person gets n of the objects ?

163. In how many ways can we group $2n$ objects in pairs if we disregard order within each pair as well as order of the pairs ?

164. Solve Problem 163 for nk objects distributed in n groups of k objects each.

165. In how many ways can 30 workers be separated into 3 groups of 10 ? Into 10 groups of 3 ?

166. In how many ways can we divide a deck of 36 cards into 2 halves if each half is to contain 2 aces ?

167. In how many ways can we package 10 books into 5 packages containing 2 books each (disregard order of the packages as well as order of the books in a package) ?

168. In how many ways can we package 9 books if 4 packages are to contain 2 books each and 1 package is to contain 1 book ?

169. Solve Problem 168 if there are to be 3 packages of 3 books each.

170. In how many ways can 3 people divide among themselves 6 apples, 1 orange, 1 plum, 1 lemon, 1 pear, 1 quince, and 1 date ?

171. What is the answer to Problem 170 if each person is to get exactly 4 fruits ?

172. Person A has 3 apples, 1 pear, 1 plum, and 1 quince. Person B has 3 apples, 1 orange, 1 lemon, and 1 date. Person C has 3 apples, 1 tangerine,

1 peach, and 1 apricot. In how many ways can they divide these fruits among themselves so that each gets 6 items?

173. In how many ways can we divide a deck of 52 cards among 13 players if each player is to get 4 cards? In how many ways can this be done if each player is to have 1 card of each suit? In how many ways can this be done if 1 player is to have cards of different suits and the remaining 12 players are each to have cards of the same suit?

174. In how many ways can 4 cards be removed from a full deck so that the cards are of exactly 3 suits? Of exactly 2 suits?

175. In how many ways can we distribute 52 cards among 4 players so that each player gets 3 cards of each of 3 suits and 4 cards of the fourth suit?

176. In how many ways can we distribute 18 different objects among 5 people so that 4 people get 4 objects each and the fifth person gets 2 objects. Solve the same problem if 3 people get 4 objects each and 2 people get 3 objects each.

177. How many selections can we make up by taking 0, 1, or 2 objects from each of 14 different pairs of objects? We disregard order in a selection.

178. In how many ways can we place 4 black balls, 4 white balls, and 4 blue balls in 6 different boxes. Any box may be left empty.

179. In how many ways can we separate 3 one-dollar bills and 10 fifty-cent coins into 4 numbered batches? Any batch may be empty.

180. Show that the number of (unordered) partitions of the number n into summands is equal to the number of (unordered) partitions of the number $2n$ into n summands.

181. Consider an ordered arrangement of n objects. In how many ways can we remove 3 of these objects without removing neighboring objects?

182. A child places black and white chess pieces (2 knights, 2 bishops, 2 rooks, 1 queen, and 1 king of each color) in the first 2 rows of a chessboard. In how many ways can he do this?

183. The setting as in Problem 182. In how many ways can these pieces be placed on the whole board?

184. Compute the number of arrangements of all the chess figures (black and white) on a chessboard.

185. In how many ways can we place 15 white checkers and 15 black checkers on 24 squares so that each square contains checkers of one color only? (This is how checkers are placed in the Eastern game of "Nardi.")

186. In how many ways can we place 20 white checkers on a chessboard so that the resulting configuration is unchanged by a rotation through 90°?

187. In how many ways can we place 20 white checkers on a chessboard so that the resulting configuration is symmetric with respect to a line which divides the chessboard in halves?

188. Solve Problem 187 with the checkers placed on black squares.

189. In how many ways can we place 12 white and 12 black checkers on black squares of a chessboard so that the resulting configuration is symmetric with respect to the center of the chessboard?

190. Solve Problem 189 if checkers corresponding under a central symmetry are to differ in color.

191. In how many ways can 20 white checkers be set out along the border lines of a chessboard so that the resulting configuration is unchanged by a rotation through 90°?

192. In how many ways can 20 white checkers be set out along the border lines of a chessboard so that the checkers on opposite sides of the board are symmetric with respect to the lines which divide the board in half?

193. In how many ways can we place 7 white balls and 2 black balls in 9 billiard pockets? We suppose the pockets different and allow empty pockets.

194. In how many ways can we place 7 white balls, 1 black ball, and 1 red ball in 9 billiard pockets?

195. In how many ways can we distribute 27 books among 3 people A, B, C, so that A and B receive twice as many books as C?

196. Eight people enter an elevator. At each of 4 floor stops at least 1 person leaves the elevator. After 4 floor stops the elevator is empty. In how many ways can this take place?

197. In how many ways can we choose 3 of the numbers from 1 to 100 so that their sum is divisible by 3?

198. In how many ways can we choose 3 out of $3n$ successive numbers so that their sum is divisible by 3?

199. Given n white balls and 1 black ball. In how many ways can some of these balls be placed in $n + 1$ billiard pockets which hold 1 ball each?

200. In how many ways can we order m white and n black balls so that the number of contacts between black and white balls is $2r - 1$? $2r$?

201. A student who passes an examination earns 3, 4, or 5 points depending on the quality of his paper. In how many ways can he earn a total of 30 points by passing examinations in 8 different subjects?

202. Show that there are $(m + n)! \, D_m/m!n!$ permutations (see p. 51) of $m + n$ objects which leave exactly n objects fixed.

203. Show that the number of ways of distributing r different things among $n + p$ people so that each of the n persons obtains at least 1 object is

$$S_r = (n + p)^r - n(n + p - 1)^r + C_n^2(n + p - 2)^r - \cdots + (-1)^n p^r.$$

204. Show that the number of partitions of $2r + x$ into $r + x$ positive summands is the same as the number of partitions of r into nonnegative summands.

205. A society of n members elects one of its members as its representative. Each member votes for just one candidate (possibly himself). The outcome of an election may be described in terms of: (a) who voted for whom; or in terms of (b) the number of votes received by various candidates. These two points of view yield different numbers of possible election outcomes. What are these numbers?

206. Show that the number of ways of placing $2n$ like objects in 3 like boxes so that the sum of the objects in any 2 of the boxes exceeds the number of objects in the third box is the same as the number of ways of placing $2n - 3$ like objects in 3 like boxes subject to the same condition.

207. Show that there are 2^{n-1} ways of choosing an odd number of objects from n objects.

208. Show that there are $3n^2 + 3n + 1$ ways of dividing $2n$ objects of one kind, $2n$ objects of a second kind, and $2n$ objects of a third kind between 2 people so that each person gets $3n$ objects.

209. In Problem 208 add $2n$ objects of a fourth kind and show that the number of ways in which the resulting $8n$ objects can be divided between 2 people so that each person gets $4n$ objects is

$$\tfrac{1}{3}(2n + 1)(8n^2 + 8n + 3).$$

210. Consider Problems 208 and 209. Show that if we distinguish between recipients, then the answers are, respectively,

$$\tfrac{1}{2}(3n^2 + 3n + 2) \quad \text{and} \quad \tfrac{1}{3}(n + 1)(8n^2 + 4n + 3).$$

211. There are $2n$ objects of each of m kinds. Show that these $2nm$ objects can be divided in half in

$$C_{mn+m-1}^{m-1} - C_m^1 C_{mn+m-2n-2}^{m-1} + C_m^2 C_{mn+m-4n-3}^{m-1} - \cdots \pm C_m^x C_{mn+m-1-x(2n+1)}^{m-1} \mp \cdots$$

ways.

212. In how many ways can we place 5 white balls, 5 black balls, and 5 red balls in 3 different boxes if each box is to contain 5 balls?

213. There are n objects of each of 3 kinds. Show that the number of ways of giving n of these objects to each of 3 people A, B, and C, is

$$C_{n+2}^2 C_{n+2}^2 - 3C_{n+3}^4 = \tfrac{1}{8}(n + 1)(n + 2)(n^2 + 3n + 4).$$

214. Three Englishmen, 3 Frenchmen, and 3 Turks are to be seated in a row so that no 3 countrymen sit together. In how many ways can this be done?

215. Solve Problem 214 if no 2 countrymen may sit together.

216. Three Englishmen, 3 Frenchmen, and 3 Turks are to be seated at a round table so that no 2 countrymen sit together. In how many ways can this be done?

217. Compute the number of rows of stamps worth 40 cents if the stamps are of the following denominations: 5, 10, 15, and 20 cents.

218. In how many ways can we change 1 dollar into 10-, 15-, 20-, and 50-cent coints?

219. Given the following row of weights: 1, 1, 2, 5, 10, 10, 20, 50 grams. In how many ways can we weigh a 78-gram object if the position of the weights in the above array is to be taken into consideration in the weighing process?

220. Given 3 black balls, 1 red ball, 1 white ball, and 1 blue ball. Compute the number of different rows of 4 of these balls.

221. Compute the number of ways of writing a natural number n as an ordered sum of 3 natural numbers.

222. How many times do we use each of the digits 0, 1,..., 9 in writing all the numbers from 1 to 999,999 inclusive? From 1 to $10^n - 1$ inclusive?

223. Compute the number of 10-digit numbers which contain only the digits 1, 2, 3 with the digit 3 appearing in each number exactly twice. How many of these numbers are divisible by 9?

224. Two numbers are said to form an inversion if the larger number precedes the smaller number. Compute the number of inversions in the set of all permutations of the numbers 1, 2,..., n.

225. Show that the number of partitions of n into 3 distinct positive summands is

$$E\left(\frac{n^2 - 6n + 12}{12}\right).$$

226. Show that the number of partitions of $12n + 5$ into 4 positive summands each $\leqslant 6n + 2$ is

$$\tfrac{1}{2}(n + 1)(12n^2 + 9n + 2).$$

227. Show that the number of partitions of $12n + 5$ into 4 distinct positive summands $\leqslant 6n + 2$ is

$$(12n^2 + 3n - 1)\frac{n}{2}.$$

228. Find the number of triples of unequal natural numbers $\leqslant 100$ which form a geometric progression.

229. Compute the number of rows of 6 Englishmen, 7 Frenchmen, and 10 Turks in which an Englishman invariably stands between a Frenchman and a Turk and in which a Frenchman and a Turk never stand side by side.

230. Solve Problem 229 for 5 Englishmen, 7 Frenchmen, and 10 Turks.

231. How many pairs of numbers are there with greatest common divisor G

and least common multiple M such that $M = Ga^\alpha b^\beta c^\gamma d^\delta$ and a, b, c, d are different primes?

232. Delete the words "greatest" and "least" in Problem 231 and solve the resulting problem.

233. Compute the number of 6-letter combinations of 20 different letters if no letter is to appear in a combination more than 2 times.

234. Of $p + q + r$ letters, p are α, q are β, and r are γ. What is the number of permutations of these letters in which α appears before β and β before γ?

235. A band 30 inches long is painted according to the scheme: red, white, blue, red, white, blue, and so on. The first stripe appearing on the band is red and the last is blue. The length of each stripe is given by an integer $\geqslant 2$. The total length of the part of the band painted a particular color is 10 inches. Compute the number of such color patterns. What is the answer if we dispense with the condition that the last stripe is blue? Show that if the length of each stripe is $\geqslant 3$, then in 153 cases the last stripe is blue, in 71 cases the last stripe is white, and in 81 cases the last stripe is red.

236. I have 6 friends. I dined with each of them 7 times, with any 2 of them 5 times, with any 3 of them 4 times, with any 4 of them 3 times, with any 5 of them 2 times, with all 6 of them 1 time, without each one of them 8 times. How many times did I dine alone?

237. Each of 12 students is taking an examination in 2 subjects. One teacher examines the students in one subject and another in the other subject, and each teacher takes 5 minutes to examine a student in a subject. In how many ways can the examinations be scheduled without a student being required to appear before both examiners at the same time?

238. In how many ways can each of 6 people select a left glove and a right glove out of a total of 6 pairs of gloves so that no person selects a matching pair of gloves?

239. The letters in $\alpha^2\beta^2\gamma^2$ are permuted so that each letter has a like neighbor on one side or on the other side. Show that the number of such permutations is 6. Show that the answer to this problem is 6 for the expression $\alpha^3\beta^3\gamma^3$, 90 for the expression $\alpha^4\beta^4\gamma^4$, and 426 for the expression $\alpha^5\beta^5\gamma^5$.

240. Each of n countries sends 4 representatives to a chess tournament. In how many ways can these people be lined up so that each person has a countryman for a neighbor?

241. The squares of a chessboard are painted 8 colors. The squares of each row are painted all the 8 colors and no 2 consecutive squares of one column are painted the same color. In how many ways can this be done?

242. Given n like things and n additional different things. In how many ways can we choose n of these $2n$ things? In how many ways can we order these $2n$ things?

243. m Frenchmen and n Englishmen stand in a row so that each person has a countryman for a neighbor. Show that the number of such arrangements is

$$m!\, n!\, [1 + (C_{m-2}^0 + C_{m-3}^1)(C_{n-2}^0 + C_{n-3}^1)$$

$$+ (C_{m-3}^1 + C_{m-4}^2)(C_{n-3}^1 + C_{n-4}^2)$$

$$+ (C_{m-4}^2 + C_{m-5}^3)(C_{n-4}^2 + C_{n-5}^3) + \cdots].$$

244. How many 6-digit numbers contain exactly 3 different digits?

245. How many m-digit numbers contain exactly k different digits?

246. Compute the number of k-samples of the digits 1, 2,..., n in which the even numbers occupy the even positions, the odd numbers occupy the odd positions, and all the numbers in the sample are arranged in increasing order.

247. Given $2n$ elements a_1, a_1, a_2, a_2,..., a_n, a_n, with $a_i \neq a_j$ for $i \neq j$. How many permutations of these $2n$ elements do not contain 2 like elements in a row?

248. Each of n selections contains q like elements. Elements belonging to different selections are different. How many permutations of these nq elements do not contain 2 like elements in a row?

249. Solve Problem 248 if the elements are arranged in a circle.

250. n books are arranged on a shelf. In how many ways can we select p of these books so that there are at least s books between any two of the selected books as well as after the pth selected book?

251. The numbers of students representing grades 5, 6, 7, 8, 9, and 10 at a mathematical contest form an arithmetical progression. The number of awards given out to each grade is equal to the difference of 2 successive terms in the progression. Show that the number of ways of distributing the awards to all the participants is the same as the number of ways of distributing all the awards to the 10th grade students in the contest.

252. A sheet of graph paper is ruled into squares which we call unit squares. We call the side of a unit square a segment. We draw a square $ABCD$ made up of 16 unit squares. We consider all shortest paths from the point A to the point C leading along the sides of the unit squares. Prove that the number of such paths is equal to 70, and that 4 segments are traversed by 35 paths, 8 by 20 paths, 4 by 18 paths, 4 by 15 paths, 4 by 12 paths, 4 by 10 paths, 4 by 5 paths, 4 by 4 paths, and 4 by 1 path. Also prove that 1 junction is crossed by 36 paths, 4 by 35 paths, 4 by 30 paths, 4 by 15 paths, 4 by 5 paths, 4 by 40 paths, and 2 by 1 path (the points A and C are excluded).

253. How many triangles can be incribed in a convex hexagon so that the vertices of each triangle coincide with some of the vertices of the hexagon?

254. What is the number of triangles each of whose sides has length 4, 5, 6, or 7?

255. What is the number of rectangular parallelepipeds such that the length of each edge is an integer from 1 to 10?

256. Compute the number of triangles determined by 4 lines (in a plane) no 2 of which are parallel and no 3 of which are concurrent.

257. Given n coplanar points such that p of them are collinear and such that no 3 of the remaining $n - p$ points are collinear. Compute the number of triangles determined by these points.

258. Given p points on a line and q points on a parallel line. Compute the number of triangles determined by these $p + q$ points.

259. Add to the points in Problem 258 r points lying on a line parallel to the other 2 lines and assume that no 3 of the $p + q + r$ points lie on a line intersecting the 3 parallel lines. How many additional triangles do we obtain?

260. Each side of a square is divided into n segments. How many triangles are determined by the division points?

261. Given n coplanar lines no 2 of which are parallel and no 3 of which are concurrent. Compute the number of points of intersection.

262. There are n coplanar lines, p of which intersect in a point A, and q in a point B. None of these lines joins the points A and B, no 2 of them are parallel, and any point C other than A or B is the point of intersection of at most 2 lines. Compute the number of points of intersection of the n lines.

263. Compute the number of regions determined by n coplanar lines no 2 of which are parallel and no 3 of which are concurrent.

264. Consider n planes such that no 4 of them have a point in common, no 3 have a line in common, and no 2 are parallel. Compute the number of regions determined by the planes.

265. Let A, B, C, D, E be 5 coplanar points such that no 2 of the lines joining these points in pairs are parallel, perpendicular, or coincident. From each point we drop perpendiculars to the lines which join the remaining points in pairs. Compute the maximal number of points of intersection of the perpendiculars.

266. Compute the number of triangles each of whose sides is an integer greater than n and not exceeding $2n$. How many of these triangles are isosceles and how many are equilateral?

267. Prove that the number of triangles each of whose sides is an integer not exceeding $2n$ is $\frac{1}{6}n(n + 1)(4n + 5)$. Also prove that if we exclude isosceles triangles, then the corresponding number is $\frac{1}{6}n(n - 1)(4n - 5)$.

268. Prove that the number of triangles each of whose sides does not exceed $2n - 1$ is $\frac{1}{6}n(n + 1)(4n - 1)$. Also show that if we exclude isosceles triangles, then the corresponding number is $\frac{1}{6}(n - 1)(n - 2)(4n - 3)$.

269. Given n coplanar lines no 3 of which are concurrent. Prove that the number of unordered groups consisting of n points of intersection no 3 of which are collinear is $\frac{1}{2}(n - 1)!$.

270. Given n coplanar points no 3 of which are collinear. Compute the number of r-sided closed polygonal lines whose vertices are among the n given points.

271. Given n points on a line and m points on a line parallel to it. Connect each point of one line to each point of the other line and assume that no 3 of the resulting lines are concurrent. Prove that the number of points of intersection of these lines is $mn(m - 1)(n - 1)/2$ (the count does not include the $m + n$ initial points).

272. Let n be the number of points in a plane such that no 3 are collinear and no 4 lie on a circle. Draw a line through every 2 of the n points and draw a circle through every 3 of the n points. Compute the maximal number of points of intersection of the set of lines with the set of circles.

273. Given n points in space such that no 4 of the points are coplanar and such that no 2 of the planes determined by triples of the given points are parallel. Compute the number of lines of intersection of the planes as well as the number of lines of intersection which do not contain any of the n given points.

274. Out of n segments of length $1, 2,..., n$, we select 4 segments in such a way that the quadrilateral on these segments can be circumscribed about a circle. Show that the number of such selections is

$$[2n(n - 2)(2n - 5) - 3 + (-1)^n]/48.$$

What is the corresponding number of selections if the sides of the quadrilateral are of equal length?

275. Given n points no 4 of which lie on a circle. What is the maximal number of points of intersection of the set of circles determined by all triples of points belonging to the given set of n points?

276. Show that n planes passing through the center of a sphere divide it into at most $n^2 - n + 2$ parts.

277. In how many geometrically different ways can we paint the faces of a cube 6 different colors? Two ways of painting a cube 6 different colors are said to be geometrically the same if they differ only by a a rigid motion of the cube.

278. In how many geometricaly different ways can we paint the faces of a tetrahedron 4 different colors?

279. In how many geometrically different ways can we paint the faces of an octahedron 8 different colors?

280. Solve the analogs of the three preceding problems for the regular dodecahedron and the regular icosahedron.

281. Modify Problems 277–280 by considering the possibility of using fewer colors than there are faces in the given figure (for example, consider the possibility of painting the faces of the cube 2, 3, 4, or 5 colors).

282. What is the number of triangles with integral sides and perimeter 40? Perimeter 43?

283. Show that there are $n + 1$ more triangles with integral sides and perimeter $4n + 3$ than there are triangles with integral sides and perimeter $4n$.

284. Prove the validity of the following table which gives the number of triangles with integral sides and perimeter N.

N	Number of triangles	N	Number of triangles
$12n$	$3n^2$	$12n + 6$	$3n^2 + 3n + 1$
$12n + 1$	$n(3n + 2)$	$12n + 7$	$(n + 1)(3n + 2)$
$12n + 2$	$n(3n + 1)$	$12n + 8$	$(n + 1)(3n + 1)$
$12n + 3$	$3n^2 + 3n + 1$	$12n + 9$	$3n^2 + 6n + 3$
$12n + 4$	$n(3n + 2)$	$12n + 10$	$(n + 1)(3n + 2)$
$12n + 5$	$(n + 1)(3n + 1)$	$12n + 11$	$3n^2 + 7n + 4$

285. A certain town has a system of bus routes satisfying the following conditions:

(1) There is a direct bus route joining any 2 bus stops;

(2) Any 2 bus routes share a unique bus stop;

(3) There are exactly n bus stops on each route.

What is the number of bus routes in the town?

286. A town has 57 bus routes. We know that:

(1) There is a direct bus route joining any 2 bus stops;

(2) Any 2 bus routes share a unique bus stop;

(3) There are at least 3 bus stops on each route.

What is the number of bus stops on each route?

287. Is it possible to have a system of 10 bus routes with bus stops distributed so that for every 8 bus routes there is a bus stop not on any of these 8 routes and every 9 bus routes pass through all of the bus stops in the system?

288. Given 3 planes and a sphere in space. What is the maximal number of spheres tangent to all 4 given figures.

289. Through each of 3 points there pass m lines such that no 2 lines from different pencils are parallel and no 3 lines from different pencils are concurrent. Compute the number of points of intersection of these lines.

290. Of n given points in space, m points belong to a plane P, and of the remaining $n - m$ points no 4 are coplanar. Compute the number of planes containing 3 of the given points.

291. Let A, B, C be 3 given points in a plane. We pass m lines through A, n lines through B, and p lines through C so that no 2 lines from different pencils are parallel and no 3 lines from different pencils are concurrent. Find the number of triangles whose vertices are different from A, B, C and belong to the set of points of intersection of the given lines.

292. Compute the number of triangles inscribed in a convex n-gon which have no points other than their vertices in common with this n-gon.

293. On each of n given lines in a plane we select p points such that none of the points is a point of intersection of the given lines and no 3 of the points lie on a line other than the given lines. Find the number of triangles whose vertices belong to this set of points.

294. Show that the number of points of intersection of the diagonals of an n-gon lying outside the n-gon is $n(n-3)(n-4)(n-5)/12$, and lying inside the n-gon is $n(n-1)(n-2)(n-3)/24$. (We assume that no 2 diagonals are parallel and no 3 diagonals are concurrent.)

295. Given n points on a circle, compute the number of polygons (convex or nonconvex) inscribed in this circle whose vertices belong to the given set of n points. How many of these polygons are convex?

296. Given $m+n$ coplanar lines such that the first m lines are parallel, none of the remaining n lines is parallel to any of the other given lines, and such that no 3 lines are concurrent. Compute the number of regions into which the given lines divide the plane.

297. Of 11 given points, 5 lie on a circle and no 4 of the remaining 6 points lie on a circle. Compute the number of circles containing at least 3 of the given points.

298. Given 10 coplanar lines such that no 2 are parallel, no 3 are concurrent, and no 4 are tangent to a circle, compute the number of circles tangent to 3 of the given 10 lines.

299. Find the number of convex k-gons whose vertices are among the vertices of a convex n-gon and whose adjoining vertices are separated by at least s vertices of the n-gon.

300. A parallelogram is cut by r lines parallel to one side and r lines parallel to its adjoining side. Compute the number of resulting parallelograms.

301. Into how many regions is a convex n-gon divided by its diagonals?

302. There is 1 card with the number 1, there are 2 cards with the number 2, 3 cards with the number 3, and so on. Show that the number of ways of withdrawing 2 cards whose numbers add up to n is $n(n^2-1)/12$ or $n(n^2-4)/12$ according as n is odd or even.

303. Of $3n+1$ objects, n are alike and the others are different. Show that there are 2^{2n} ways of withdrawing n of these objects.

304. In how many ways can we withdraw 3 of the numbers $1, 2, 3,..., 2n$ to obtain an arithmetical progression? Answer the same question for the sequence $1, 2,..., 2n+1$.

305. Consider a number of simple closed curves in the plane such that each curve intersects each of the remaining curves in at least 2 points. Let n_r denote the number of points of intersection of the curves in which exactly r of the curves meet. Then the number of regions bounded by arcs of the curves and containing no such arcs in their interiors is

$$1 + n_2 + 2n_3 + \cdots + rn_{r+1} + \cdots.$$

306. Consider 2 coplanar pencils of m and n lines, respectively. Into how many regions do these lines divide the plane if no 2 of the lines are parallel and no line joins the centers A and B of the pencils?

307. Can each of 77 telephones be connected to exactly 15 of these telephones?

308. Compute the sum of the coefficients of the polynomial obtained by expanding the expression

$$(7x^3 - 13y^2 + 5z^2)^{1964} \, (y^3 - 8y^2 + 6y + z)^2 + (2x^2 + 18y^3 - 21)^{1965}.$$

309. A box contains 100 colored balls. Of these, 28 are red, 20 green, 12 yellow, 20 blue, 10 white, and 10 black. What is the minimal number of balls which one must withdraw to be sure to have withdrawn 15 balls of one color?

310. We can paint all the faces of a cube white, or all of them black, or some white and some black. Find the number of geometrically different ways of painting the cube (see Problem 277).

311. Solve Problem 310 if instead of painting the faces of a cube we paint its vertices.

312. Models of polyhedra can be made of plane paper cutouts which are bent along the edges. In the case of a regular tetrahedron, it is possible to make 2 different cutouts. How many different cutouts are there for a cube?

313. A regular dodecahedron can be painted 4 colors so that any 2 adjoining faces are painted different colors. In how many geometrically different ways can this be done?

314. It is possible to select 4 out of the 6 edges of a tetrahedron such that the selected edges form a closed quadrilateral in three-space. The vertices of this quadrilateral include all of the vertices of the tetrahedron. Can the same be done with an octahedron? A dodecahedron? An icosahedron? How many solutions are there to each of these problems?

315. A particle is located at the origin. In a unit of time it splits into 2 particles of which one moves a unit to the left and the other a unit to the right. This process is repeated after each unit of time. Also, if 2 particles end up in the same spot, they annihilate each other (so that, for example, after 2 units of time there will be only 2 particles left). How many particles will there be after 129 units of time? After n units of time?

316. A certain alphabet uses 6 letters encoded for telegraphic transmission as follows:

$$\bullet; \quad -; \quad \bullet\,\bullet; \quad -\,-; \quad \bullet\,-; \quad -\,\bullet\,.$$

A certain word was transmitted without breaks between groups of symbols denoting different letters. The result was an unbroken chain of 12 dots and dashes. In how many ways can this chain of dots and dashes be decoded?

317. Compare the size of the set of all numbers between 1 and 10,000,000 inclusive which do not contain the digit 1 with the set of the remaining numbers in the same range.

318. Consider 7-symbol "words," where a symbol is either a dot or a dash. What is the largest set of words such that any 2 words differ in at least 3 places?

319. A circular loop is divided into p parts, p prime. In how many ways can we paint the loop n colors if we identify patterns which differ by a rotation of the loop?

320. Each of the numbers $1, 2,..., n^2$ is placed in one of n^2 squares of an n by n ruled sheet in such a way that the numbers in each row and in each column form an arithmetic progression. In how many ways can this be done?

321. A person has no more than 300,000 hairs on his head. Show that there are at least 10 people in Moscow (population about 6 million) with the same number of hairs.

322. Given $2n + 1$ objects. Show that there are as many ways of choosing an odd number of these objects as there are ways of choosing an even number of these objects.

323. Show that there are more ways of changing 1 dollar into 2- and 5-cent coins than there are ways of changing 1 dollar into 3– and 5–cent coins.

324. In how many was can we change a 20–cent coin into 1–, 2–, and 5–cent coins?

325. Show that a selection of weights containing 1 mg (mg stands for milligram), 2 mg, 2 mg, 5 mg, 10 mg, 20 mg, 20 mg, 50 mg, 100 mg, 200 mg, 200 mg, 500 mg, 1 g (g stands for gram), and so on, suffices to give any integral number of milligrams.

326. Find the sum of all even 4-digit numbers composed of the digits 0, 1, 2, 3, 4, or 5.

327. $2n$ cards are shuffled as follows: First the cards are divided into 2 batches of n cards each. Then we alternate the cards of the 2 batches so that the card numbered $n + 1$ becomes first, the card numbered 1 becomes second, the card numbered $n + 2$ becomes third, the card numbered 2 becomes fourth, and so on. Show that after r shuffles, the card originally occupying the position p occupies the position x, where x is the remainder in the division of $p \cdot 2^r$ by $2n + 1$.

328. Show that if in Problem 327 the deck contains $6m + 2$ cards, then the cards numbered $2m + 1$ and $4m + 2$ exchange positions after each shuffle.

329. Show that if the deck in Problem 327 contains $14m + 6$ cards, then after 3 shuffles the cards numbered $2m + 1$, $2(2m + 1)$, $3(2m + 1)$, $4(2m + 1)$, $5(2m + 1)$, $6(2m + 1)$ return to their original positions.

330. If in Problem 327, $2^x - 1$ is divisible by $2n + 1$, then after x shuffles all the cards in a deck of $2n$ cards return to their original positions.

331. A deck of cards is shuffled as follows: We take the first card, put the second card on top of the first card, put the third card under the first card, and so on. Show that if the deck consists of $6n - 2$ cards, then the position of the $2n$th card remains unchanged.

332. 22 cards are shuffled as in Problem 331. Show that card 8 stays fixed, cards 5 and 14 exchange locations, and cards 3, 13, 18 are permuted cyclically.

333. Retain the setting of Problem 331. Show that the cards in a deck of 16, 32, 42, 28, 36, 12, 20, 46, 22, 52, 14, 18, 26, 30, 50 cards return to their original positions after 5, 6, 8, 9, 9, 10, 10, 10, 12, 12, 14, 18, 26, 30, 50 shuffles, respectively.

334. A square is subdivided into 16 squares which are painted white, black, red, and blue. In how many ways can the painting be done if the squares in any row or column are to be painted different colors?

335. Fifteen students about to go for a walk are lined up in 5 rows of 3. How many times can the students be lined up so that 2 students who are in the same row on one occasion are not in the same row on another occasion?

336. Show that if n is a natural number, then $(n^2)!/(n!)^{n+1}$ is a natural number; if m and n are odd natural numbers, then $(mn)!/[(m!)^{(n+1)/2}(n!)^{(m+1)/2}]$ is a natural number.

337. n objects are arranged on a circle. Let f_n be the number of permutations of these objects in which each object follows a different object from the one it followed originally. Show that

$$f_n + f_{n+1} = D_n.$$

(For the definition of D_n see p. 51.)

338. Find the number of integral solutions of the equation

$$x_1 + x_2 + \cdots + x_p = m,$$

with $0 \leqslant l \leqslant x_k \leqslant n$, $k = 1,\dots, p$.

339. There are 7 copies of one book, 8 copies of another book, and 9 copies of a third book. In how many ways can each of 2 people be given 12 of these books?

340. Consider all n-combinations with repetitions of n letters. Show that each letter appears in these combinations C_{2n-1}^n times.

341. The distance from A to B is 999 miles. Along the way, at one mile intervals, are signs indicating the distances to A and B, respectively: $(0; 999)$, $(1; 998),\dots, (999; 0)$. How many of these signs involve just 2 different digits?

342. Show that the number of all possible permutations with repetitions of up to m white balls and up to n black balls is $P(m + 1, n + 1) - 2$.

343. Show that the sum of the white balls in all posible permutations with repetitions of up to m white balls and up to n black balls is

$$1 + \frac{mn + m - 1}{n + 2} P(m + 1, n + 1),$$

and that the corresponding number for black balls is

$$1 + \frac{mn + n - 1}{m + 2} P(m + 1, n + 1).$$

Test your answer using the letters in "gaaga."

344. Show that the number of all possible permutations with repetitions of up to m white balls, up to n black balls, and 1 red ball which include the red ball is

$$1 + \frac{mn + m + n}{m + n + 4} P(m + 2, n + 2).$$

345. Show that the number of all possible permutations with repetitions of up to m white balls, up to n black balls, and 1 red ball is

$$\frac{(m + 1)(n + 1)}{m + n + 3} P(m + 2, n + 2) - 1.$$

Check your answer using the letters in the word "Ararat."

346. I have 7 friends. In how many ways can I invite them for supper 3 at a time on 7 occasions if no 2 of them are to meet in my house twice?

347. In how many ways can I invite my 7 friends 3 at a time on 7 occasions if the triples are to be different and no friend is to be left out?

348. In how many ways can I invite my 7 friends 3 at a time on 7 occasions if the triples are to be different and no friend is to be present on all 7 occasions?

349. Show that the number of k-samples, $k = 1,..., n$ of n objects, $n \geqslant 2$, is $E(en! - 1)$.

350. Show that each object appears in the totality of samples in Problem 349 $E[e(n - 1)(n - 1)!]$ times.

351. Consider a series of $2n$ throws of a coin. Show that there are

$$1 + (C_n^1)^2 + \cdots + (C_n^n)^2 = C_{2n}^n$$

series in which the number of heads at no time exceeds the number of tails.

352. In how many ways can we distribute $3n$ books among 3 people so that the shares form an arithmetical progression?

353. There are n different pairs of like letters. These $2n$ letters are ordered so that no 2 like letters appear together. Show that the number of such arrangements is

$$\frac{1}{2^n} \left[(2n)! - \frac{n}{1} 2(2n - 1)! + \frac{n(n - 1)}{1 \cdot 2} 2^2 (2n - 2)! - \cdots \right].$$

354. *r* different things are divided among $n + p$ people so that at least n of them get 1 or more things. Show that this can be done in

$$(n + p)^r - n(n + p - 1)^r + \frac{n(n - 1)}{1 \cdot 2} (n + p - 2)^r - \cdots$$

ways.

355. Let Π_n^k denote the number of ways of putting n different things into k like boxes with no box left empty. Show that for $n > 1$

$$1 - \Pi_n^2 + 2! \, \Pi_n^3 - 3! \, \Pi_n^4 + \cdots = 0.$$

356. There are m boxes containing $n, 2n, \ldots, mn$ different objects, respectively. In how many ways can we withdraw mn objects by withdrawing n objects from each of the boxes?

357. A basket contains $2n + r$ different apples and $2n - r$ different pears. Show that for a given n the number of ways of selecting n apples and n pears is largest when $r = 0$.

358. A convex 1000-gon contains in its interior 500 points. Of the 1500 points (1000 vertices and 500 interior points) no 3 are collinear. The polygon is divided into triangles in such a way that all the 1500 points are vertices of triangles and no triangle has any other point as a vertex. Find the number of triangles.

359. Five people play a number of games of domino (two against two) in such a way that each player is once the partner and twice the opponent of any other player. Compute the number of games played and the possible distributions of the players.

360. Find the number of closed plane polygonal lines issuing from a point 0 and consisting of $2n$ (possibly repeated) horizontal and vertical strokes of unit length.

361. Each of n horizontal lines intersects each of n vertical lines. Compute the number of closed $2n$-sided polygonal lines whose sides consist of portions of each of the n horizontal lines and each of the n vertical lines.

362. A factory produces rattles in the form of a ring with 3 red and 7 blue balls. How many different rattles can be made (two rattles which differ by a cyclic permutation or a flip are not regarded as different)?

363. Consider a group of n people. Some of them know one another. Each pair of strangers has exactly 2 common acquaintances and each pair of acquaintances has no common acquaintances. Show that each person has the same number of acquaintances among the people in the group.

364. $a + b$ points on a circle divide it into $a + b$ arcs. We mark a of the points with an A and the remaining b points with a B. If both end points of an arc are marked with an A, then we associate with it the number 2. If both end points of an arc are marked with a B, then we associate with it the number $\frac{1}{2}$. Finally, if the end points of an arc are marked with different letters, then we associate with it the number 1. Show that the value of the product of these numbers is 2^{a-b}.

365. We number the rows of a chessboard 1 to 8 and mark the columns a, b, c, d, e, f, g, h. Next we assign arbitrary numerical values to the letters a, b, c, d, e, f, g, h. Finally, we write on each square the product of its coordinates and place 8 rooks on the board so that they cannot attack each other. Find the product of the numbers associated with the covered squares.

366. The organizing committee of a mathematical competition consists of 11 people. The materials for the competition are locked in a safe. How many locks should be placed on the safe and how many keys should be given to each member of the organizing committee if each group of 6 committee members should be able to open the safe and no smaller group of committee members should be able to open it?

367. A chain consists of 60 links each of which weighs 1 oz. What is the smallest number of links which must be opened so that the resulting pieces of chain (including the opened links) can be grouped to give any weight from 1 oz to 60 oz.? Solve this problem if the pieces of chain are intended as weights for a two-arm scale.

368. Find the number of pairs of integers x, y between 1 and 1000 such that $x^2 + y^2$ is divisible by 49.

369. The sum of a 2-digit number and of the number obtained by reversing the order of its digits is a perfect square. How many such numbers are there?

370. Find the sum of all 4-digit numbers composed of the digits 1, 2, 3, 4, 5, 6 and divisible by 3.

371. Find the sum of all even 4-digit numbers composed of the digits 0, 1, 2, 3, 4, 5.

372. Find the number of integral solutions of the inequality

$$|x| + |y| \leqslant 1000.$$

373. Given points $A_1, A_2, ..., A_{16}$ on a circle. We construct all possible convex polygons whose vertices are among the points $A_1, A_2, ..., A_{16}$. We separate these polygons into two classes one of which consists of the polygons for which A_1 is a vertex, and the other of the remaining polygons. Which class contains more polygons?

374. A knight is placed on an infinite chessboard. Find the number of positions which he can reach after $2n$ moves.

375. Given 1955 points. What is the largest number of triples of these points with the property that any 2 triples have a point in common?

376. Show that the number of digits in the number 123456 $\cdots$ 100,000,000 obtained by writing in succession the numbers from 1 to 100,000,000 is equal to the number of zeros in the sequence 1, 2, 3,..., 10^9.

377. How many 4-digit numbers in the range from 0001 to 9999 (inclusive) have the property that the sum of the first 2 digits is equal to the sum of the last 2 digits?

378. A school teaches $2n$ subjects. All the students receive grades of "very good" or "good" in all the subjects, no two students have identical records, and no student can be said to be superior to another student (we say of one student that he is superior to another if there is no subject in which the grade of the first student is poorer than the grade of the second student and there is at least one subject in which the first student has a higher grade than the second student). Show that the number of students in the school does not exceed C_{2n}^n .

379. Put $M_r = A_m^r$, the number of r-samples of m objects, and $N_r = A_n^r$. Show that the value of A_{m+n}^r is given by the sum obtained by expanding $(M + N)^r$ and changing exponents to indices.

380. Find the coefficient of x^8 in the expansion of

$$(1 + x^2 - x^3)^9.$$

381. Find the coefficient of x^m in the expansion of

$$(1 + x)^k + (1 + x)^{k+1} + \cdots + (1 + x)^n.$$

Discuss the case $m < k$ and the case $m \geqslant k$.

382. Find the coefficients of x^{17} and x^{18} in the expansion of $(1 + x^5 + x^7)^{20}$.

383. Compare the coefficients of x^{17} in the expansions of $(1 + x^2 - x^3)^{1000}$ and $(1 - x^2 + x^3)^{1000}$.

384. Let a_0 , a_1 , a_2 ,... be the coefficients in the expansion of $(1 + x + x^2)^n$ arranged in increasing powers of x. Prove that

(a) $a_0 a_1 - a_1 a_2 + a_2 a_3 - \cdots - a_{2n-1} a_{2n} = 0,$

(b) $a_0^2 - a_1^2 + a_2^2 - \cdots + (-1)^{n-1} a_{n-1}^2 = \frac{1}{2} a_n + \frac{1}{2}(-1)^{n-1} a_n^2 ,$

(c) $a_r - n a_{r-1} + C_n^2 a_{r-2} - \cdots + (-1)^r C_n^r a_0 = 0,$ if r is not divisible by 3.

(d) $a_0 + a_2 + a_4 + \cdots = \frac{1}{2}(3^n + 1),$
    $a_1 + a_3 + a_5 + \cdots = \frac{1}{2}(3^n - 1).$

385. Find the number of different terms in the expansion of

$$(x_1 + x_2 + \cdots + x_n)^3.$$

386. Find the coefficient of x^k in the expansion of

$$(1 + x + x^2 + \cdots + x^{n-1})^2.$$

387. Show that

$$\frac{[C_{n+1}^{r+1} - C_n^r]C_{n-1}^{r-1}}{(C_n^r)^2 - C_{n+1}^{r+1}C_{n-1}^{r-1}} = r.$$

388. Show that

$$C_n^1 + 6C_n^2 + 6C_n^3 = n^3,$$

$$1 + 7C_n^1 + 12C_n^2 + 6C_n^3 = (n+1)^3.$$

389. Show that

$$1 + 14C_n^1 + 36C_n^2 + 24C_n^3 = (n+1)^4 - n^4,$$

$$C_n^1 + 14C_n^2 + 36C_n^3 + 24C_n^4 = n^4.$$

390. Show that

$$1 - 3C_n^2 + 9C_n^4 - 27C_n^6 + \cdots = (-1)^n 2^n \cos\frac{2n\pi}{3}$$

$$C_n^1 - 3C_n^3 + 9C_n^5 - \cdots = \frac{(-1)^{n+1}2^{n+1}}{\sqrt{3}}\sin\frac{2n\pi}{3}.$$

391. Show that

(a) $\quad C_n^0 + C_n^3 + C_n^6 + \cdots = \frac{1}{3}\left(2^n + 2\cos\frac{n\pi}{3}\right),$

(b) $\quad C_n^1 + C_n^4 + C_n^7 + \cdots = \frac{1}{3}\left(2^n + 2\cos\frac{(n-2)\pi}{3}\right),$

(c) $\quad C_n^2 + C_n^5 + C_n^8 + \cdots = \frac{1}{3}\left(2^n + 2\cos\frac{(n+2)\pi}{3}\right),$

(d) $\quad C_n^0 + C_n^4 + C_n^8 + \cdots = \frac{1}{2}\left(2^{n-1} + 2^{n/2}\cos\frac{n\pi}{4}\right).$

392. Show that for $n \geqslant 2$ and $|x| < 1$ we have

$$(1 + x)^n + (1 - x)^n \leqslant 2^n.$$

393. Show that for $m > n$

$$\sum_{x=0}^{n} \frac{n(n-1)\cdots(n-x+1)}{m(m-1)\cdots(m-x+1)} = \frac{m+1}{m-n+1}$$

and

$$\sum_{x=0}^{n} \frac{C_n^x C_n^r}{C_{2n}^{x+r}} = \frac{2n+1}{n+1}$$

394. Show that

$$\frac{m}{1} + \frac{m(m+1)}{1\cdot 2} + \cdots + \frac{m(m+1)\cdots(m+n-1)}{1\cdot 2\cdots n}$$

$$= \frac{n}{1} + \frac{n(n+1)}{1\cdot 2} + \cdots + \frac{n(n+1)\cdots(n+m+1)}{1\cdot 2\cdots m}.$$

395. Show that

$$\sum_{x=1}^{n} \frac{C_{n-1}^{x-1}}{C_{2n-1}^{x}} = \frac{2}{n+1}.$$

396. Show that

$$\sum_{x=1}^{n} \frac{C_{n-1}^{x-1}}{C_{n+q}^{x}} = \frac{n+q+1}{(q+1)(q+2)}.$$

397. Show that

$$\sum_{x=1}^{n} \frac{C_{n-2}^{x-2}}{C_{n+q}^{x}} = \frac{2(n+q+1)}{(q+1)(q+2)(q+3)}.$$

398. Show that

$$(C_n^1)^2 + 2(C_n^2)^2 + 3(C_n^3)^2 + \cdots + n(C_n^n)^2 = \frac{(2n-1)!}{[(n-1)!]^2}.$$

399. Show that

$$\frac{1}{[(n-1)!]^2} + \frac{1}{1!\,2!}\frac{1}{[(n-2)!]^2} + \frac{1}{2!\,3!}\frac{1}{[(n-3)!]^2} + \cdots = \frac{(2n-1)!}{[n!\,(n-1)!]^2}.$$

400. Show that

$$\frac{(n+r-1)!}{r!} - \frac{n}{1}\frac{(n+r-3)!}{(r-2)!} + \frac{n(n-1)}{1\cdot 2}\frac{(n+r-5)!}{(r-4)!} - \cdots = \frac{n!\,(n-1)!}{r!\,(n-r)!}.$$

401. Compute the following sums:

(a) $C_n^1 + 2C_n^2 + 3C_n^3 + \cdots + nC_n^n$,

(b) $C_n^0 + 2C_n^1 + 3C_n^2 + \cdots + (n+1)C_n^n$,

(c) $C_n^2 + 2C_n^3 + 3C_n^4 + \cdots + (n-1)C_n^n$,

(d) $C_n^0 + 3C_n^1 + 5C_n^2 + \cdots + (2n-1)C_n^n$,

(e) $C_n^0 - 2C_n^1 + 3C_n^2 - \cdots + (-1)^n(n+1)C_n^n$,

(f) $3C_n^1 + 7C_n^2 + 11C_n^3 + \cdots + (4n-1)C_n^n$,

(g) $C_n^1 - 2C_n^2 + 3C_n^3 - \cdots + (-1)^{n-1}nC_n^n$,

(h) $\dfrac{C_n^0}{1} + \dfrac{C_n^1}{2} + \dfrac{C_n^2}{3} + \cdots + \dfrac{C_n^n}{n+1}$,

(i) $\dfrac{C_n^0}{2} + \dfrac{C_n^1}{3} + \dfrac{C_n^2}{4} + \cdots + \dfrac{C_n^n}{n+2}$,

(j) $\dfrac{C_n^0}{1} - \dfrac{C_n^1}{2} + \dfrac{C_n^2}{3} - \cdots + (-1)^n\dfrac{C_n^n}{n+1}$,

(k) $(C_n^0)^2 - (C_n^1)^2 + (C_n^2)^2 - \cdots + (-1)^n(C_n^n)^2$.

402. Find the largest coefficient in the expansions of

$$(a+b+c)^{10}, \qquad (a+b+c+d)^{14}.$$

403. Express the coefficients Y_1, Y_2,... in the expansion

$$(1-4x)^{-1/2} = 1 + Y_1 x + Y_2 x^2 + \cdots$$

in terms of the binomial coefficients. Expand $(1-4x)^{1/2}$.

404. Show that the Y_n in Problem 403 satisfy the relations

(a) $Y_n + \frac{1}{2}Y_1 Y_{n-1} + \frac{1}{3}Y_2 Y_{n-2} + \cdots + \dfrac{1}{n+1}Y_n = \frac{1}{2}Y_{n+1}$,

(b) $Y_0 Y_n + Y_1 Y_{n-1} + Y_2 Y_{n-2} + \cdots + Y_n Y_0 = 4^n$,

(c) $\dfrac{Y_0 Y_n}{1(n+1)} + \dfrac{Y_1 Y_{n-1}}{2 \cdot n} + \dfrac{Y_2 Y_{n-2}}{3(n-1)} + \cdots + \dfrac{Y_n Y_0}{(n+1) \cdot 1} = \dfrac{Y_{n+1}}{n+2}$.

405.   Each number in the number triangle

$$1$$
$$1 \ 1 \ 1$$
$$1 \ 2 \ 3 \ 2 \ 1$$
$$1 \ 3 \ 6 \ 7 \ 6 \ 3 \ 1$$
$$\cdot \ \cdot \ \cdot \ \cdot \ \cdot \ \cdot$$

is equal to the sum of the number directly above it and its two immediate neighbors (locations not occupied by numbers should be thought of as being occupied by zeros). Show that beginning with the third row each row contains an even number.

406.   The first row of the number triangle

$$0 \ 1 \ 2 \ 3 \ \cdot \ \cdot \ \cdot \ \cdot \ \cdot \ 1957 \qquad 1958$$
$$1 \ 3 \ 5 \ \cdot \ \cdot \ \cdot \ \cdot \ \cdot \qquad \qquad 3915$$
$$\cdot \ \cdot \ \cdot \ \cdot \ \cdot \ \cdot \ \cdot \ \cdot \ \cdot \ \cdot \ \cdot \ \cdot \ \cdot \ \cdot \ \cdot \ \cdot \ \cdot$$

consists of the numbers 0, 1,..., 1958. An element in any other row is the sum of its 2 closest neighbors in the row above it. Show that the element in the last row of the triangle is divisible by 1958.

407.   Consider the Fibonacci numbers $u_n : u_0 = 0, \ u_1 = 1, \ u_2 = 1, \ u_3 = 2,$ $u_4 = 3, u_5 = 5,$ and so on (note that we started the sequence with 0 and 1 and not with 1 and 2 as in Chap. VI). Show that:

(a)   For all m and n, $u_{n+m} = u_{n-1}u_m + u_n u_{m+1}$.

(b)   For all m and $n = km$, the number u_n is divisible by u_m .

(c)   Two neighboring Fibonacci numbers are relatively prime.

408.   Find the greatest common divisor of the 1000th and 770th terms in the Fibonacci sequence.

409.   Does the set of the first 100,000,001 Fibonacci numbers contain a number which ends in 4 zeros?

410.   Show that the sum of 8 successive Fibonacci numbers is not a Fibonacci number.

411.   Show that:

(a) $u_2 + u_4 + \cdots + u_{2n} = u_{2n+1} - 1,$

(b) $u_1 + u_3 + \cdots + u_{2n-1} = u_{2n} ,$

(c) $u_1^2 + u_2^2 + \cdots + u_n^2 = u_n u_{n+1} ,$

(d) $u_{n+1}^2 = u_n u_{n+2} + (-1)^n$,

(e) $u_1 u_2 + u_2 u_3 + \cdots + u_{2n-1} u_{2n} = u_{2n}^2$,

(f) $u_1 u_2 + u_2 u_3 + \cdots + u_{2n} u_{2n+1} = u_{2n+1}^2 - 1$,

(g) $n u_1 + (n-1)u_2 + (n-2)u_3 + \cdots + 2u_{n-1} + u_n = u_{n+4} - (n+3)$,

(h) $u_3 + u_6 + \cdots + u_{3n} = \dfrac{u_{3n+2} - 1}{2}$,

(i) $u_{3n} = u_{n+1}^3 + u_n^3 - u_{n-1}^3$.

412. Show that every natural number N can be written as the sum of different Fibonacci numbers such that no 2 summands are neighbors in the Fibonacci sequence.

413. Let $p \geqslant q \geqslant r$ be integers such that $p < q + r$ and $p + q + r = 2s$. Show that the number of ways of dividing p black balls, q white balls, and r red balls between 2 people so that each person gets s balls is equal to

$$s^2 + s + 1 - \tfrac{1}{2}(p^2 + q^2 + r^2).$$

414. Show that if $q + r < p$, then the answer in Problem 413 is increased by $\tfrac{1}{2}(p - s)(p - s - 1)$.

415. $pq + r$ different objects, $0 \leqslant r < p$, are divided among p people as evenly as possible (this means that each person gets q or $q + 1$ objects). Show that this can be done in

$$C_p^r \frac{(pq + r)!}{(q + 1)^r (q!)^p}.$$

ways.

416. Compute the sum

$$\sum_{i_n=1}^{m} \sum_{i_{n-1}=1}^{i_n} \cdots \sum_{i_1=1}^{i_2} \sum_{i_0=1}^{i_1} 1.$$

417. Prove the identity

$$C_{n+m}^m = \sum P(k_1, \ldots, k_m, n - k_1 - \cdots - k_m + 1),$$

where the sum extends over all nonnegative integral solutions of the equation $k_1 + 2k_2 + \cdots + mk_m = m$.

418. Find the general solution of each of the following recurrence relations:

(a) $a_{n+2} - 7a_{n+1} + 12a_n = 0$,

(b) $a_{n+2} + 3a_{n+1} - 10a_n = 0$,

(c) $a_{n+2} - 4a_{n+1} + 13a_n = 0$,

(d) $a_{n+2} + 9a_n = 0,$

(e) $a_{n+2} + 4a_{n+1} + 4a_n = 0,$

(f) $a_{n+3} - 9a_{n+2} + 26a_{n+1} - 24a_n = 0,$

(g) $a_{n+3} + 3a_{n+2} + 3a_{n+1} + a_n = 0,$

(h) $a_{n+4} + 4a_n = 0.$

419. Find a_n , given the recurrence relation and initial terms

(a) $a_{n+2} - 5a_{n+1} + 6a_n = 0,$ $a_1 = 1,$ $a_2 = -7,$

(b) $a_{n+2} - 4a_{n+1} + 4a_n = 0,$ $a_1 = 2,$ $a_2 = 4,$

(c) $a_{n+2} + a_{n+1} + a_n = 0,$ $a_1 = -\frac{1}{4},$ $a_2 = -\frac{1}{2},$

(d) $a_{n+3} - 9a_{n+2} + 26a_{n+1} - 24a_n = 0,$ $a_1 = 1,$ $a_2 = -3,$ $a_3 = -29.$

420. Find a sequence such that $a_1 = \cos \alpha,$ $a_2 = \cos 2\alpha,$ and

$$a_{n+2} - 2 \cos \alpha a_{n+1} + a_n = 0.$$

421. Show that the sequence with general term $a_n = n^k$ satisfies the relation

$$a_{n+k} - C_k^1 a_{n+k-1} + C_k^2 a_{n+k-2} + \cdots + (-1)^k C_k^k a_n = 0.$$

422. Find a sequence such that

$$a_{n+2} + 2a_{n+1} - 8a_n = 2^n.$$

423. Use the identity $(1 + x)^p (1 + x)^{-k-1} = (1 + x)^{p-k-1}$ to show that

$$\sum_{s=0} (-1)^s C_{k+s}^s C_p^{n-s} = C_{p-k-1}^n .^*$$

424. Use the identity $(1 - x)^{-m-1} (1 - x)^{-q-1} = (1 - x)^{-m-q-2}$ to show that

$$\sum_{s=0} C_{p-s}^m C_{q-s}^q = C_{p+q+1}^{p-m} .$$

425. Use the identity $(1 + x)^n = (1 - x^2)^n (1 - x)^{-n}$ to show that

$$\sum_{s=0} (-1)^s C_{n+k-2s}^n C_{n+1}^s = C_{n+1}^k .$$

426. Use the identiy $(1 + x)^n (1 - x^2)^{-n} = (1 - x)^{-n}$ to show that

$$\sum_{s=0} C_n^{k-2s} C_{n+s-1}^s = C_{n+k-1}^k .$$

---

* Here and in the sequel, the summation extends over all nonnegative values of s for which the left-hand side of the inequality is defined.

427. Use the identiy $(1 - x^2)^{-p-1} = (1 + x)^{-p-1}(1 - x)^{-p-1}$ to show that

$$\sum_{s=0} (-1)^s C_{p+2k-s}^p C_{p+s}^p = C_{p+k}^k .$$

428. Use the identities

$$(1 - x)^{-2p} \left[1 - \left(\frac{x}{1 - x} \right)^2 \right]^{-p} = (1 - 2x)^{-p}$$

and

$$(1 - x)^{2p} \left[1 - \left(\frac{x}{1 - x} \right)^2 \right]^{p} = (1 - 2x)^p$$

to show that

$$\sum_{s=0} C_{p+s}^s C_{2p+m}^{2p+2s+1} = 2^{m-1} C_{m+p-1}^p$$

and

$$\sum_{s=0} (-1)^s C_p^s C_{2p-2s}^{m-2s} = 2^p C_p^m .$$

429. Show that

$$\sum_{s=0} C_{p+s}^s C_{2p+m}^{2p+2s} = 2^{m-1} \frac{2p + m}{m} C_{m+p-1}^p .$$

430. Use the identities

$$(1 - x)^{\pm 2p} \left[1 + \frac{2x}{(1 - x)^2} \right]^{\pm p} = (1 + x^2)^{\pm p}$$

to show that

$$\sum_{s=0} (-1)^s C_{p+s-1}^s C_{2m+2p+s}^{2m+1-s} 2^s = 0,$$

$$\sum_{s=0} (-1)^s C_{p+s-1}^s C_{2m+2p+s-1}^{2m-s} 2^s = (-1)^m C_{p+m-1}^m ,$$

$$\sum_{s=0} (-1)^s C_p^s C_{2p-2s}^{2m+1-s} 2^s = 0,$$

$$\sum_{s=0} (-1)^s C_p^s C_{2p-2s}^{2m-s} 2^s = C_p^m ,$$

and then show that

$$\sum_{s=0} C_{2p+2m}^{2s} C_{p+m-s}^{p} = 2^{2m}(p+m)\frac{(p+2m-1)!}{p!\,(2m)!},$$

$$\sum_{s=0} C_{2p+2m+1}^{2s+1} C_{p+m-s}^{p} = 2^{2m}(2p+2m+1)\frac{(p+2m)!}{p!\,(2m+1)!},$$

$$\sum_{s=1} C_{2p+2m}^{2s-1} C_{p+m-s}^{p} = 2^{2m-1} C_{p+2m-1}^{p},$$

$$\sum_{s=0} C_{2p+2m+1}^{2s} C_{p+m-s}^{p} = 2^{2m} C_{p+2m}^{p}.$$

431. Consider the formulas

$$[(1+x)^p \pm (1-x)^p]^2 = (1+x)^{2p} + (1-x)^{2p} \pm 2(1-x^2)^p,$$

$$[(1+x)^p + (1-x)^p][(1+x)^p - (1-x)^p] = (1+x)^{2p} - (1-x)^{2p},$$

for positive and negative values of p and show that

$$2\sum_{s=0} C_p^{2s} C_p^{2m-2s} = C_{2p}^{2m} + (-1)^m C_p^m,$$

$$2\sum_{s=0} C_p^{2s+1} C_p^{2m-2s+1} = C_{2p}^{2m+2} + (-1)^m C_p^{m+1},$$

$$2\sum_{s=0} C_p^{2s} C_p^{2m-2s+1} = C_{2p}^{2m+1},$$

$$2\sum_{s=0} C_{p+2s}^{p} C_{p+2m-2s}^{p} = C_{2p+2m+1}^{2p+1} + C_{p+m}^{p},$$

$$2\sum_{s=0} C_{p+2s}^{p-1} C_{p+2m-2s}^{p-1} = C_{2p+2m+1}^{2p-1} - C_{p+m}^{p},$$

$$2\sum_{s=0} C_{p+2s}^{p} C_{p+2m-2s+1}^{p} = C_{2p+2m+2}^{2p+1}.$$

432. Consider the expression

$$[(1+x)^{p+1} \pm (1-x)^{p+1}][(1+x)^p \pm (1-x)^p]$$

for all combinations of signs and show that

$$2 \sum_{s=0} C_{p+1}^{2s} C_p^{2m-2s} = C_{2p+1}^{2m} + (-1)^m C_p^m,$$

$$2 \sum_{s=0} C_{p+1}^{2s} C_p^{2m-2s+1} = C_{2p+1}^{2m+1} - (-1)^m C_p^m,$$

$$2 \sum_{s=0} C_{p+1}^{2s+1} C_p^{2m-2s} = C_{2p+1}^{2m+1} + (-1)^m C_p^m,$$

$$2 \sum_{s=0} C_{p+1}^{2s+1} C_p^{2m-2s+1} = C_{2p+1}^{2m+2} + (-1)^m C_p^{m+1},$$

$$2 \sum_{s=0} C_{p+2s-1}^{p-1} C_{p+2m-2s}^{p} = C_{2p+2m}^{2p} + C_{p+m}^{p},$$

$$2 \sum_{s=0} C_{p+2s-1}^{p-1} C_{p+2m-2s+1}^{p} = C_{2p+2m+1}^{2p} + C_{p+m}^{p},$$

$$2 \sum_{s=0} C_{p+2s}^{p-1} C_{p+2m-2s}^{p} = C_{2p+2m+1}^{2p} - C_{p+m}^{p}$$

$$2 \sum_{s=0} C_{p+2s}^{p-1} C_{p+2m-2s+1}^{p} = C_{2p+2m+2}^{2p} - C_{p+m+1}^{p}.$$

433. Use the relation

$$\left(1 - \frac{1}{x}\right)^m (1 - x)^{-n-1} = \frac{(-1)^m}{x^m} (1 - x)^{m-n-1}$$

to show that

$$\sum_{s=0} (-1)^s C_m^{m-k+s} C_{n+s}^n = C_{m-n-1}^k.$$

434. Show that

$$\sum_{s=0} (-1)^s C_m^s C_s^n = \begin{cases} 0, & \text{for } m \neq n, \\ (-1)^n, & \text{for } m = n. \end{cases}$$

435. Use the relation

$$(1 - x)^{-n} (1 - x^h)^n = (1 + x + \cdots + x^{h-1})^n$$

to show that

$$\sum_{s=0} (-1)^s C_{m-sh}^{n-1} C_n^s = \begin{cases} 0, & \text{for } m > hn - 1, \\ 1, & \text{for } m = hn - 1. \end{cases}$$

436. Use the relation

$$(1 - x)^{-n-1}(1 - x^h)^n = \frac{(1 + x + \cdots x^{h-1})^n}{1 - x}$$

to show that for $m \geqslant hn$

$$\sum_{s=0} (-1)^s C^n_{m-sh} C^s_n = h^n.$$

437. Use the relation

$$(1 + x)^{\pm p} (1 - x)^{\pm p} = (1 - x^2)^{\pm p}$$

to show that

$$\sum_{s=0} (-1)^s C^{m-s}_p C^s_p = \begin{cases} (-1)^{m/2} C^{m/2}_p, & \text{for } m \text{ even,} \\ 0, & \text{for } m \text{ odd.} \end{cases}$$

$$\sum_{s=0} (-1)^s C^p_{p+m-s} C^p_{p+s} = \begin{cases} (-1)^{m/2} C^{m/2}_{p+(m/2)}, & \text{for } m \text{ even,} \\ 0, & \text{for } m \text{ odd.} \end{cases}$$

438. Show that

$$\sum_{s=0} (-1)^s [C^s_m]^2 = \begin{cases} (-1)^{m/2} C^{m/2}_m, & \text{for } m \text{ even,} \\ 0, & \text{for } m \text{ odd.} \end{cases}$$

439. Denote the expression

$$a(a + 1)(a + 2) \cdots (a + n - 1)$$

by $(a)_n$; show that

$$(a + b)_n = \sum_{m=0}^{n} C^m_n (a + m)_{n-m} (b - m + 1)_m .$$

Answers and Hints

1. By the rule of product, the number of roads is $5 \cdot 3 = 15$.

2. By the rule of product, the number of choices is $100^2 = 10,000$.

3. 20.

4. 6.

5. 4.

6. 48.

7. 25; 20.

8. 480; 437.

9. 1024; 4032.

10. We can choose the black square in 32 ways. Then we delete the row and column which share this square. We are then left with 24 white squares to choose from. Thus the pair of squares can be chosen in $32 \cdot 24 = 768$ ways.

11. By the rule of product, the number of ways is $12 \cdot 9 \cdot 10 = 1080$.

12. $6 \cdot 5 = 30$.

13. $3 \cdot 7 \cdot 7 = 147$.

14. We can buy a selection of 3 "one-novel" books or a selection of 1 "two-novel" book and 1 "one-novel" book. The first can be done in $6 \cdot 3 \cdot 4$ ways. The second can be done in $5 \cdot 4 + 7 \cdot 6$ ways. It follows that the purchase can be effected in $6 \cdot 3 \cdot 4 + 5 \cdot 4 + 7 \cdot 6 = 134$ ways. (Our argument makes use of the rule of product and the rule of sum.)

15. We have the additional possibility of buying a copy of the combined edition of "Rudin" and "Fathers and Sons" and a copy of "Nobleman's Nest." This can be done in $3 \cdot 3 = 9$ ways. Hence the present total of choices is $134 + 9 = 143$.

16. In the first case: for then Jill can make $11 \cdot 10 = 110$ choices. In the second case, she can make only $12 \cdot 9 = 108$ choices.

17. $6 \cdot 8 \cdot 10 = 480$. If the first 2 tops fall on the side marked with a one, then the third top can fall in 10 ways. If the first and third tops fall on the side marked with a one, then the second top can fall in 8 ways. Finally, if the last 2 tops fall on the side marked with a one, then the first top can fall in 6 ways. The apparent total of $6 + 8 + 10$ possibilities must be diminished by 2, for the possibility of all 3 tops falling on the side marked with a one has been counted 3 times. It follows that the number of different outcomes is 22.

18. Since the order of the colors is immaterial, the number of choices is $C_5^3 = 10$.

19. Here the order of the colors matters, so that the number of flags is $A_5^3 = 60$. If one of the stripes is to be red, then the number of flags is $3 \cdot A_4^2 = 36$.

20. $A_5^2 = 20$ dictionaries.

21. $A_{10}^2 - A_5^2 = 70$.

22. This is a matter of counting the number of 4-arrangements with repetitions of elements of 13 kinds. The number of such arrangements is $13^4 = 28,561$. If no two of the cards can have the same value, then our problem reduces to counting 4-arrangements without repetitions of elements of 13 kinds. The number of such arrangements is $A_{13}^4 = 17,160$.

23. In this case, an arrangement is determined as soon as we have selected, say, a diamonds-card and a spades-card. Thus the number of possible arrangements is $13^2 = 169$.

24. The number of choices is $300 + 300 \cdot 299 + 300 \cdot 299 \cdot 298 = 26,820,600$.

25. The relation "is the neighbor of" is preserved by cyclic permutations and reflections. In the case of 4 people, we have $2 \cdot 4 = 8$ such transformations. Since the number of permutations of 4 people is $4! = 24$, the number of different seating arrangements is $24/8 = 3$. For 7 people, the corresponding number is $7!/14 = 360$. For n people, the corresponding number is $(n-1)!/2$.

The number of ways of seating 7 people so that certain 2 people are neighbors is twice the number of ways of seating 6 people (since the neighbors can exchange seats), that is, $5! = 120$. Similarly, the number of ways of seating 7 people so that a certain person has certain two neighbors is $4! = 24$.

26. Consider the number of ways of making up, say, the team consisting of 1 boy and 3 girls. The boy can be chosen in $C_3^1 = 3$ ways and the 3 girls can be chosen in $C_5^3 = 10$ ways. By the rule of product, a team of this kind can be made up in $3 \cdot 10 = 30$ ways.

27. There are n^k ways of distributing n different objects in k different boxes. In this case, the number of ways is $3^6 = 729$.

28. By the rule of product, the number of ways is $7 \cdot 9 = 63$.

29. One person can choose 2 books in $C_7^2 = 21$ ways, and the other in $C_9^2 = 36$ ways. Hence the number of possible exchanges is $21 \cdot 36 = 756$.

30. We group the $5!$ unrestricted arrangements of speakers in pairs which differ only in the order of speakers A and B. In each pair there is exactly 1 acceptable arrangement. Hence the number of acceptable arrangements is $5!/2 = 60$.

31. In this case, the pair A, B can be regarded as a single speaker, so that the number of arrangements is $4! = 24$.

32. There are 2 ways of assigning chairs for men and women. Following such an assignment, the men can be seated in $5!$ ways and, subsequently, the women can be seated in $5!$ ways. This gives $2(5!)^2 = 28,800$ ways.

33. The number of arrangements is reduced by a factor of 10 and so is equal to 2880.

34. The number of ways of choosing 10 cards out of 52 is C_{52}^{10}. The number of selections without an ace is C_{48}^{10}. Hence the number of selections with at least 1 ace is $C_{52}^{10} - C_{48}^{10}$. The number of selections with exactly 1 ace is $C_4^1 C_{48}^9$. The number of selections with at least 2 aces is $C_{52}^{10} - C_{48}^{10} - 4C_{48}^9$. The number of selections with exactly 2 aces is $C_4^2 C_{48}^8$ (the 2 aces can be selected in C_4^2 ways and the remaining 8 cards can be selected in C_{48}^8 ways).

35. The number of signals is 3^m (see Problem 27).

36. We associate with every set of teeth a sequence of 32 zeros and ones (the presence or absence of a tooth in a certain position is denoted by a 1 or a 0 in that position). Since the number of such sequences is 2^{32}, this is the maximal number of inhabitants in the country.

37. First we choose 1 of the 3 "no preference" passengers. This can be done in 3 ways. We seat this passenger to face the locomotive. This determines a seating arrangement of the passengers up to a permutation of each of the 2 groups of 5 passengers. It follows that the number of seating arrangements is $3(5!)^2 = 43{,}200$.

38. $A_9^4 = 3024$.

39. $C_{52}^5 = 2{,}598{,}960$.

40. There are $26 \cdot 10^4$ licenses with 1 letter, $26^2 \cdot 10^4$ licenses with 2 letters, and $26^3 \cdot 10^4$ licenses with 3 letters. By the rule of sum, there are at most $(26 + 26^2 + 26^3) \cdot 10^4 = 18{,}278 \cdot 10^4$ licenses.

41. The 2 "apple days" can be chosen out of 5 days in $C_5^2 = 10$ ways.

42. C_{m+n}^n.

43. $P(2, 3, 4) = 1260$.

44. Since the oranges are different, the number of ways is $A_8^5 = 6720$.

45. Each orange can be given to each of the 8 sons. Hence the number of ways is $8^5 = 32{,}768$.

46. $P(2, 2, 2, 1, 1, 1, 1, 1)$; $P(3, 1, 1, 1, 1, 1)$; $P(2, 2, 2, 1, 1, 1)$.

47. $C_{30}^4 = 27{,}405$; $A_{30}^4 = 657{,}720$.

48. $P(2, 2, 2, 1, 1) = 5040$.

49. The 6 subscribers can be selected in C_n^6 ways. There are 6! permutations of 6 people and each such permutation determines 3 ordered pairs with the members of each pair ordered. Since neither type of order is relevant, we must divide $C_n^6 \cdot 6!$ by $2^3 \cdot 3! = 48$. This yields the answer $n!/[48(n-6)!]$.

50. $C_{10}^{12} = C_{21}^{12}$; $C_{10}^8 = C_{17}^8$; C_{10}^8.

51. A selection can include 2, 3, or 4 women. The number of selections corresponding to each of these three possibilities is $C_4^2 C_7^4$, $C_4^3 C_7^3$, and $C_4^4 C_7^2$. Therefore the required number of ways is equal to $C_4^2 C_7^4 + C_4^3 C_7^3 + C_4^4 C_7^2 = 371$.

52. Our numbers must end in 12, 24, 32, 44, or 52. Each of the first 2 digits can be chosen in 5 ways. This gives a total of $5^2 \cdot 5 = 125$ numbers.

53. Each of the n passengers has m choices. Hence the number of ways is m^n. If we are concerned only with the *number* of passengers who get off at each stop, then the number of ways is C_{n+m-1}^{m-1}.

54. Since a and b appear together, we can think of the pair as a single symbol. Since we can interchange the positions of a and b, we obtain $2(n-1)!$ permutations with a and b together. It follows that there are $n! - 2(n - 1)!$ permutations in which a and b do not appear together. Similarly, there are $n! - 6(n - 2)!$ permutations in which a, b, and c do not appear together. Finally, by the principle of inclusion and exclusion, the number of permutations in which no two of the three symbols a, b, c appear together is $n! - 6(n - 1)! + 6(n - 2)!$.

55. The 3 judges can name 10^3 triples of winners. $A_{10}^3 = 720$ triples will include 3 different athletes. It follows that in 280 cases the triples will include the same athlete at least twice. The required fraction is therefore 0.28.

56. $3^4 = 81$.

57. We can turn a necklace over and permute its beads cyclically without producing a different necklace. Hence the number of different necklaces is $7!/14 = 360$.

58. The necklaces differ by the number of small beads contained between the two large beads. Hence the number of different necklaces is 3.

59. The number of possible pairs of letters is $26^2 = 726 < 2000$.

60. $A_{10}^7 = 604{,}800$; $C_{10}^3 = 120$. The 2 girls can choose partners in A_7^2 ways. The remaining 5 boys can choose partners in A_8^5 ways, so that the number of possible outcomes is $A_7^2 A_8^5 = 282{,}240$. The number of different trios without partners is now $C_8^3 = 56$.

61. $C_3^1 C_6^2 C_{60}^{20}$; $C_5^1 C_{60}^{20}$.

62. The 4 girls can be chosen in C_{12}^4 ways. Then the boys can be chosen in A_{15}^4 ways (now order matters!). Hence the number of ways is

$$C_{12}^4 A_{15}^4 = 17{,}417{,}400.$$

63. If we ignore restrictions, then each chicken may be included in the selection or left out. Of the $2^3 = 8$ possible ways of assigning the chickens, 1 is not acceptable (no chicken included in the selection). This leaves 7 acceptable ways of assigning the chickens. Similarly, there are $2^4 - 1 = 15$ acceptable ways of assigning the ducks, and $2^2 - 1 = 3$ acceptable ways of assigning the geese. It follows that there are $7 \cdot 15 \cdot 3 = 315$ acceptable selections.

64. This can be done in $P(m, n, p) = (m + n + p)!/m!\, n!\, p!$ ways.

65. The black-bound books can be permuted in $m!$ ways and the red-bound books can be permuted in $n!$ ways. This gives $m!\, n!$ arrangements in which the black-bound books come first. If the black-bound books are together, then we can place them in $n + 1$ ways between red-bound books. This gives a total of $m!\, n!(n + 1) = m!(n + 1)!$ ways of placing the set of black-bound books.

66. Each person can be included in the detachment or not. If we exclude the empty selection, we have $2^{15} - 1 = 32{,}767$ possibilities.

67. Each prime p_k can enter a divisor of q with one of the $\alpha_k + 1$ exponents $0, 1,..., \alpha_k$. By the rule of product, the number of divisors is $(\alpha_1 + 1) \cdots (\alpha_n + 1)$. To compute their sum, we consider the expression

$$(1 + p_1 + \cdots + p_1^{\alpha_1}) \cdots (1 + p_n + \cdots + p_n^{\alpha_n}).$$

If we remove brackets in this product, then we obtain a sum in which each divisor of q appears exactly once. If we use the formula for the sum of a geometric progression, then the above product, and therefore also the required sum of the divisors of q, is seen to have the value

$$\frac{p_1^{\alpha_1+1} - 1}{p_1 - 1} \ldots \frac{p_n^{\alpha_n+1} - 1}{p_n - 1}.$$

68. First we place a single half-dollar coin in each batch. The remaining 7 half-dollar coins can be distributed in $C_{11}^4 = 330$ ways (see p. 171).

69. We add 4 dividers to the 20 books and permute these 24 objects. The number of resulting permutations is 24!/4!. Each of these permutations determines a particular arrangement of the books.

70. Reasoning as in the preceding problem, we obtain the answer 8!/3! = 6720.

71. Since we are only concerned with the number of votes cast in favor of each issue, we need only compute the number of ways of distributing 30 like objects in 5 different boxes. This number is $P(30, 4) = 46{,}376$.

72. There are 3^{12} ways of binding 12 books in 3 colors of cloth. In $3 \cdot 2^{12}$ cases, the books are bound in at most 2 colors of cloth, and in 3 cases the books are bound in just 1 color of cloth. By the principle of inclusion and exclusion, there are $3^{12} - 3 \cdot 2^{12} + 3$ ways of binding the books in all 3 colors of cloth.

73. We separate the 26 letters into 6 nonempty batches by means of 5 dividers. The 5 dividers can be placed in the 25 spaces between the 26 letters. This gives C_{25}^5 ways of placing the dividers and so C_{25}^5 ways of making a particular arrangement of the letters into 6 words. Since the number of permutations of 26 letters is 26! we obtain $26!C_{25}^5$ ordered arrangements of 6 words. Since the order of the words is immaterial, there are $26!C_{25}^5/6!$ ways of making up 6 words.

74. Twelve people can be selected out of 17 in C_{17}^{12} ways. A certain 2 people enter C_{15}^{10} of these selections. Hence the number of admissible selections is $C_{17}^{12} - C_{15}^{10}$.

75. There are $P(5, 6, 7)$ permutations of the precious stones. A bracelet is not affected by cyclic permutations or by being turned over. Hence the required number is $P(5, 6, 7)/36 = 18!/(36 \cdot 5! \cdot 6! \cdot 7!)$.

76. There are 3 ways of choosing stones of 1 kind, $3 \cdot 2 = 6$ ways of choosing stones of 2 kinds, and 1 way of choosing stones of 3 kinds. This gives a total of 10 ways.

77. The cups can be set in A_4^3 ways, the saucers in A_5^3 ways, and the teaspoons can be set in A_6^3 ways. This gives $A_4^3 A_5^3 A_6^3 = 172,800$ ways of setting the table.

78. If the man invites k women, then he must invite $6 - k$ men. But then his wife must invite $6 - k$ women and k men. By the rules of sum and product, the number of ways is

$$\sum_{k=0}^{5} (C_5^k)^2 (C_7^{6-k})^2 = 267,148.$$

79. Of the people who have no seating preference, 0, 1, 2, 3, or 4 can be seated on port side. If the selected crew contains k "no preference" people, then we can select $4 - k$ people who prefer port seats. Then we are left with $12 + (9 - k)$ people out of whose number we must choose 4 starboard members of the crew. This gives $C_9^k C_{10}^{4-k} C_{21-k}^4$ choices. Summing over k, we obtain the answer

$$\sum_{k=0}^{4} C_9^k C_{10}^{4-k} C_{21-k}^4 = \frac{9! \, 10!}{4!} \sum_{k=0}^{4} \frac{(21 - k)!}{k! \, (9 - k)! \, (4 - k)! \, (6 + k)! \, (17 - k)!} \, .$$

80. The number 9 can be partitioned into 3 different summands in 3 ways: $9 = 1 + 2 + 6 = 1 + 3 + 5 = 2 + 3 + 4$. Four drawings will yield a sum less than 9: $1 + 2 + 3 = 6$, $1 + 2 + 4 = 7$, $1 + 2 + 5 = 1 + 3 + 4 = 8$. Since the number of different drawings is C_{10}^3, there are $C_{10}^3 - 4$ drawings with sum $\geqslant 9$.

81. First we select 4 cards of different suits. This can be done in 13^4 ways. Two additional cards of different suits can be selected in $C_4^2 \cdot 12^2 = 864$ ways. Combining these ways with the ways of selecting the first 4 cards and taking into consideration the possibility of interchanging cards of the same suit, we obtain $216 \cdot 13^4$ ways. Two cards of the same suit can be selected in $4C_{12}^2 = 264$ ways. Again combining these ways with the ways of selecting the first 4 cards and again bearing in mind the possibility of interchanging cards of the same suit, we obtain $88 \cdot 13^4$ ways. Hence the answer to our problem is $304 \cdot 13^4$.

82. On the first day we have $C_{10}^6 = 210$ choices, on the second day we have $C_{10}^6 - 1 = 209$ choices, and on the third day we have $C_{10}^6 - 2 = 208$ choices. This gives a total of $210 \cdot 209 \cdot 208 = 9,129,120$ choices.

83. Since $C_6^3 = 20$, each set of 3 friends must be invited exactly once in the 20-day period, that is, two ways can differ only in the order of the various trios. It follows that the number of ways is 20!.

84. Each man can choose his place of employment in 5 ways and each woman can choose hers in 4 ways. It follows that the people in question have $5^3 \cdot 4^2 = 2000$ choices.

85. In the first position we can write one of the 26 letters and in each successive position we can write one of 25 letters. This gives $26 \cdot 25^4 = 10,156,250$ words.

86. In the first case award winners can be chosen in C_{20}^6 ways, and the prizes can be awarded to a group of 6 in $P(3, 2, 1)$ ways. By the rule of product, the number of ways of awarding the prizes in this case is $C_{20}^6 P(3, 2, 1)$. In the second case, the first book can be awarded in C_{20}^3 ways, the second book in C_{20}^2 ways, and the third book in C_{20}^1 ways. Hence in this case the prizes can be awarded in $C_{20}^3 C_{20}^2 C_{20}^1$ ways.

87. We associate with the domino (p, q) the domino $(n - p, n - q)$. If $p + q = n - r$, then $(n - p) + (n - q) = n + r$, which shows that there are as many dominos with face-sum $n - r$ as there are dominos with face-sum $n + r$. The number of dominos with face-sum $n - r$ is $(n - r + 1)/2$ for $n - r$ odd, and $(n - r + 2)/2$ for $n - r$ even; in both cases the number of dominos is $E[(n - r + 2)/2]$, where $E(x)$ is the integral part of x. There are altogether C_{n+1}^2 dominos.

88. The statement of the problem implies that men and women alternate. Hence the number of seating arrangements is $2(7!)^2$.

89. There are 8 ways of choosing 3 horses, one each from the pairs A, A'; B, B'; C, C'. There are $C_{10}^3 = 120$ ways of choosing the additional horses. Finally, there are 6! ways of harnessing a set of 6 horses. This yields the answer $8 \cdot 120 \cdot 6! = 691,200$.

90. The consonants can be selected in C_9^4 ways and the vowels can be selected in C_7^3 ways. The 7 letters can be permuted in 7! ways. Hence the number of words is $C_9^4 C_7^3 \cdot 7!$. If no two consonants can appear together, then the word-pattern is CVCVCVC and the number of permutations is $3!4!$. Hence in this case the number of words is $C_9^4 C_7^3 \cdot 3!\, 4!$.

91. By the principle of inclusion and exclusion, the number of workers is $6 + 6 + 7 - 4 - 3 - 2 + 1 = 11$. The number of people who know only Russian is $6 - 4 - 2 + 1 = 1$. The number of people who know only French is $7 - 3 - 2 + 1 = 3$.

92.   By the principle of inclusion and exclusion, the number of people who took along pies was $92 - 47 - 38 - 42 + 28 + 31 + 26 - 25 = 25$.

93.   The 5 pairs of men can be selected in $10!/[(2!)^5 \cdot 5!]$ ways (order within a pair as well as the order of the pairs are irrelevant). The 5 pairs of women can be subsequently selected in $10!/(2!)^5$ ways (now the order of the pairs matters). This gives $(10!)^2/(2^{10} \cdot 5!)$ ways.

94.   Consider the boat with the married couple. The other 2 occupants of this boat can be selected in 9^2 ways. The occupants of the other 4 boats can be selected in $(8!)^2/(2^8 \cdot 4!)$ ways. Hence the required number of cases is $(9!)^2/(2^8 \cdot 4!)$.

95.   There arise two possibilities. If both married couples are in the same boat, then the remaining people can occupy the remaining 4 boats in $(8!)^2/(2^8 \cdot 4!)$ ways. If the 2 married couples are in different boats, then these 2 boats can be filled in $(A_8^2)^2$ ways and the remaining 3 boats can be filled in $6!/(2^6 \cdot 3!)$ ways. The total number of cases is $17(8!)^2/(2^8 \cdot 4!)$.

96. $7^4 - 7^3 = 2058$.

97.   Let the number formed by the first 3 digits be x. Then the number formed by the last 3 digits can take on the values $0, 1,..., 999 - x$, that is, $1000\text{-}x$ different values. Since x varies from 100 to 999, it follows that in order to obtain the answer to our problem we must add all the numbers from 1 to 900. The result is 405,450.

98.   The squares for the white checkers can be selected in C_{32}^{12} ways. Then the squares for the black checkers can be selected in C_{20}^{12} ways. The required number of ways is $C_{32}^{12}C_{20}^{12}$.

99.   We separate the P_7 permutations of the word "Jupiter" into classes such that two permutations in the same class differ only in the order of the vowels. Each of these classes contains exactly 1 permutation in which the 3 vowels appear in alphabetical order. Since the number of classes is $P_7/P_3 = 840$, this is also the number of required permutations.

100.   The 4 letters "a" appear together in $4 \cdot 3! = 4!$ permutations. Since the number of all permutations is $P(4, 1, 1, 1)$, the answer is $P(4, 1, 1, 1) - 4! = 186$.

101.   We can think of the pair of letters "op" as a single symbol. Hence the number of required permutations is $P(2, 1, 1, 1, 1) = 360$.

102.   The letters other than "i" can be permuted in $P(1, 4, 2)$ ways. The 4 letters "i" can be placed in 8 locations. This implies C_8^4 choices. It follows that the number of required permutations is $P(1, 4, 2)\, C_8^4$.

103. In this case, the vowels as well as the consonants can be independently permuted in $P(2, 1, 1) = 12$ ways. For every arrangement of the 4 consonants there are 5 places for the 4 vowels. Hence there are $C_5^4 = 5$ sets of locations for the vowels. Hence the number of required permutations is $5 \cdot 12^2 = 720$.

104. We write the vowels in the order in which they appear. Then there are 5 places for the letter "b." Once this letter has been written down, there are 6 places for the letter "v." After that there are 7 places for the letter "c." Hence the total number of ways is $5 \cdot 6 \cdot 7 = 210$.

105. Reasoning as in the preceding problem, we obtain the answer $A_{11}^7/P_3 = 277,200$ (one must bear in mind the fact that the letter "l" appears in our word 3 times).

106. First we fix the order of the 2 vowels (2 ways). Then we place 2 of the consonants between the vowels ($A_4^2 = 12$ ways). The first of the 2 remaining consonants can be placed before or after the vowels (2 ways). After that there are 3 positions for the second consonant. In all there are $2 \cdot 12 \cdot 2 \cdot 3 = 144$ ways.

107. We select 3 of the 5 consonants and place them in the required positions (A_5^3 ways). The remaining 5 letters are placed arbitrarily in the remaining 5 positions (5! ways). This gives $5!A_5^3 = 7200$ ways.

108. $C_5^2 C_3^1 = 30$; $C_4^1 C_3^1 = 12$.

109. $P(3, 1, 1, 1) - 4! = 96$ ways (see Problem 100).

110. First we place the consonants (3! ways). This leaves 4 locations for the 3 letters "o," and so 4 ways of placing these 3 letters. In all there are 24 ways.

111. A selection can include 0, 1,..., 4 letters "a," that is, "a" can enter a selection in 5 ways. Similarly, "n" can enter in 3 ways, "e" in 5 ways, "y" in 3 ways, "f" in 3 ways, "o" in 7 ways, "r" in 3 ways, "t" in 5 ways, and "h" in 3 ways. This gives a total of $5 \cdot 3 \cdot 5 \cdot 3 \cdot 3 \cdot 7 \cdot 3 \cdot 5 \cdot 3 = 212,625$ selections.

112. The number of selections in which all 3 letters are distinct is $C_9^3 = 84$. The number of selections in which exactly 2 letters are distinct is $9 \cdot 8 = 72$. The number of selections in which all 3 letters are the same is 4. This gives a total of $84 + 72 + 4 = 160$ selections.

113. If order is taken into consideration, then there are $A_9^3 + 3A_9^2 + 4 = 724$ selections.

114. Since the order of the vowels and consonants is fixed, we can only choose, say, the 3 positions of the vowels. This can be done in C_7^3 ways.

115. In case of the word "triannual" we must start and end with a consonant. The consonants can be permuted in $P(2, 1, 1, 1)$ ways and the vowels in $P(2, 1, 1)$

ways. In all, there are $P(2, 1, 1, 1) P(2, 1, 1) = 720$ ways. In case of the word "samovar," the corresponding count is $P_4 \cdot P(2, 1) = 72$.

116. The 3 letters "a" can occupy any 3 of 6 locations. This gives C_6^3 acceptable permutations. If no two letters "a" can appear together, then our problem is that of putting 3 like things in 4 (different) boxes (formed by the consonants). The latter can be done in $C_4^3 = 4$ ways.

117. There are 180 permutations of the word "zigzag." The letters "z" are together in 60 permutations. The letters "g" are together in 60 permutations. The letters "z" appear together and the letters "g" appear together in 24 permutations. By the principle of inclusion and exclusion the number of admissible permutations is $180 - 60 - 60 + 24 = 84$. In case of the word "Tartar," the number of admissible permutations is

$$90 - 30 - 30 - 30 + 12 + 12 + 12 - 6 = 30.$$

118. There are 3 selections which include the letters "t," "a," and "r." Likewise, there are 3 selections which have exactly 2 distinct letters. This gives a total of 6 selections. The number of 4-digit numbers composed of the digits in the number 123,123 is $3P(2, 1, 1) + 3P(2, 2) = 54$.

119. By the principle of inclusion and exclusion,

$$10^6 - 4 \cdot 9^6 + 6 \cdot 8^6 - 4 \cdot 7^6 + 6^6 = 23{,}160$$

of our numbers include all the digits 1, 2, 3, 4. Also,

$$4 + 4^2 + 4^3 + 4^4 + 4^5 + 4^6 = (4^7 - 4)/3 = 5460$$

of our numbers contain only the digits 1, 2, 3, 4.

120. Each of the digits 1, 2, 3, 4 appears in each column $P_4/4 = 6$ times. Hence the sum of the units is $6(1 + 2 + 3 + 4) = 60$, the sum of the tens is 600, and so on. Hence the required sum is $60 + 600 + 6000 + 60{,}000 = 66{,}660$.

121. Here the number of permutations is 12, the digits 1 and 5 appear in each column 3 times, and the digit 2 appears in each column 6 times. Hence the sum of the units is $3 \cdot 1 + 3 \cdot 5 + 6 \cdot 2 = 30$. But then the required sum is $30 + 300 + 3000 + 30{,}000 = 33{,}330$.

122. The sum is 11,110.

123. The sum is 16,665.

124. If we set aside the restriction that no number starts with 0, then we obtain the sum 2,666,640. The sum of the numbers which begin with a 0 is

66,660 (see solution of Problem 120). Hence the answer to our problem is given by the difference $2,666,640 - 66,660 = 2,599,980$.

125. Using the digits 8 and 9, we can make 2^k k-digit numbers. It follows that the number of required numbers is $\sum_{k=1}^{6} 2^k = 126$.

126. Reasoning as in Problem 125, we obtain the answer $\sum_{k=1}^{6} 3^k = 1092$.

127. Since the first digit must not be 0, we obtain $2 \sum_{k=0}^{5} 3^k = 728$ numbers.

128. Each of the digits 1, 2, 3, 4 appears in each column $4^2 = 16$ times. Hence the sum of the units is $16(1 + 2 + 3 + 4) = 160$, the sum of the tens is 1600, and the sum of the hundreds is 16,000. The required sum is 17,760.

129. The first required sum is 3,999,960. To compute the second required sum, note that each digit appears in each column A_8^4 times. Hence the sum of the units is $A_8^4(1 + 2 + \cdots + 9) = 75,600$ and the required sum is 839,991,600.

130. The last digit can be 3 or 9. The remaining 3 digits can be permuted in 3! ways. This gives a total of 12 odd numbers. Similarly, the total of even numbers is also 12.

131. There are $C_6^3 = 20$ sets of places for the 3 odd digits. Once the positions of the odd digits are fixed, the positions of the even digits are also fixed. Then each of these positions can be filled in 5 ways (for the time being the digit 0 is allowed in the first position). This gives a total of $20 \cdot 5^6$ numbers. Of these, $10 \cdot 5^5$ begin with the digit 0. Hence there are $20 \cdot 5^6 - 10 \cdot 5^5 = 281,250$ required numbers.

132. $C_6^3 \cdot 5^6 = 312,500$ numbers.

133. The first digit can be chosen in 9 ways, the next 4 digits can be chosen in 10 ways, and the last digit (necessarily even) can be chosen in 5 ways. This gives $9 \cdot 10^4 \cdot 5 = 450,000$ numbers. There are 499,999 admissible numbers from 1 to 999,999.

134. Apart from the digit 0, the remaining digits yield one of the following sequences: 3; 2, 1; 1, 2; 1, 1, 1. It remains to distribute zeros so that the first digit is not a zero. For 3, this can be done in 1 way. For each of the sequences 2, 1 and 1, 2 this can be done in 9 ways. For 1, 1, 1 this can be done in $C_9^2 = 36$ ways. Hence the required total of numbers is $1 + 9 + 9 + 36 = 55$. In solving this problem for the numbers from 1 to 9,999,999 we assign locations to the nonzero digits. For 3 this can be done in C_{10}^1 ways, for each of the pairs 1, 2 and 2, 1 this can be done in C_{10}^2 ways, and for 1, 1, 1 this can be done in C_{10}^3 ways. Hence in this case we obtain $C_{10}^1 + 2C_{10}^2 + C_{10}^3 = 340$ admissible numbers.

135. The first position can be filled in 9 ways, the second in 9 ways, the third in 8 ways, and so on. Hence there are $9 \cdot 9!$ required numbers.

136. There are $E(1000/5)$ numbers from 0 to 999 which are divisible by 5 ($E(x)$ is the integral part of x), $1 + E(1000/7)$ numbers which are divisible by 7, and $1 + E(1000/35)$ numbers which are divisible by 35. By the principle of inclusion and exclusion there are

$$1000 - E\left(\frac{1000}{5}\right) - E\left(\frac{1000}{7}\right) + E\left(\frac{1000}{35}\right) = 686$$

numbers in the indicated range which are not divisible by either 5 or 7.

137. Reasoning as in Problem 136, we find that there are 228 admissible numbers.

138. There are 729 numbers in this range which do not include the digit 9, so that $1000 - 729 = 271$ numbers include the digit 9. There are exactly 27 numbers in which the digit 9 enters exactly twice (099, 909, 990, 199, and so on). The digit 0 appears in 1 one-digit number, in 9 two-digit numbers, and in 171 three-digit numbers for a total of 181 numbers. The digit 0 appears twice in 9 numbers. Both of the digits 0 and 9 appear in 36 numbers ($8 + 8 + 16 = 32$ three-digit numbers in which the third digit is not 0 or 9, and the 4 numbers 90, 900, 909, 990). The digits 8 and 7 appear in 54 numbers. The number of n-digit numbers which do not have the same digit twice in a row is 9^n for $n > 1$, and 10 for $n = 1$. Hence the number of numbers from 0 to 999,999 with this property is $10 + 9^2 + 9^3 + 9^4 + 9^5 + 9^6 = 597{,}871$.

139. A 4-digit number may contain 4 different digits (1, 2, 3, 5), just 3 different digits, or just 2 different digits. Hence the total of required numbers is

$$P_4 + 2C_3^2 P(2, 1, 1) + P(2, 2) = 24 + 6 \cdot 12 + 6 = 102.$$

140. Reasoning as in Problem 139, we obtain the answer

$$2P(2, 1, 1, 1) + 3P(3, 1, 1) + 2P(2, 2, 1) + 3P(4, 1) + P(3, 2) = 265.$$

141. In a 6-digit number there may appear 1, 2, or 3 pairs of equal digits. One pair can be selected in C_5^1 ways. The number of permutations of 4 different and 2 equal digits is $P(2, 1, 1, 1, 1) = 6!/2! = 360$. In $5! = 120$ of these permutations 2 equal digits appear together. Hence in this case we obtain $5(360 - 120) = 1200$ admissible numbers. Two pairs of equal digits can be selected in $C_5^2 = 10$ ways and, then, 2 additional digits can be selected in C_3^2 ways. The number of permutations of these digits is $P(2, 2, 1, 1) = 180$. In $2 \cdot (5!/2!) = 120$ of these permutations there is (at least) 1 pair of equal digits in a row and in $4! = 24$ of these permutations there are 2 such pairs. By the principle of inclusion and exclusion, this case contributes

$$10 \cdot 3(180 - 120 + 24) = 2520$$

admissible numbers. Similarly, the case of 3 pairs of equal digits contributes

$$C_5^3 \left(\frac{6!}{(2!)^3} - 3 \cdot \frac{5!}{(2!)^2} + 3 \cdot \frac{4!}{2!} - 3! \right) = 300$$

admissible numbers. In all, we obtain 4020 admissible numbers.

142. The number of 5-digit numbers which can be composed of the given digits is

$$3 \cdot \frac{5!}{2!} + C_3^2 C_2^1 \frac{5!}{(2!)^2} + C_3^1 \cdot \frac{5!}{3!} + C_2^1 \cdot \frac{5!}{3!\,2!} = 440.$$

In $3P_3 + 2(P_3/2!) = 24$ of these numbers the digit 3 appears 3 times in a row. This gives 416 admissible numbers.

143. The number of permutations of the given digits is $P(2, 2, 2, 2)$. In $P(2, 2, 2, 1)$ of these permutations 1 of the given digits appears twice in a row, in $P(2, 2, 1, 1)$ of these permutations 2 of the given digits appear twice in a row, in $P(2, 1, 1, 1)$ of these permutations 3 of the given digits appear twice in a row, and in $P(1, 1, 1, 1)$ of these permutations 4 of the given digits appear twice in a row. By the principle of inclusion and exclusion, the required number of permutations (in which no digit appears twice in a row) is

$$P(2, 2, 2, 2) - 4P(2, 2, 2, 1) + 6P(2, 2, 1, 1)$$
$$- 4P(2, 1, 1, 1) + P(1, 1, 1, 1) = 864.$$

144. Reasoning as before, we obtain the answer

$$\frac{8!}{(2!)^3} - 3 \cdot \frac{7!}{(2!)^2} + 3 \cdot \frac{6!}{2!} - 5! = 2220.$$

145. Reasoning as before, we obtain the answer

$$\frac{10!}{(3!)^2} - 2 \cdot \frac{8!}{3!} + 6! = 88,080.$$

146. Reasoning as before, we obtain the answer 20,040.

147. The first number can be chosen in 20 ways and the second in 10 ways (its parity is determined). Since the order of the numbers is immaterial, the number of ways is $20 \cdot 10/2 = 100$.

148. Either all 3 numbers are even or 1 is even and 2 are odd. Hence the number of choices is $C_{15}^3 + C_{15}^2 C_{15}^1 = 2030$.

149. There are 2 choices at each of 11 junctions. Hence the number of admissible routes is $2^{11} = 2048$.

150. Since 1 choice is made at the point of departure, there remain $2^{10} = 1024$ possibilities.

151. Reasoning as before we obtain the answer $3^5 = 243$.

152. If we take p 10-cent coins, then we can take $0, 1, ..., 20 - p$ 15-cent coins, that is, we have $21 - p$ choices. Since p varies from 0 to 20, we obtain the answer $1 + 2 + 3 + \cdots + 21 = 231$.

153. The total number of selections of the 5 coins is $C_{13}^5 = 1287$. Hence the maximal number of false guesses is 1286.

154. The number of 5-digit numbers is 90,000. Of these, all the digits are even in $4 \cdot 5^4 = 2500$ cases and all the digits are odd in $5^5 = 3125$ cases. In $4^5 = 1024$ cases, the numbers contain no digits smaller than 6, and in $3 \cdot 4^4 = 768$ cases the numbers contain no digits larger than 3. $5! = 120$ numbers contain all the digits 1, 2, 3, 4, 5; $4 \cdot 4! = 96$ numbers contain all the digits 0, 2, 4, 6, 8.

155. It is clear from the conditions of the problem that two throws yield the same sum if and only if the numbers turned up in one throw are a permutation of the numbers turned up in the second throw. Hence the number of different sums is $C_6^2 + 6 = 21$.

156. Reasoning as in Problem 155, we obtain the answer $C_6^3 + 2C_6^2 + 6 = 56$.

157. The number of throws in which the dice turn up the same number is 6. The throws in which the dice turn up 2 different numbers can arise in the following 3 ways: 1 die turns up 1 number and the remaining 5 dice turn up a different number, 2 dice turn up 1 number and the remaining 4 dice turn up a different number, 3 dice turn up 1 number and the remaining 3 dice turn up a different number. In all these cases the 2 different numbers turned up by the 6 dice can be selected in A_6^2 ways. In the first case the die which turns up the "one-die" number can be selected in 6 ways. It follows that this case contributes $6A_6^2 = 180$ throws. Similarly, the second case contributes $A_6^2 P(2, 4) = 450$ throws, and the third case contributes $C_6^3 P(3, 3) = (1/2!) A_6^2 P(3, 3) = 300$ throws. Hence the number of throws in which the dice turn up 2 different numbers is $180 + 450 + 300 = 930$. To compute the number of throws in which the dice turn up 3 different numbers we find all partitions of the number 6 into 3 summands: $6 = 1 + 1 + 4 = 1 + 2 + 3 = 2 + 2 + 2$. We obtain, accordingly:

$$\frac{1}{2!} A_6^3 P(1, 1, 4) = 1800,$$

$$A_6^3 P(1, 2, 3) = 7200,$$

$$\frac{1}{3!} A_6^3 P(2, 2, 2) = 1800,$$

that is, there are 10,800 cases in which the dice turn up 3 different numbers. The partitions of 6 into 4 summands are:

$$6 = 1 + 1 + 1 + 3 = 1 + 1 + 2 + 2.$$

Hence the number of throws in which the dice turn up 4 different numbers is

$$\frac{1}{3!} A_6^4 P(1, 1, 1, 3) + \frac{1}{(2!)^2} A_6^4 P(1, 1, 2, 2) = 7200 + 16,200 = 23,400.$$

The number of throws in which the dice turn up 5 different numbers is $(1/4!) A_6^5 P(1, 1, 1, 1, 2)$ and the number of throws in which the dice turn up 6 different numbers is $6! = 720$. Observe that

$$6 + 930 + 10,800 + 23,400 + 10,800 + 720 = 6^6.$$

158. For a given throw, the n dice can be grouped in classes depending on the number turned up by each die. Hence our problem is to find the number of distributions of n like elements in 6 different boxes and this number is C_{n+5}^5 (see p. 171).

159. Since $1,000,000 = 2^6 \cdot 5^6$, every decomposition of 1,000,000 into 3 factors is of the form

$$1,000,000 = (2^{\alpha_1} \cdot 5^{\beta_1})(2^{\alpha_2} \cdot 5^{\beta_2})(2^{\alpha_3} \cdot 5^{\beta_3}),$$

where $\alpha_1, \alpha_2, \alpha_3, \beta_1, \beta_2, \beta_3$ are nonnegative integers such that $\alpha_1 + \alpha_2 + \alpha_3 = \beta_1 + \beta_2 + \beta_3 = 6$. Now, the number of partitions of 6 into 3 nonnegative summands is $C_8^2 = 28$. Hence the number of decompositions of 1,000,000 into 3 ordered factors is $28^2 = 784$.

160. The decompositions obtained in the preceding problem can be divided into 3 classes according as all 3 factors are equal, 2 factors are equal and the third factor is different, all three factors are different. The first class consists of 1 decomposition only, namely, $1,000,000 = 100 \cdot 100 \cdot 100$. Next we find the number of decompositions in the second class. If each of the 2 equal factors is of the form $2^\alpha \cdot 5^\beta$, then $2\alpha + \alpha_3 = 2\beta + \beta_3 = 6$. Now, the equation $2x + y = 6$

has 4 nonnegative integer solutions: $x = 0$, $y = 6$; $x = 1$, $y = 4$; $x = 2$, $y = 2$; $x = 3$, $y = 0$. Since we can combine any α with any β, we get 16 possible ways of selecting a pair α, β. One of these pairs, namely the pair $\alpha = 2$, $\beta = 2$ has to be rejected since it leads to the decomposition of 1,000,000 into 3 equal factors already accounted for. We are thus left with 15 pairs, each leading to 3 ordered decompositions. Hence the second class consists of 45 ordered decomposition. If we disregard order, the number of decompositions is 15. Finally, the number of ordered decompositions in the third class is $784 - 1 - 45 = 738$. These can be arranged in groups, the decompositions in each group differing from each other in the order of their factors. It follows that each group contains 6 decompositions. In all, we obtain $1 + 15 + 123 = 139$ unordered decompositions of the number 1,000,000 into 3 factors.

161. There are 2 choices for each of the 9 coins. Hence the number of distributions is 2^9.

162. We order the $3n$ objects and give the first n objects to the first person, the second n objects to the second person, and the remaining n objects to the third person. Since order of the objects in a share is immaterial, we obtain $(3n)!/(n!)^3 = C_{3n}^n C_{2n}^n$ ways.

163. Reasoning as before, we find that the number of groupings is $(2n)!/(2^n n!)$.

164. Reasoning as before we obtain the answer $(nk)!/(k!)^n n!$.

165. $\dfrac{30!}{(10!)^3 \, 3!}$; $\dfrac{30!}{(3!)^{10} \, 10!}$.

166. The 4 aces can be arranged in pairs in $4!/(2!)^3 = 3$ ways. The remaining 32 cards can be arranged in batches of 16 in $32!/[(16!)^2 \, 2!]$ ways. Since the 2 arrangements can be combined in 2 ways, we obtain the answer $3 \cdot 32!(16!)^2$.

167. The number of ways is $10!/(2^5 \cdot 5!) = 945$.

168. 945.

169. $9!/(3!)^4 = 280$.

170. Six apples can be divided among 3 people in C_8^2 ways. The remaining 6 fruits can be divided in 3^6 ways since each fruit can be given to any of the 3 people. This implies a total of $3^6 C_8^2 = 20,412$ ways.

171. First we distribute the apples. Since each person gets at most 4 apples, distribution of the apples, apart from permutations, takes place in accordance with one of the following schemes:

$$6 = 4 + 2 + 0 = 4 + 1 + 1 = 3 + 3 + 0 = 3 + 2 + 1 = 2 + 2 + 2.$$

If the distribution of the apples follows the scheme $4 + 2 + 0$, then it remains to select 2 out of 6 fruits for the second person and to give the remaining fruits to the third person. This can be done in C_6^2 ways. If we bear in mind the possibility of permuting the 3 persons, then we see that this case contributes $3!C_6^2$ admissible distributions. In case of the scheme $4 + 1 + 1$, the second person must obtain 3 out of 6 fruits (C_6^3 ways). Since 2 people now have the same number of apples, the number of relevant permutations of the 3 people is $P(2, 1) = 3$. Hence this case contributes $3C_6^3$ admissible distributions. In case of the scheme $3 + 3 + 0$, we must select 1 of 6 fruits for the first person and 1 of 5 fruits for the second person. The number of relevant permutations of the 3 people is again 3. Hence this case contributes $3C_6^1C_5^1$ distributions. The remaining schemes are analyzed in much the same way. In all, we obtain

$$6C_6^2 + 3C_6^3 + 3C_6^1C_5^1 + 6C_6^1C_5^2 + C_6^2C_4^2 = 690$$

admissible distributions.

172. Since $9 = 6 + 3 + 0 = 6 + 2 + 1 = 5 + 4 + 0 = 5 + 3 + 1 = 4 + 3 + 2 = 5 + 2 + 2 = 4 + 4 + 1 = 3 + 3 + 3$, we obtain, reasoning as in the preceding problem,

$$6[C_9^3 + C_9^4 + C_9^1C_8^2 + C_9^1C_8^3 + C_9^2C_7^3] + 3(C_9^1C_8^4 + C_9^2C_7^2) + C_9^3C_6^3 = 19{,}068$$

admissible distributions.

173. The cards in the deck can be distributed among the 13 players in $52!/(4!)^{13}$ ways (see Problem 164). If each player is to obtain 1 card of each suit, then the number of distributions is 13^4 (13 choices for each of the suits). In the third case one can choose 1 card of each suit in 13^4 ways. After that, the remaining 12 cards of one suit can be distributed in batches of 4 in $12!/[(4!)^3\,3!]$ ways, and all the remaining cards (48 of them) can be distributed in batches of 4 in $(12!)^4/[(4!)^{12}\,(3!)^4]$ ways. These batches can be distributed among the 12 players in 12! ways. If we bear in mind the fact that the player with cards of all suits can be selected in 13 ways, we obtain in the present case the answer $(13!)^5/[(4!)^{12}(3!)^4]$.

174. The suit that is absent from the selection and the suit that is repeated can be selected in A_4^2 ways, the 2 cards of the repeated suit can be selected in C_{13}^2 ways, and the 2 cards of the remaining 2 suits can be selected in $(C_{13}^1)^2$ ways. Hence the answer $A_4^2(C_{13}^1)^2\,C_{13}^2 = 518{,}184$.

If the selection is to contain cards of exactly 2 suits, then we can have either 2 cards of each suit or 1 card of one suit and 3 cards of the other suit. In the first case we select 2 suits and then 2 cards from each suit. In the second case we select the first and second suit (here the order of the suits is important) and then select 3 cards of the first suit and 1 card of the second suit. Hence we get

$$C_4^2(C_{13}^2)^2 + A_4^2 C_{13}^3 C_{13}^1 = 81{,}120$$

cases.

175. We divide the 13 cards of each suit in accordance with the scheme $3 + 3 + 3 + 4$. This can be done in $13!/[4!(3!)^4]$ ways. The 4-card batches can be distributed among the players in 4! ways and the 3-card batches of each suit in 3! ways. Thus there are $(3!)^4 4!$ ways of distributing the different batches and, therefore, the cards can be distributed in

$$\left(\frac{13!}{4! \, (3!)^4}\right)^4 4! \, (3!)^4 = \frac{(13!)^4}{(4!)^3 (3!)^{12}}$$

ways.

176. We arrange the 5 people in a certain order. Next we order the 18 objects and divide them into 4 groups of 4 objects each and 1 group of 2 objects. We give the group of 2 objects to 1 of the 5 people, and each of the groups of 4 objects to each of the remaining people (the first group to the first of the remaining people, the second to the second, and so on). Since order within the groups is irrelevant, we obtain $5 \cdot 18!/[(4!)^4 \, 2!]$ distributions. In the second case we obtain, likewise, $18! C_5^2/[(4!)^3 \, (3!)^2]$ distributions.

177. There are 3 possibilities for each pair: we select no object from the pair, we select 1 object from the pair, or we select 2 objects from the pair. Hence the number of selections is $3^{14} = 4{,}782{,}969$.

178. The 4 black balls can be placed in 6 different boxes in C_9^5 ways. The same is true for the white and blue balls. By the rule of product we obtain $(C_9^5)^3 = 2{,}000{,}376$ ways of placing the balls in 6 different boxes.

179. Reasoning as in Problem 178 we obtain the answer $C_6^3 C_{13}^3 = 5720$.

180. Represent each partition of the number n into summands by means of a graph. If we add to the graph a column of n dots, then we get a graph representing a partition of the number $2n$ into n summands.

181. Choose 3 numbers from the range 1 to $n - 2$. Increase the largest number by 2 and the next-to-largest number by 1. The result is a triple of numbers of the required kind. Also, all the required triples can be obtained in this way. Since 3 numbers can be chosen from the range 1 to $n - 2$ in C_{n-2}^3 ways, this is the required answer.

182. $P(2, 2, 2, 2, 2, 2, 1, 1, 1, 1) = 16!/2^6$ ways.

183. Place identical checkers on the 48 unoccupied squares of the chessboard and compute the number of permutations of the resulting arrangement. This number is $P(48, 2, 2, 2, 2, 2, 2, 1, 1, 1, 1) = 64!/(2^6 \cdot 48!)$.

184. Reasoning as before, we obtain the answer

$$P(32, 8, 8, 2, 2, 2, 2, 2, 2, 1, 1, 1, 1).$$

185. Suppose that p squares are taken up with white checkers and q squares are taken up with black checkers. There are C_{14}^{p-1} ways of setting out 15 white checkers on p squares so that no square is empty. Similarly, there are C_{14}^{q-1} ways of setting out 15 black checkers on q squares so that no square is empty. The p squares for the white checkers and the q squares for the black checkers can be selected in $P(p, q, 24 - p - q)$ ways. Hence the number of ways of setting out the pieces is

$$\sum_{p,q} P(p, q, 24 - p - q) C_{14}^{p-1} C_{14}^{q-1},$$

where we sum over all p and q such that

$$1 \leqslant p \leqslant 15, \quad 1 \leqslant q \leqslant 15, \quad p + q \leqslant 24.$$

186. The 64 squares can be divided into 16 groups of 4 such that the squares in a group are permuted by a rotation through 90^0. Our problem is to select 5 such groups. This can be done in $C_{16}^5 = 4368$ ways.

187. Reasoning as in Problem 186, we obtain the answer C_{32}^{10}.

188. Since we have only half as many squares as in Problem 187, the number of ways is C_{16}^{10}.

189. Six white and 6 black checkers must be set out on 16 black squares which belong to half the chessboard. This can be done in $P(6, 6, 4) = 16!/(6!6!4!)$ ways.

190. We must select 12 out of 16 black squares belonging to one half of the chessboard and settle these squares with arbitrary checkers. This can be done in $2^{12}C_{16}^{12} = 7,454,720$ ways. (It is clear that a square in the remaining half of the board which is centrally symmetric to a square occupied by, say, a black checker must be settled with a white checker.)

191. The problem comes down to choosing 5 out of 7 squares in a row. This can be done in $C_7^5 = 21$ ways.

192. The corner squares may or may not be occupied. Consider the case of an occupied corner square. The 12 inside squares of the row and column determined by an occupied corner square contain 8 checkers. These can be set out in $C_{12}^8 = 495$ ways. Now consider the case of an unoccupied corner square. The 12 inside squares of the row and column determined by an unoccupied corner square contain 10 checkers. These can be set out in $C_{12}^{10} = 66$ ways. This gives a total of 561 ways.

193. The 7 white balls can be placed in 9 different pockets in C_{15}^8 ways and the 2 black balls in C_{10}^8 ways. Hence the number of ways is $C_{15}^8 C_{10}^8 = 289{,}575$.

194. Reasoning is before we obtain the answer $C_{15}^8 (C_9^8)^2 = 521{,}235$.

195. First we give 9 books to person C. This can be done in C_{27}^9 ways. The remaining 18 books can be distributed between A and B in 2^{18} ways. Hence the answer $2^{18} C_{27}^9$.

196. There are 4^8 ways in which the 8 people can leave the elevator at the 4 floor stops. In 3^8 cases no passenger leaves at a preassigned floor. In 2^8 cases no passenger leaves at 2 preassigned floors. In 1 case no passenger leaves at 3 preassigned floors. The principle of inclusion and exclusion yields the answer $4^8 - 4 \cdot 3^8 + 6 \cdot 2^8 - 4 = 40{,}824$.

197. The following cases arise: all of the summands are divisible by 3, just 1 summand is divisible by 3, no summand is divisible by 3. In the first case the summands may be selected in C_{33}^3 ways. In the second case, upon division by 3, one summand leaves the remainder 1 and the other leaves the remainder 2. Consider the numbers from 1 to 100. Upon division by 3, 34 of these numbers leave the remainder 1, 33 leave the remainder 2, and 33 leave the remainder 0. It follows that in the second case the summands may be selected in $C_{34}^1 (C_{33}^1)^2$ ways. In the third case all 3 summands leave the remainder 1 or all 3 summands leave the remainder 2. These possibilities give rise to C_{34}^3 and C_{33}^3 possibilities respectively. In all we have $2C_{33}^3 + C_{34}^3 + C_{34}^1 (C_{33}^1)^2 = 53{,}922$ ways.

198. Reasoning as in the preceding problem, we obtain the answer

$$3C_n^3 + (C_n^1)^3 = \frac{n}{2}(3n^2 - 3n + 2).$$

199. Pockets for p white balls can be chosen in C_{n+1}^p ways. With p pockets occupied by white balls, there are $n - p + 2$ choices for the black ball (we may place it in one of the empty $n + 1 - p$ pockets or we may decide to leave it out). Hence the answer

$$\sum_{p=0}^{n} (n - p + 2)C_{n+1}^p = \sum_{s=1}^{q} sC_q^s + \sum_{p=0}^{q-1} C_q^p.$$

Since

$$\sum_{s=1}^{q} sC_q^s = q2^{q-1}$$

and

$$\sum_{p=0}^{q-1} C_q^p = 2^q - 1$$

(see Problem 401a), our answer takes the simple form $(q + 2) \, 2^{q-1} - 1$.

200. Let W denote a batch of white balls and B a batch of black balls. Our problem implies that the balls are arranged in accordance with one or another of the schemes $BWBW \cdots BW$ or $WBWB \cdots WB$, where the number of pairs of letters in each scheme is r. Now, m white balls can be separated into r nonempty batches in C_{m-1}^{r-1} ways, and n black balls can be separated into r nonempty batches in C_{n-1}^{r-1} ways. Hence in this case $(2r - 1$ contacts) the number of ways is $2C_{n-1}^{r-1}C_{m-1}^{r-1}$. The corresponding number in the second case $(2r$ contacts) is $C_{m-1}^r C_{n-1}^{r-1} + C_{m-1}^{r-1} C_{n-1}^r$.

201. Let $A(m, n)$ denote the number of ways of earning m points as a result of passing n examinations. Then it is clear that

$$A(30, 8) = A(25, 7) + A(26, 7) + A(27, 7),$$

and so on. By successive reduction of m, we obtain the answer 784.

202. First we choose the n fixed objects. This can be done in C_{m+n}^n ways. The remaining m objects are permuted so that none stay fixed. This can be done in D_m ways (see p. 51). Hence the number of admissible permutations is $[(m + n)!/(m!n!)]D_m$.

203. There are $(n + p)^r$ ways of distributing r things among $n + p$ people. In $(n + p - 1)^r$ cases, a certain person will not get anything. In $(n + p - 2)^r$ cases, a certain 2 people will not get anything, and so on. Now application of the principle of inclusion and exclusion yields the required result.

204. The first column of the graph of a partition of $2r + x$ into $r + x$ positive summands contains $r + x$ points. Deletion of this column yields the graph of a partition of r into nonnegative summands.

205. (a) Since each of the n members can vote for each of the n candidates, the number of outcomes is n^n.

(b) Here n votes are distributed among n candidates. This can happen in C_{2n-1}^{n-1} ways.

206. Consider an admissible partition of $2n$ into summands a, b, c, $a + b + c = 2n$, $a \leqslant b \leqslant c$. We claim that $a \neq 1$; indeed, $a = 1$ implies $b + c = 2n - 1$, and so $b < c$ contradicting $b + 1 > c$. Also, $a + b > c$, and $a + b$ and c have the same parity. This implies that $a + b \geqslant c + 2$. But then $a - 1$, $b - 1$, $c - 1$ form a partition of $2n - 3$ such that $(a - 1) + (b - 1) > c - 1$. This establishes a one-to-one correspondence between the partitions of $2n$ and $2n - 3$.

207. The result follows from the equality

$$C_n^1 + C_n^3 + C_n^5 + \cdots = 2^{n-1}.$$

208. Suppose the first person obtained x objects of the first kind, y objects of the second kind, and z objects of the third kind. Then $x + y + z = 3n$, $0 \leqslant x, y, z \leqslant 2n$. In other words, our problem is to find the number of nonnegative solutions of the equation $x + y + z = 3n$ with each entry in a solution $\leqslant 2n$. If we ignore this restriction, then our problem reduces to the problem of finding the number of ways of dividing $3n$ like objects among 3 people. This number is C_{3n+2}^2. Now we find the number of nonnegative solutions of $x + y + z = 3n$ with $x > 2n$. This number is equal to the number of nonnegative integral solutions of the equations $y + z = k$, $0 \leqslant k < n$ (taken one at a time), that is, $1 + 2 + \cdots + n = n(n + 1)/2$. Also, there are $n(n + 1)/2$ nonnegative solutions of $x + y + z = 3n$ with $y > 2n$, and $n(n + 1)/2$ nonnegative solutions of $x + y + z = 3n$ with $z > 2n$. If we reject the $\frac{3}{2}n(n + 1)$ "bad" solutions, we obtain the answer $3n^2 + 3n + 1$.

209. Reasoning as before, we obtain

$$C_{4n+3}^3 - 4 \sum_{k=0}^{2n-1} C_{k+2}^2 = C_{4n+3}^3 - 4C_{2n+2}^3 = \tfrac{1}{3}(2n + 1)(8n^2 + 8n + 3).$$

210. The mathematical counterpart of removing distinctions of kind is to identify the solutions x, y, z and $2n - x$, $2n - y$, $2n - z$ of the equation $x + y + z = 3n$ in Problem 208. With one exception (that of the solution $x = n, y = n, z = n$) corresponding solutions are different. This means that the reduced number of solution is $(3n^2 + 3n)/2 + 1$.

211. Here we are required to find the number of nonnegative solutions of the equation $x_1 + x_2 + \cdots + x_m = mn$, with $0 \leqslant x_k \leqslant 2n$, $1 \leqslant k \leqslant m$. If we drop the restriction $x_1 \leqslant 2n$ (but keep $0 \leqslant x_1$), then we obtain C_{mn+m-1}^{m-1} solutions. Now we find the number of nonnegative solutions of $x_1 + x_2 + \cdots + x_m = mn$

for which $x_1 > 2n$. This number is equal to the number of nonnegative solutions of the equations $x_2 + x_3 + \cdots + x_m = k$, $0 \leqslant k \leqslant mn - 2n - 1$, that is,

$$\sum_{k=0}^{mn-2n-1} C_{k+m-2}^{m-2} = C_{mn-2n+m-2}^{m-1}.$$

Each of the possibilities $x_2 > 2n, \ldots, x_m > 2n$ contributes again as many "bad" solutions. It would appear that the number of solutions to be rejected is $C_m^1 C_{mn+m-2n-2}^{m-1}$. But it must be remembered that in rejecting this number of solutions we are rejecting certain solutions (say, those with $x_1 > 2n$, $x_2 > 2n$) twice. The correct number of solutions of our problem is obtained by applying the principle of inclusion and exclusion.

212.   There are 231 ways. This problem is a special case of the next problem.

213.   Let x_1, x_2, x_3 denote the shares of objects of the first kind and y_1, y_2, y_3 the shares of objects of the second kind. Then we must solve the system of equations, $x_1 + x_2 + x_3 = n$, $y_1 + y_2 + y_3 = n$, with $0 \leqslant x_k + y_k \leqslant n$, $1 \leqslant k \leqslant 3$. If we remove the restrictions $x_k + y_k \leqslant n$, $1 \leqslant k \leqslant 3$, then each equation yields C_{n+2}^2 solutions, and we obtain $(C_{n+2}^2)^2$ "unrestricted" solutions of our problem. The number of unrestricted solutions which violate the condition $x_1 + y_1 \leqslant n$ is equal to the number of nonnegative solutions of the systems of equations $x_2 + x_3 = r$, $y_2 + y_3 = s$, with $0 \leqslant r < n$, $0 \leqslant s < n$, and $r + s < n$. The number of nonnegative solutions of a system $x_2 + x_3 = r$, $y_2 + y_3 = s$ is $(r + 1)(s + 1)$. Hence the total number of solutions of our system is

$$\sum_{s=0}^{n-1} \sum_{r=0}^{n-s-1} (r + 1)(s + 1) = \frac{1}{2} \sum_{s=0}^{n-1} (s + 1)(n - s)(n - s + 1)$$

$$= \sum_{s=0}^{n-1} C_{s+1}^1 C_{n-s+1}^2 = C_{n+3}^4$$

(see p. 38). There are again as many nonrestricted solutions which violate the condition $x_2 + y_2 \leqslant n$, and again as many which violate the condition $x_3 + y_3 \leqslant n$. If we reject all of these solutions, we obtain the answer $(C_{n+2}^2)^2 - 3C_{n+3}^4$. For $n = 5$, we obtain 231 as the answer to Problem 212.

214.   There are 9! permutations of 9 people. Let us compute in how many of these permutations 3 Englishmen sit together. All such permutations can be obtained from one by permuting the group of Englishmen (3! ways) and by permuting the group of 3 Frenchmen, 3 Turks, and the "block" of 3 Englishmen (7! ways). In this way we obtain 3!7! permutations. There are again as many permutations in which 3 Frenchmen sit together, and again as many in which

3 Turks sit together. Next we see that there are $(3!)^2 \, 5!$ permutations in which the Englishmen sit together and the Frenchmen sit together, and that there are $(3!)^4$ in which all countrymen sit together in 3 groups of 3. By the principle of inclusion and exclusion, the answer to our problem is

$$9! - 3 \cdot 3!7! + 3(3!)^2 \, 5! - (3!)^4 = 283{,}824.$$

215. The total number of permutations is $9!$. Let us compute the number of permutations in which a certain 2 Englishmen sit together. The remaining 7 people and the block of 2 Englishmen account for $8!$ such permutations. Since we can permute the 2 Englishmen within the block (in 2 ways), the number of permutations in which a certain 2 Englishmen sit together is $2!8!$. Since the 2 Englishmen can be selected in C_3^2 ways and there are 3 nationalities to consider, the appropriate term in the inclusion–exclusion formula is $3C_3^2 2!8!$. Now we compute the number of permutations in which a certain 2 Englishmen as well as a certain 2 Frenchmen sit together. The 2 two-man blocks and the remaining 5 people form a group of 7 "objects." Considering permutations within each block, we arrive at the figure of $(2!)^2 \, 7!$ permutations. Also, our 2 pairs can be selected in $(C_3^2)^2$ ways. This means that the corresponding term in the inclusion–exclusion formula is $(C_3^2)^3 \, (2!)^2 \, 7!$. It remains to consider the following cases of "togetherness":

   (a) 1 triple of countrymen,
   (b) 3 pairs of countrymen,
   (c) 1 triple and 1 pair of countrymen,
   (d) 2 triples of countrymen,
   (e) 1 triple and 2 pairs of countrymen,
   (f) 2 triples and 1 pair of countrymen,
   (g) 3 triples of countrymen.

By the principle of inclusion and exclusion we obtain the answer

$$9! - 9 \cdot 2!8! + 27(2!)^2 \, 7! + 3 \cdot 3!7! - (2!)^3 \, 6! - 18 \cdot 3!2!6!$$
$$+ \, 3(3!)^2 \, 5! + 27 \cdot 3!(2!)^2 5! - 9(3!)^2 \, 2!4! + (3!)^4.$$

216. This problem is similar to Problem 217 except that the individual figures are now computed differently. A certain 2 Englishmen can sit together in $2!9$ ways, and then the others can be seated in $7!$ ways. This accounts for $2!9 \cdot 7!$ permutations. To compute the number of cases in which a pair of Englishmen sit together and a pair of Frenchmen sit together, we note that, apart from permutation of countrymen within a pair, the pair of Englishmen can be seated in 9 ways, and then the block of 2 Frenchmen and the 5 other people can be permuted in $6!$ ways. With permutations within a pair taken into consideration, the required number is $(2!)^2 \, 9 \cdot 6!$ Similar arguments apply in the remaining cases. By the principle of inclusion and exclusion we arrive at the answer

$$9! - 9 \cdot 2!9 \cdot 7! + 27(2!)^2 9 \cdot 6! + 3 \cdot 3!9 \cdot 6!$$
$$- (2!)^3 9 \cdot 5! - 18 \cdot 3!2!9 \cdot 5! + 3(3!)^2 9 \cdot 4!$$
$$+ 27 \cdot 3!(2!)^2 9 \cdot 4! - 9(3!)^2 2!9 \cdot 3! + (3!)^3 9 \cdot 2!.$$

217. Let $F(N)$ denote the number of rows of stamps worth N cents. We separate the rows into classes according to the denomination of the last stamp. Then we obtain the recurrence relation

$$F(N) = F(N - 5) + F(N - 10) + F(N - 15) + F(N - 20).$$

Using this relation and the fact that $F(5) = 1$, we find that $F(40) = 108$. (Note: $F(N) = 0$, for $N < 0$ and $F(0) = 1$.)

218. Let $F(n_1, ..., n_m ; N)$ denote the number of ways of paying N cents with coins worth n_1, n_2, ..., n_m cents, respectively. Then we have the recurrence relation

$$F(n_1, ..., n_m ; N) = F(n_1, ..., n_{m-1} ; N) + F(n_1, ..., n_m ; N - n_m)$$

(see p. 78). It is now easy to obtain the answer $F(10, 15, 20, 50; 100) = 20$. (Note: Include analog of condition in brackets given in the solution of Problem 217.)

219. Using the previous recurrence relation, we obtain the answer 4.

220. A row can contain 3, 2, or 1 black balls. If a row contains 3 black balls, then the fourth ball can be selected in 3 ways, and the 4 balls can be permuted in $P(3, 1) = 4$ ways. Hence this case contributes 12 possibilities. Similarly, the case of 2 black balls contributes $C_3^2 P(2, 1, 1) = 36$ possibilities, and the case of 1 black ball contributes $4!$ possibilities. In all, the number of different rows is $12 + 36 + 24 = 72$.

221. The number of such ordered sums is equal to the number of ways of placing n identical balls in 3 different boxes, that is, C_{n-1}^2.

222. First we compute the number of zeros used in writing all the numbers from 1 to 999,999 inclusive. The digit 0 appears last in 99,999 numbers (10, 20,..., 999,990), second in 99,990 numbers, third in 99,900 numbers, and so on. In all, the digit 0 appears

$$99,999 + 99,990 + 99,900 + 99,000 + 90,000 = 488,889$$

many times. The total number of digits in our numbers is

$$9 + 2 \cdot 90 + 3 \cdot 900 + 4 \cdot 9000 + 5 \cdot 90,000 + 6 \cdot 900,000 = 5,888,889.$$

Since all the digits other than 0 appear in our numbers the same number of times, each of them appears $(5{,}888{,}889 - 4{,}888{,}889)/9 = 600{,}000$ times.

223. The positions of the pair of digits 3 can be selected in C_{10}^2 ways. The remaining 8 positions can be filled with digits 1 and 2 in 2^8 ways. Hence the answer is $2^8 C_{10}^2 = 11{,}520$.

The sum of the digits in each of our numbers lies between $8 \cdot 1 + 2 \cdot 3 = 14$ and $8 \cdot 2 + 2 \cdot 3 = 22$. If one of our numbers is to be divisible by 9, then the sum of its digits must be 18. Hence the sum of the 1's and 2's in our number must be 12. This sum can be obtained by taking 4 ones and 4 twos. Since the number of permutations of 4 ones, 4 twos, and 2 threes is

$$P(4, 4, 2) = \frac{10!}{4! \, 4! \, 2!} = 3150,$$

this is also the answer to this part of the problem.

224. If 2 digits a and b form an inversion, then by interchanging their order we obtain a permutation in which a and b no longer form an inversion. The number of permutations of the numbers 1 to n is $n!$. Since the number of ways of choosing 2 out of n digits is C_n^2 and there are as many inversions as noninversions, it follows that the number of inversions is $n! C_n^2 / 2$.

225. The number of ordered partitions of n into 3 positive summands is $C_{n-1}^2 = (n^2 - 3n + 2)/2$. If n is even, then $(n - 2)/2$ of these ordered partitions have 2 equal summands. If n is odd, then $(n - 1)/2$ of these ordered partitions have equal summands. Also, if n is divisible by 3, then there is 1 partition of n with 3 equal summands. Applying the principle of inclusion and exclusion, we find easily that the number of ordered partitions of n into distinct summands is

$$\frac{n^2 - 3n + 2}{2} - \frac{3}{2}(n - 2) + 2 = \frac{n^2 - 6n + 12}{2}, \qquad \text{if} \quad n = 6k,$$

$$\frac{n^2 - 3n + 2}{2} - \frac{3}{2}(n - 1) = \frac{n^2 - 6n + 5}{2}, \qquad \text{if} \quad n = 6k + 1,$$

$$\frac{n^2 - 3n + 2}{2} - \frac{3}{2}(n - 2) = \frac{n^2 - 6n + 8}{2}, \qquad \text{if} \quad n = 6k + 2,$$

$$\frac{n^2 - 3n + 2}{2} - \frac{3}{2}(n - 1) + 2 = \frac{n^2 - 6n + 9}{2}, \qquad \text{if} \quad n = 6k + 3,$$

$$\frac{n^2 - 3n + 2}{2} - \frac{3}{2}(n - 2) = \frac{n^2 - 6n + 8}{2}, \qquad \text{if} \quad n = 6k + 4,$$

$$\frac{n^2 - 3n + 2}{2} - \frac{3}{2}(n - 1) = \frac{n^2 - 6n + 5}{2}, \qquad \text{if} \quad n = 6k + 5.$$

If we disregard order of the summands in our partitions, then we obtain 6 times fewer partitions. It is not difficult to see that in all cases the number of (unordered) partitions of n into 3 distinct positive summands is equal to the integral part of $(n^2 - 6n + 12)/12$.

226. The idea of the proof is to compute the number of all *ordered* partitions of $12n + 5$ which otherwise satisfy the restrictions of the problem and have 4, 3, and 2 different summands, respectively. It is then easy to find the number of (unordered) partitions in each of the three categories. Adding these three numbers, we obtain the answer to our problem. The computations follow.

The number of ordered partitions of $12n + 5$ into 4 positive summands x, y, z, t is C_{12n+4}^3. The number of such partitions with $x = y$ is equal to the number of positive integral solutions of $2x + z + t = 12n + 5$. To compute this number, note that the number of positive solutions of $z + t = 12n - 2k + 5$ is $12n - 2k + 4$. It follows that the number of positive solutions of $2x + z + t = 12n + 5$ is

$$\sum_{k=1}^{6n+1} (12n - 2k + 4) = (6n + 1)(6n + 2) = 2C_{6n+2}^2.$$

The number of positive solutions of $x + y + z + t = 12n + 5$ with $x = y = z$ is equal to the number of positive solutions of $3x + t = 12n + 5$, that is, $4n + 1$.

Now we compute the number of positive solutions of $x + y + z + t = 12n + 5$ in which all the entries are $\leqslant 6n - 2$. To do this, we first compute the number of solutions with $x = k \geqslant 6n + 3$. We rewrite our equation in the form $y + z + t = 12n + 5 - k$ and note that for a fixed k this equation has $C_{12n+4-k}^2$ positive solutions. This means that the number of solutions with $x \geqslant 6n + 3$ is

$$\sum_{k=6n+3}^{12n+2} C_{12n+4-k}^2 = C_{6n+2}^3.$$

Since x is only 1 of 4 variables, we see that the number of positive solutions of $x + y + z + t = 2n + 5$ in which all the entries are $\leqslant 6n + 2$ is $C_{12n+4}^3 - 4C_{6n+2}^3$.

Now we compute the number of positive solutions of $x + y + z + t = 12n + 5$ in which 2 entries are equal and all entries are $\leqslant 6n + 2$. We note that the number of positive solutions of $2x + z + t = 12n + 5$ with $z \geqslant 6n + 3$ is $3n(3n + 1) = 2C_{3n+1}^2$. It follows readily that the number of positive solutions of our equation with $x = y$ and all the entries $\leqslant 6n + 2$ is $2(C_{6n+2}^2 - 2C_{3n+1}^2)$. Since x and y can be replaced by other pairs of the variables x, y, z, t, we see that the number of positive solutions of $x + y + z + t = 12n + 5$ in which 2 entries are equal and all the entries are $\leqslant 6n + 2$ is

$$2C_4^2(C_{6n+2}^2 - 2C_{3n+1}^2).$$

Observe that the number of positive solutions of $3x + t = 12n + 5$ in which $t \geqslant 6n + 3$ is $2n$. Hence the number of positive solutions of $x + y + z + t = 12n + 5$ in which just 3 entries are equal and all entries are $\leqslant 6n + 2$ is $4(2n + 1)$.

To complete our computations, we make use of the principle of inclusion and exclusion. Before we do so, however, we note that in rejecting the positive solutions with 2 equal entries we are rejecting the positive solutions with 3 equal entries 3 times.

We can now claim that the number of positive solutions of $x + y + z + t = 12n + 5$ with 4 different entries all $\leqslant 6n + 2$ is

$$[C_{12n+4}^3 - 4C_{6n+2}^3] - 2C_4^2[C_{6n+2}^2 - 2C_{3n+1}^2] + 8(2n + 1) = 12n(12n^2 + 3n - 1),$$

the number of positive solutions with exactly 3 different entries and all entries $\leqslant 6n + 2$ is

$$2C_4^2[C_{6n+2}^2 - 2C_{3n+1}^2] - 12(2n + 1) = 12n(9n + 4),$$

and the number of positive solutions with exactly 2 different entries and all entries $\leqslant 6n + 2$ is $4(2n + 1)$.

In order to go from ordered to unordered partitions of $12n + 5$ into 4 positive summands $\leqslant 6n + 2$, we must divide the number of ordered partitions in the first of our 3 classes by $4! = 24$, in the second by $P(2, 1, 1) = 12$, and in the third by $P(3, 1) = 4$. The sum of the resulting numbers is the answer to our problem. The answer is

$$\frac{n}{2}(12n^2 + 3n - 1) + n(9n + 4) + 2n + 1 = \frac{n+1}{2}(12n^2 + 9n + 2).$$

227. The answer is $(12n^2 + 3n - 1)\, n/2$, the first summand in the answer to Problem 226.

228. A geometric progression $a, aq, aq^2,\ldots$ is determined by its first term a and the multiplier q. In an increasing progression, we must have $aq^2 \leqslant 100$, that is, $a \leqslant 100/q^2$. Hence the number of increasing geometric progressions with multiplier q is $E(100/q^2)$. The number of all geometric progressions of triples of different natural numbers is

$$2\left[E\left(\frac{100}{4}\right) + E\left(\frac{100}{9}\right) + E\left(\frac{100}{16}\right) + \cdots + E\left(\frac{100}{100}\right)\right] = 102.$$

(The presence of the factor 2 is explained by the fact that the admissible decreasing geometric progressions are obtained by reversing order in the admissible increasing geometric progressions.)

229. Let F denote a row of Frenchmen, T a row of Turks, and a an Englishman. Our problem allows 2 types of arrangements: $FaTaFaTaFaTaF$ and $TaFaTaFaTaFaT$. In the first scheme we separate the 7 Frenchmen into 4 nonempty groups (C_6^3 ways) and the 10 Turks into 3 nonempty groups (C_9^2 ways). In any specific arrangement, we can permute the nationals of each group among themselves. Hence the number of arrangements of the first type is $6!7!10!C_6^3C_9^2$. Similarly, the number of arrangements of the second type is $6!7!10!C_6^2C_9^3$. The answer to our problem is

$$6! \ 7! \ 10! \ [C_6^3C_9^2 + C_6^2C_9^3] = 6! \ 7! \ 10! \ 1980.$$

230. Reasoning as in Problem 229, we obtain the answer $5!7!10!1080$.

231. Consider a solution pair (P, Q). Then P is a product of G and some (possibly none) of the numbers a^α, b^β, c^γ, d^δ, and Q is the product of G and the remaining numbers (if any) in the set a^α, b^β, c^γ, d^δ. Since there are $2^4 = 16$ ways of dividing 4 numbers in 2 groups, it follows that there are 16 solution pairs (P, Q). If order in a solution pair is disregarded, then there are 8 solution pairs.

232. Each solution pair is of the form (GA, GB), where A and B are divisors of the number $a^\alpha b^\beta c^\gamma d^\delta$. This number has $N = (\alpha + 1)(\beta + 1)(\gamma + 1)(\delta + 1)$ divisors (see Problem 67). If we identify the pairs (GA, GB) and (GB, GA), then the number of solution pairs is $N + N(N - 1)/2 = C_{N+1}^2$. Otherwise the number of solution pairs is N^2.

233. There are C_{20}^6 combinations in which all 6 letters are different, $C_{20}^1C_{19}^4$ combinations with just 1 pair of like letters, and so on. The total of admissible combinations is

$$C_{20}^6 + C_{20}^1C_{19}^4 + C_{20}^2C_{18}^4 + C_{20}^3 = 146,400.$$

234. An admissible permutation starts with a block of, say, k letters α followed by a single β. The remaining letters can be any permutation of $p - k$ letters α, $q - 1$ letters β, and r letters γ. The number of the latter permutations is $P(p - k, q - 1, r)$. Summing on k from 1 to p, we obtain the answer

$$\sum_{k=1}^{p} \frac{(p + q + r - k - 1)!}{(p - k)! \ (q - 1)! \ r!} = C_{q+r-1}^r \sum_{k=1}^{p} C_{p-k+q+r-1}^{p-k}.$$

Since

$$\sum_{k=1}^{p} C_{p-k+q+r-1}^{p-k} = C_{p+q+r-1}^{p-1} \,,$$

the number of admissible permutations is $C_{q+r-1}^{r} C_{p+q+r-1}^{p-1}$.

235. Concentrate on a particular color. The numbers which give the lengths of successive stripes of this color in a pattern form an ordered partition of 10 into summands from 2 to 10. The number of such partitions of 10 into k summands is the coefficient of x^{10} in the expansion

$$(x^2 + x^3 + \cdots + x^{10})^k$$

$$= \left(\frac{x^2 - x^{11}}{1 - x}\right)^k$$

$$= x^{2k}(1 - x^9)^k(1 - x)^{-k}$$

$$= x^{2k}\left(1 - kx^9 - \frac{k(k-1)}{1 \cdot 2} x^{18} - \cdots\right)\left(1 + kx + \frac{k(k+1)}{1 \cdot 2} x^2\right.$$

$$\left. + \frac{k(k+1)(k+2)}{1 \cdot 2 \cdot 3} x^3 + \cdots + \frac{k(k+1) \cdots (k+9)}{10!} x^{10} + \cdots\right).$$

It is clear that the values of this coefficient for $k = 1, 2, 3, 4, 5$ are, respectively, 1, 7, 15, 10, 1. Since in any pattern the value of k is the same for all 3 colors and the lengths of the stripes of different color can be combined independently, we see that the number of admissible patterns is $1^3 + 7^3 + 15^3 + 10^3 + 1^3 = 4720$.

Now we drop the restriction that the patterns must end in blue. If the pattern ends in red, then there are $k + 1$ stripes of red to k stripes of white and k stripes of blue. Hence in this case we have $1 + 7 \cdot 1^2 + 15 \cdot 7^2 + 10 \cdot 15^2 + 1 \cdot 10^2 = 3093$ patterns. Similarly, if the last stripe is white, then we have $1^2 + 7^2 + 15^2 \cdot 7 + 10^2 \cdot 15 + 1^2 \cdot 10 = 3135$ patterns. The total number of patterns is 10,948.

If the minimal length of a stripe is 3 inches, then the problem reduces to determining the number of ordered partitions of k summands from 3 to 10. If $k = 1, 2, 3$, then we have, respectively, 1, 5, 3 ordered partitions. Hence there are $1^3 + 5^3 + 3^3 = 153$ patterns which end in blue, $1 + 5 \cdot 1^2 + 3 \cdot 5^2 = 81$ patterns which end in red, and $1^2 + 5^2 \cdot 1 + 3^2 \cdot 5 = 71$ patterns which end in white.

236. Since I dined just once with all 6 and twice with each group of 5, I must have dined just once with each group of just 5. This and the data in the problem show that I never dined with just 4 or just 3 or just 2 of my friends. Since I met with each of my friends 7 times and the 6 different dinners account for 6 meetings, I dined once with just 1 of my friends. Each friend was absent 6 times from the dinners mentioned so far (5 times from dinners with just 1 and once from a dinner with just 5). Since each friend was absent from 8 dinners, it follows that I dined twice alone.

237.   The schedule for each teacher can be made up in 12! ways. Hence the schedules for both teachers can be made up in $(12!)^2$ ways. The number of schedules with at least 1 snag is $C_{12}^1 A_{12}^1 (11!)^2$. The number of schedules with at least 2 snags is $C_{12}^2 A_{12}^2 (10!)^2$, and so on. By the principle of inclusion and exclusion, the number of reasonable schedules is

$$(12!)^2 \left[1 - 1 + \frac{1}{2!} - \frac{1}{3!} + \frac{1}{4!} - \cdots + \frac{1}{12!} \right] = 12! \cdot 176{,}214{,}841.*$$

238.   Reasoning as before, we obtain the answers

$$(6!)^2 \left[1 - 1 + \frac{1}{2!} - \frac{1}{3!} + \cdots + \frac{1}{6!} \right] = 190{,}800,$$

and

$$(A_9^6)^2 - C_6^1 A_9^1 (A_8^5)^2 + C_6^2 A_9^2 (A_7^4)^2 - C_6^3 A_9^3 (A_6^3)^2 + C_6^4 A_9^4 (A_5^2)^2 - C_6^5 A_9^5 (A_4^1)^2 + C_6^6 A_9^6 .$$

239.   We are to count permutations of the letters in which each letter has a like letter for a neighbor. This means that we are really concerned with permutations of $a = \alpha^2$, $b = \beta^2$, $c = \gamma^2$, and the number of such permutations is 6. The same answer holds for the expression $\alpha^3 \beta^3 \gamma^3$. In the admissible permutations of $\alpha^4 \beta^4 \gamma^4$, each letter also appears in a block of like letters of even length. We put $\alpha^2 = a_1$, $\alpha^2 = a_2$, $\beta^2 = b_1$, $\beta^2 = b_2$, $\gamma^2 = c_1$, $\gamma^2 = c_2$ and disregard temporarily the equalities $a_1 = a_2$, $b_1 = b_2$, $c_1 = c_2$. Then we obtain 720 permutations of the letters a_1, a_2, b_1, b_2, c_1, c_2. These permutations form classes which differ only by permutations of a_1 and a_2, b_1 and b_2, c_1 and c_2. Each of these groups contains 8 permutations of a_1, a_2, b_1, b_2, c_1, c_2 which reduce to a single permutation of $\alpha^4 \beta^4 \gamma^4$. Hence the number of admissible permutations of the letters in $\alpha^4 \beta^4 \gamma^4$ is $720/8 = 90$.

Finally we consider permutations of $\alpha^5 \beta^5 \gamma^5$. If we put temporarily $\alpha^2 = \alpha_1$, $\alpha^3 = a_2$, $\beta^2 = b_1$, $\beta^3 = b_2$, $\gamma^2 = c_1$, $\gamma^3 = c_2$, then the admissible permutations of the letters in $\alpha^5 \beta^5 \gamma^5$ are permutations of the letters a_1, a_2, b_1, b_2, c_1, c_2. However, some permutations of a_1, a_2, b_1, b_2, c_1, c_2 determine the same permutation of α, β, γ in $\alpha^5 \beta^5 \gamma^5$; for example, $a_1 a_2 b_1 c_2 b_2 c_1$ and $a_2 a_1 b_1 c_2 b_2 c_1$ give $\alpha^5 \beta^2 \gamma^3 \beta^3 \gamma^2$. This happens whenever 2 letters in one of the pairs (a_1 , a_2), (b_1 , b_2), or (c_1 , c_2) appear together. The letters a_1, a_2 appear together in $2 \cdot 5!$ permuta-

---

* A reasonable schedule is a permutation without fixed points of $1, \ldots, 12$. There are D_{12} such permutations, and each can be permuted in 12! ways. Hence the answer $12! D_{12}$ (Translators).

tions (the same is true for the letters b_1, b_2 and for the letters c_1, c_2). The letters in each of the pairs (a_1, a_2) and (b_1, b_2) are together in $(2!)^2 4!$ permutations (the same is true of the letters in the pairs (a_1, a_2), (c_1, c_2) as well as in the pairs (b_1, b_2), (c_1, c_2)). Finally, all 3 pairs appear together in $(2!)^3 3! = 48$ permutations. By the principle of inclusion and exclusion, we see that in

$$6! - 6 \cdot 5! + 3(2!)^2 \cdot 4! - (2!)^3 3! = 240$$

permutations no 2 letters of a pair are together; in

$$3[2 \cdot 5! - 2(2!)^2 4! + (2!)^3 3!] = 288$$

permutations the letters in just 1 of the pairs (a_1, a_2), (b_1, b_2), (c_1, c_2) appear together; in

$$3[(2!)^2 4! - (2!)^3 3!] = 144$$

permutations the letters in just 2 of these pairs appear together. It follows that the number of admissible permutations of α, β, γ in $\alpha^5\beta^5\gamma^5$ is

$$240 + \frac{288}{2!} + \frac{144}{(2!)^2} + \frac{48}{(2!)^3} = 426$$

(if, say, a_1 and a_2 are together, then interchanging their positions has no effect on the order of α, β, γ).

240. First we arrange the players of each nationality into ordered pairs without considering the order of the pairs themselves. This can be done for each nationality in $4!/2 = 12$ ways. These arrangements for the individual nationalities can be combined in 12^n ways. Each combination of this kind consists of $2n$ pairs, and these pairs can be permuted in $(2n)!$ ways. Hence the total of admissible permutations is $12^n(2n)!$.

241. The first row can be painted in $8!$ ways. Each row after the first can be painted in

$$D_8 = 8! \left[1 - 1 + \frac{1}{2!} - \frac{1}{3!} + \cdots + \frac{1}{8!} \right] = 14{,}833$$

ways. Hence the number of ways is $8!(14{,}833)^7$.

242. There are 2^n subsets of a set of n different objects (a subset may contain $0, 1, ..., n$ objects). By possibly supplementing these subsets with some (including none or all) of the n like objects, we obtain all selections of n of our $2n$ objects. Hence the number of such selections is 2^n. The number of permutations of these $2n$ objects is $(2n)!/n!$

243. In each admissible permutation, blocks of 2 or more Frenchmen alternate with blocks of 2 or more Englishmen. The difference between the number of blocks of Frenchmen and blocks of Englishmen is at most 1. We compute the number of ways of separating n Englishmen into p ordered groups of 2 or more people. The Englishmen can be permuted in $n!$ ways. A particular one of these permutations can be broken up into p groups of 2 or more people in $P(n - 2p, p - 1) = C_{n-p-1}^{p-1}$ ways (see p. 68). Hence the required number of ways is $n!C_{n-p-1}^{p-1}$. The m Frenchmen can be separated into s groups of the indicated type in $m!C_{m-s-1}^{s-1}$ ways. These arrangements can be combined into admissible arrangements of Frenchmen and Englishmen in one of the following ways:

(a) p groups of Englishmen and $p - 1$ groups of Frenchmen;

(b) p groups of Englishmen and p groups of Frenchmen, with Frenchmen coming first;

(c) p groups of Englishmen, p groups of Frenchmen, with Englishmen coming first;

(d) p groups of Englishmen and $p + 1$ groups of Frenchmen.

Hence the answer to our problem is given by the formula

$$m!\, n!\, [2(C_{m-2}^0 C_{n-2}^0 + C_{m-3}^1 C_{n-3}^1 + C_{m-4}^2 C_{n-4}^2 + \cdots)$$
$$+ (C_{m-2}^0 C_{n-3}^1 + C_{m-3}^1 C_{n-4}^2 + \cdots)$$
$$+ (C_{m-3}^1 C_{n-2}^0 + C_{m-4}^2 C_{n-3}^1 + \cdots)].$$

If we remove brackets in the formula in the statement of our problem (p. 189), then we obtain the expression just given.

244. First we compute how many of our numbers do not contain the digit 0. We note that 3 nonzero digits can be selected in C_9^3 ways. With certain 3 nonzero digits we can make 3^6 six-digit numbers, with certain 2 nonzero digits we can make 2^6 six-digit numbers, and with a certain nonzero digit we can make 1^6 six-digit numbers. By the principle of inclusion and exclusion there are

$$3^6 - C_3^1 2^6 + C_3^2 1^6 = 540$$

nonzero six-digit numbers which contain all 3 selected nonzero digits. Hence there are $84 \cdot 540 = 45{,}360$ six-digit numbers containing exactly 3 nonzero digits.

If a number contains the digit 0, then it must include 2 additional digits. These can be selected in $C_9^2 = 36$ ways. Take one such triple, say, the triple 0, 1, 2. Then the first digit is 1 or 2. Suppose the first digit is 1. Then the remaining 5 digits can be 0, 1, or 2 provided that each of the digits 0, 2 is included at least once. By the principle of inclusion and exclusion, we see that the 5 digits in question can be selected in

$$3^5 - C_2^1 2^5 + 1^6 = 180$$

ways. But then the number of 6-digit numbers composed of 0, 1, 2 and containing all of these digits is $2 \cdot 180 = 360$. Hence the total of 6-digit numbers composed of 3 digits including the digit 0 is $45{,}360 + 12{,}960 = 58{,}320$.

245. Reasoning as in Problem 244, we obtain the answer

$$C_9^k[k^m - C_k^1(k-1)^m + C_k^2(k-2)^m - \cdots + (-1)^{k-1}C_k^{k-1}1^m]$$

$$+ (k-1)C_9^{k-1}[k^{m-1} - C_{k-1}^1(k-1)^{m-1} + C_{k-1}^2(k-2)^{m-1} - \cdots$$

$$+ (-1)^{k-2}C_{k-1}^{k-2}1^{m-1}].$$

246. Let $\Gamma_n^{(k)}$ denote the number of admissible samples. These samples start with a 1 or do not start with a 1. If a sample starts with a 1, then we decrease each of its digits by 1 and delete the initial zero (for example, 14,589 is first reduced to 03,478 and then becomes 3478). The resulting number is an admissible $(k-1)$-sample of the digits $1, 2,..., n-1$. Hence the number of admissible k-samples of the digits $1, 2,..., n$ which begin with the digit 1 is $\Gamma_{n-1}^{(k-1)}$. The remaining k-samples start with an odd digit >1. If we decrease the digits in each of these k-samples by 2, then we obtain an admissible k-sample of the digits $1,..., n-2$, so that the number of our k-samples which do not start with the digit 1 is $\Gamma_{n-2}^{(k)}$. Hence the recurrence relation

$$\Gamma_n^{(k)} = \Gamma_{n-1}^{(k-1)} + \Gamma_{n-2}^{(k)}.$$

If we put $F_n^{(k)} = C_N^k$, where $N = E[(n+k)/2]$, then we have

$$F_{n-1}^{(k-1)} + F_{n-2}^{(k)} = C_{N-1}^{k-1} + C_{N-1}^k = C_N^k = F_n^{(k)}.$$

This means that the numbers $F_n^{(k)}$ satisfy the same recurrence relation as the numbers $\Gamma_n^{(k)}$.

Next we show that $F_n^{(n)} = \Gamma_n^{(n)}$ and $F_{n+1}^{(n)} = \Gamma_{n+1}^{(n)}$. Note that there is exactly one way of arranging the numbers $1,..., n$ in increasing order. Therefore $\Gamma_n^{(n)} = 1 = C_n^n = F_n^{(n)}$. Again, there is exactly one way of selecting n out of the $n+1$ integers $1,..., n+1$ in agreement with the conditions of our problem. Therefore $\Gamma_{n+1}^{(n)} = 1 = C_n^n = F_{n+1}^{(n)}$. It follows that for all n and k

$$\Gamma_n^{(k)} = F_n^{(k)} = C_N^k,$$

where $N = E[(n+k)/2]$.

247. The number of permutations of the given elements is

$$P(2, 2,..., 2) = \frac{(2n)!}{2^n}.$$

We compute in how many permutations the elements in a certain k pairs appear together. In such a permutation we can think of the elements in each of these k pairs as a single object. Then we obtain a permutation of k different elements and of the elements of $n - k$ pairs. The number of such permutations is $(2n - k)!/2^{n-k}$. The k pairs can be selected in C_n^k ways. Applying the principle of inclusion and exclusion, we see that the number of admissible permutations is

$$\frac{(2n)!}{2^n} - C_n^1 \frac{(2n - 1)!}{2^{n-1}} + C_n^2 \frac{(2n - 2)!}{2^{n-2}} - \cdots + (-1)^n C_n^n n!.$$

248. Reasoning as in the preceding problem, we obtain the answer

$$\frac{(qn)!}{(q!)^n} - C_n^1 \frac{(qn - q + 1)!}{(q!)^{n-1}} + C_n^2 \frac{(qn - 2q + 2)!}{(q!)^{n-2}} - \cdots.$$

249. The given elements can be permuted in $(qn)!/(q!)^n$ ways. We compute the number of permutations in which elements from certain k of our q-selections appear together. Choose a definite q-selection. Without separating these elements we can arrange them in a circle in nq ways. The number of permutations of the remaining $k - 1$ q-selections, thought of as $k - 1$ different objects, and of the $(n - k)q$ remaining elements is $(qn - qk + k - 1)!/(q!)^{n-k}$. It is clear that each of these permutations determines a unique arrangement of our nq objects in a circle. Hence the number of arrangements of our nq objects in a circle such that elements from a certain k q-selections are together is $qn(qn - qk + k - 1)!/(q!)^{n-k}$. Since the k q-selections can be chosen in C_n^k ways, it follows from the principle of inclusion and exclusion that the number of admissible permutations is

$$qn \left[\frac{(qn - 1)!}{(q!)^n} - C_n^1 \frac{(qn - q)!}{(q!)^{n-1}} + C_n^2 \frac{(qn - 2q + 1)!}{(q!)^{n-2}} - \cdots + (-1)^n C_n^n (n - 1)! \right].$$

250. We adjoin to each selected book s books which follow it. Then our problem reduces to the problem of selecting p out of $n - ps$ objects, and this can be done in C_{n-ps}^p ways.

251. Let a be the number of 5th graders and let d be the difference between the number of students in two consecutive grades. The number of ways of dividing the $6d$ prizes among the children in grades 5, 6, 7, 8, 9 is

$$A_a^d A_{a+d}^d A_{a+2d}^d \cdots A_{a+5d}^d.$$

The number of ways of dividing the $6d$ prizes among the children in grade 10 is A_{a+5d}^{6d}. The equality

$$A_a^d A_{a+d}^d A_{a+2d}^d \cdots A_{a+5d}^d = A_{a+5d}^{6d}$$

follows from the obvious fact that $A_n^m A_{n+k}^k = A_{n+k}^{m+k}$.

252. The number of paths is $C_8^4 = 70$. To answer the remaining questions, we compute the number of paths through each of the segments of the square and each of its junctions. For instance, the segment EF (Fig. 36) is traversed by 18 paths ($C_3^1 = 3$ paths lead from A to E and $C_4^2 = 6$ paths lead from F to C). The point E is crossed by 30 paths (3 paths lead from A to E and $C_5^2 = 10$ paths lead from E to C). Analogous computations apply to the other segments and junctions.

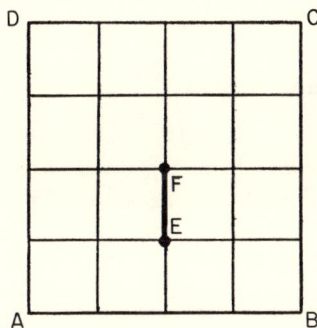

FIG. 36

253. $C_6^3 = 20$.

254. We are dealing here with 3-combinations with repetitions of objects of 4 kinds, that is,

$$\bar{C}_4^3 = C_6^3 = 20.$$

255. $\bar{C}_{10}^3 = C_{12}^3 = 220$.

256. We obtain 4 triangles.

257. If no 3 of the n points are collinear, then the n points determine C_n^3 triangles. Since, however, p of the points are collinear, we have to reject C_p^3 triangles. We are left with $C_n^3 - C_p^3$ triangles.

258. Any 2 points of one line and any point of the other line determine a triangle. Hence the number of triangles is $C_p^2 C_q^1 + C_p^1 C_q^2 = (pq/2)(p + q - 2)$.

259. We obtain

$$C_r^2(C_p^1 + C_q^1) + C_r^1(C_p^2 + C_q^2) + C_r^1 C_p^1 C_q^1 = \frac{r}{2}(p + q)(p + q + r - 2)$$

additional triangles.

260. The triangles can be of two kinds: either each vertex lies on a different side of the square, or 2 vertices lie on one side and the third on another. To compute the number of triangles in the first case, we select 3 of the four sides of the square (this can be done in $C_4^3 = 4$ ways) and then select 1 out of $n - 1$ points on each of the sides. Hence there are $4(C_{n-1}^1)^3$ triangles in the first case. In the second case we select 1 side of the square (4 ways) and 2 of the $n - 1$ division points on this side (C_{n-1}^2 ways), and then we select 1 of the remaining 3 sides (3 ways) and 1 of the $n - 1$ division points (C_{n-1}^1 ways). Hence, in the second case we obtain $12C_{n-1}^1 C_{n-1}^2$ triangles. The total number of triangles is

$$4(C_{n-1}^1)^3 + 12C_{n-1}^1 C_{n-1}^2 = 2(n - 1)^2(5n - 8).$$

261. There are C_n^2 points of intersection.

262. In general, n lines can have C_n^2 points of intersection. However, the p lines intersecting at A contribute only 1 point of intersection instead of C_p^2, and the q lines intersecting at B contribute again only 1 point of intersection instead of C_q^2. Hence the number of points of intersection is $C_n^2 - C_p^2 - C_q^2 + 2$.

263. Let the number of regions determined by $k - 1$ lines be N_{k-1}. We draw an additional line. This line is divided by the $k - 1$ lines into k parts which determine k additional regions. Hence $N_k = N_{k-1} + k$, $k = 1, 2,..., n$. For $k = n$, we obtain $1 + 1 + 2 + \cdots + n = \frac{1}{2}(n^2 + n + 2)$ regions.

264. Suppose there are already $k - 1$ planes. We introduce an additional plane. This plane intersects the $k - 1$ planes along $k - 1$ lines. By Problem 263, the $k - 1$ lines determine in the plane $(k^2 - k + 2)/2$ plane regions. In turn, each of the plane regions gives rise to a new region in space. Hence, if we denote by R_k the number of regions determined by k planes, then $R_k = R_{k-1} + (k^2 - k + 2)/2$. It follows that the n planes determine

$$1 + \frac{1}{2} \sum_{k=1}^{n} (k^2 - k + 2) = \frac{1}{6}(n + 1)(n^2 - n + 6)$$

regions.

265. There are $C_5^2 = 10$ lines joining the points A, B, C, D, E in pairs. There are 4 lines passing through each point. It follows that we can drop 6 perpendiculars from each point. Consider any 2 points, say, B and C. The perpendiculars dropped from B to the lines passing through C intersect all perpendiculars from C. There are 3 lines issuing from C which do not pass through B. Hence we can drop perpendiculars from B to these 3 lines. These perpendiculars intersect the perpendiculars from C in $3 \cdot 6 = 18$ points. Each of the perpendiculars from B to the remaining 3 lines (not passing through C) intersects only 5 of the perpendiculars from C; this is so because each of the perpendiculars from B in question is parallel to a perpendicular from C. As a result, we get additional 15 points. Hence the perpendiculars issuing from 2 points intersect in $18 + 15 = 33$

points. Since there are 10 pairs of points, there are $33 \cdot 10 = 330$ points of intersection. However, some of these points coincide. Specifically, the points A, B, C, D, E determine $C_5^3 = 10$ triangles. The altitudes of such a triangle belong to our set of perpendiculars, but they intersect in 1 rather than in 3 points. This means a loss of 2 points per triangle, and so a loss of $10 \cdot 2 = 20$ points. It follows that our perpendiculars intersect in 310 points.

266. We can choose for the sides of the triangles any 3 numbers x, y, z such that $n + 1 \leqslant x, y, z \leqslant 2n$. Hence the number of triangles is $\bar{C}_n^3 = C_{n+2}^3$. To compute the number of isosceles triangles, note that for any given base we have n triangles. Therefore the number of isosceles triangles is n^2. The number of equilateral triangles is n.

267. Our problem reduces to finding triples of numbers x, y, z such that $x \leqslant y \leqslant z \leqslant 2n$ and $x + y > z$. Let $x = p$. Then y can assume values from p to $2n$. If y assumes values from p to $2n - p + 1$, then to each value of y there correspond p values of z such that $y \leqslant z < y + p$, $z \leqslant 2n$. If y assumes values from $2n - p + 2$ to $2n$, then to each value of y there correspond $2n - y + 1$ values of z. In all, for fixed $x = p$, we obtain

$$2p(n - p + 1) + \sum_{y=2n-p+2}^{2n} (2n - y + 1) = 2pn - \tfrac{3}{2}p^2 + \tfrac{3}{2}p$$

pairs (y, z) such that x, y, z satisfy the necessary conditions. Hence, the number of triangles such that $1 \leqslant x \leqslant n$ and $1 \leqslant y, z \leqslant 2n$ is

$$\sum_{p=1}^{n} \left(2pn - \frac{3}{2}p^2 + \frac{3}{2}p \right) = \frac{n}{2}(n + 1)^2.$$

By Problem 266, there are C_{n+2}^3 triangles for which $x \geqslant n + 1$. Therefore, the total number of triangles satisfying the conditions of our problem is

$$\frac{n}{2}(n + 1)^2 + \frac{n(n + 1)(n + 2)}{6} = \frac{n(n + 1)(4n + 5)}{6}.$$

As for the second part of our problem, note that the number of isosceles triangles with base $x = 2k$ is $2n - k$, and the number of isosceles triangles with base $x = 2k + 1$ is also $2n - k$. The total number of isosceles triangles is therefore

$$\sum_{k=1}^{n} (2n - k) + \sum_{k=0}^{n-1} (2n - k) = 3n^2.$$

Excluding these triangles, we obtain the answer

$$\frac{n(n+1)(4n+5)}{6} - 3n^2 = \frac{n(n-1)(4n-5)}{6}.$$

268. The solution of this problem is analogous to the solution of Problem 267. The number of triangles for a given $x = p \leqslant n - 1$ is $2np - 3p^2/2 + p/2$. If we let x vary from 1 to $n - 1$, then we obtain

$$\sum_{p=1}^{n-1} \left(2np - \frac{3}{2}p^2 + \frac{p}{2}\right) = \frac{n(n+1)(n-1)}{2}.$$

triangles. Since the number of triangles for $x \geqslant n$ is C_{n+2}^3, the total number of triangles is

$$\frac{n(n+1)(n-1)}{2} + \frac{n(n+1)(n+2)}{6} = \frac{n(n+1)(4n-1)}{6}.$$

The number of isosceles triangles is

$$\sum_{k=1}^{n-1} (2n - k - 1) + \sum_{k=0}^{n-1} (2n - k - 1) = 3n^2 - 3n + 1,$$

and the number of equilateral triangles is

$$\frac{n(n+1)(4n-1)}{6} - 3n^2 + 3n - 1 = \frac{1}{6}(n-1)(n-2)(4n-3).$$

269. Since every group consists of n points and no 3 of these points are collinear, it follows that each line contains exactly 2 points belonging to a given group. Therefore, if we number the lines and pick on the first line its point of intersection with the second line, on the second line its point of intersection with the third line,..., on the nth line its point of intersection with the first line, we obtain an admissible group. All the admissible groups can be obtained in this manner. Cyclic permutation of the numbers assigned to the lines as well as reversal of their order yield the same group of n points. Hence the number of admissible groups is $P_n/2n = (n-1)!/2$.

270. We can select r vertices in stated order in A_n^r ways. Since cyclic permutation of the vertices as well as reversal of their order yield the same polygon, the number of required polygons is $(1/2r) A_n^r$.

271. Pick 2 points on one line and 2 points on the other line. The lines joining these 4 points in pairs intersect in 2 points (the points of intersection of the diagonals of the trapezoid and of two of its sides; we are disregarding the points on the parallel lines). Since on the first line we can select C_n^2 pairs of points and on the second line C_m^2 pairs of points, the number of required points is $2C_n^2C_m^2$.

272. n points determine C_n^3 circles. Of these circles, C_{n-1}^2 pass through 1 given point and C_{n-2}^1 pass through 2 given points. Hence a line joining 2 given points intersects the circles in at most $2C_{n-2}^3 + (2C_{n-1}^2 - C_{n-2}^1) + 2$ points. Since n points determine C_n^2 lines, the number of points of intersection is at most

$$C_n^2[2C_{n-2}^3 + 2C_{n-1}^2 - C_{n-2}^1 + 2].$$

273. Each line of intersection is determined by 2 planes and each plane by 3 given points. We divide the lines into classes according as the planes determining each line have no given points in common, have just 1 given point in common, or have 2 given points in common. In the first case, the number of lines is $C_n^3C_{n-3}^3/2$ (we select 3 point out of n, and then 3 more points out of the remaining $n-3$ points; order is immaterial). In the second case, the number of lines is $\frac{3}{2}C_n^3C_{n-3}^2$. In the last case, the number of lines is $\frac{3}{2}C_n^3C_{n-1}^1$. The total number of lines is

$$\tfrac{1}{2}C_n^3(C_{n-3}^3 + 3C_{n-3}^2 + 3C_{n-3}^1) = \frac{n(n-1)(n-2)(n-3)(n^2+2)}{72}.$$

Of these lines

$$\tfrac{1}{2}C_n^3C_{n-3}^3 = \frac{n(n-1)(n-2)(n-3)(n-4)(n-5)}{72}$$

do not contain any of the given points.

274. Let a, b, c, d denote the sides of the quadrilateral. Without loss of generality, we may assume that a is the smallest of its sides, c is opposite to a, and $b < d$. Then $a < b < d$ and $a < c$. Also, since our quadrilateral is circumscribed about a circle, we must have $a + c = b + d$. It follows that $a + c > 2b$. But then, for given a and b, c varies from $2b - a + 1$ to n, and we must have $2b - a \leqslant n - 1$.

We have shown that $b \leqslant (a + n - 1)/2$ and that $2b - a + 1 \leqslant c \leqslant n$. The last inequality shows that for given a and b, c takes on $n + a - 2b$ values. Also, b varies from $a + 1$ to $E[(a + n - 1)/2] = s$. It follows that for a fixed a, the number of quadrilaterals is

$$\sum_{b=a+1}^{s} (n + a - 2b) = (s - a)(n - s - 1).$$

Let n be even, $n = 2m$. Then for odd values of a, $a = 2k - 1$, we have $s = E[(n + a - 1)/2] = m + k - 1$, and, therefore, $(m - k)^2$ quadrilaterals; for even values of a, $a = 2k$, we have $s = E[(n + a - 1)/2] = m + k - 1$, and, therefore, $(m - k - 1)(m - k)$ quadrilaterals. Summing on a, we find that the number of quadrilaterals (for n even) is

$$\sum_{k=1}^{m} (m - k)^2 + \sum_{k=1}^{m} (m - k)(m - k + 1)$$

$$= \frac{m(m - 1)(4m - 5)}{6} = \frac{n(n - 2)(2n - 5)}{24}$$

The case when n is odd is solved analogously.

If the sides of the quadrilaterals are of equal length, then $a \leqslant b \leqslant d \leqslant n$, $a \leqslant c$ and $a + c = b + d$. It follows that $b < (a + n)/2$ and $2b - a \leqslant c \leqslant n$. Hence, for fixed a, the number of quadrilaterals is $(n - s + 1)(s - a + 1)$, where $s = E[(a + n)/2]$.

For n even, the number of quadrilaterals is $n(n + 2)(2n + 5)/24$; for n odd, the corresponding number is $(n + 1)(2n^2 + 7n + 3)/24$.

275. The number of circles is C_n^3. The number of circles passing through 1 given point is C_{n-1}^2, and the number of circles passing through 2 given points is C_{n-2}^1. Let A, B, C be 3 of the n given points and consider the circle determined by these points. There are $C_n^3 - 3C_{n-1}^2 + 3C_{n-2}^1 - 1$ circles which do not pass through A, B, C. Each of these circles intersects our circle in 2 points. Furthermore, there are $3(C_{n-1}^2 - 2C_{n-2}^1 + 1)$ circles passing through exactly 1 of the points A, B, C. Each of these circles intersects our circle in 1 point different from A, B, C. The remaining circles intersect our circle in 2 of the points A, B. C. Hence our circle has at most

$$2(C_n^3 - 3C_{n-1}^2 + 3C_{n-2}^1 - 1) + 3(C_{n-1}^2 - 2C_{n-2}^1 + 1) = \frac{(n - 3)(n - 4)(2n - 1)}{6}$$

points of intersection different from A, B, C. Considering all the circles, the maximal number of points of intersection different from the n given ones is

$$\frac{1}{2} C_n^3 \frac{(n - 3)(n - 4)(2n - 1)}{6} = \frac{5(2n - 1)}{3} C_n^5$$

and if we add the n given points, the corresponding number is

$$\frac{5(2n - 1)}{3} C_n^5 + n.$$

276. Suppose we have already k planes of the required type. Introduction of the $(k + 1)$th plane results in a division of the sphere into at most $2k$ additional parts. Therefore, n planes of the required type divide the sphere into at most $2 + 2 + 4 + 6 + \cdots + 2(n-1) = n^2 - n + 2$ parts.

277. The total number of ways of painting the faces of a cube 6 different colors is $6! = 720$. We separate these ways of painting into classes of geometrically identical ways. Since there are 24 rigid motions of a cube (6 motions take each face into a preassigned face and 4 motions carry it into itself), each class consists of 24 geometrically identical painting schemes. Hence the required number of ways is $720/24 = 30$.

278. This problem is solved in the manner of the preceding problem. The number of ways is $4!/12 = 2$.

279. There are $8!/24 = 1680$ geometrically different ways.

280. There are $12!/60$ ways for the dodecahedron and $20!/60$ ways for the icosahedron.

282. We must find all triples x, y, z of natural numbers such that $x \leqslant y \leqslant z$, $x + y + z = 40$, $x + y > z$. These data imply that $14 \leqslant z \leqslant 19$. If $z = 19$, then $x + y = 21$ with $x \leqslant y \leqslant 19$. Hence $11 \leqslant y \leqslant 19$ and we have 9 triangles corresponding to $z = 19$. Similarly, for $z = 18, 17, 16, 15, 14$ we have, respectively, 8, 6, 5, 3, 2 triangles. In all there are 33 triangles. The number of triangles with perimeter 43 is 44.

283. Consider a triangle with perimeter $4n$. Let x, y, z denote the sides of this triangle. By adding 1 to each of the sides of our triangle, we obtain a triangle with sides $x + 1$, $y + 1$, $z + 1$, that is, a triangle with perimeter $4n + 3$. In addition, we have the $n + 1$ triangles whose sides are, respectively,

$$(1, 2n + 1, 2n + 1), \qquad (2, 2n, 2n + 1), \ldots, (n + 1, n + 1, 2n + 1).$$

284. Let $N = 12n$. We must find the number of triples of natural numbers x, y, z such that $x \leqslant y \leqslant z$, $x + y + z = 12n$, $x + y > z$. These data imply that $4n \leqslant z \leqslant 6n - 1$. If $z = 2k$, then $x + y = 12n - 2k$ and the number of solutions of this equation with $x \leqslant y \leqslant z = 2k$ is $3k - 6n + 1$. If $z = 2k + 1$, then the corresponding number of solutions is $3k - 6n + 2$. Hence the number of triangles is

$$\sum_{k=2n}^{3n-1} (3k - 6n + 1) + \sum_{k=2n}^{3n-1} (3k - 6n + 2) = 3n^2.$$

The remaining cases are analyzed in a similar manner. In going from N to $N + 3$, we can argue as in the solution of Problem 283.

285. We show that there are exactly n routes passing through every stop.

Let l be one of the routes and let B be a stop not on the route l (Fig. 37). In view of condition (1), we can go by direct route from B to each of the n stops A_1 ,..., A_n of the route l. Also, in view of condition (2), each of the routes through B passes through one of the stops A_1 ,..., A_n (otherwise there would be at least one route

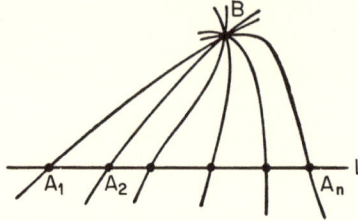

Fig. 37

through B from which one could not transfer to the route l) and only through one (otherwise there would be a route through B from which one could transfer to the route l at two stops). Again, no two of the routes through B pass through the same stop on the route l (otherwise the two routes would have two stops in common and one could transfer from one of those routes to the other at two stops). It follows that the number of routes through B is equal to the number of stops on l, that is, n.

It remains to show that there are exactly n routes through each of the stops A_1 ,..., A_n on the route l. For this it suffices to show that for each of these stops there is a route l' not passing through it (then the stop in question would be related to the route l' in the same way in which the stop B is related to the route l). Since there are at least two routes, it follows that in addition to the route l there is some route l'' which intersects the route l in a unique point, say A_1 (Fig. 38).

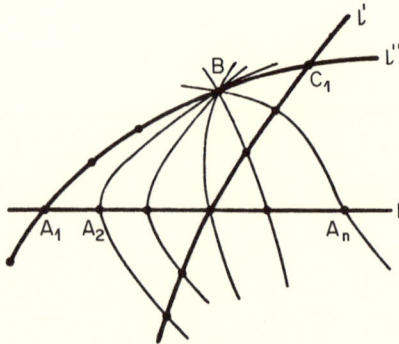

Fig. 38

Then the stops $A_2, ..., A_n$ are not on the route l'', and we may therefore conclude that there are exactly n routes passing through each of them. Suppose B is a stop on l'' different from A_1. The route connecting B and A_2 does not pass through A_1. This implies that there are exactly n routes passing through A_1.

Since there is a route which does not pass through any preassigned stop and there are exactly n stops on each route, there are exactly n routes passing through each stop. Consider a particular stop l. Through each of its n stops there pass $n-1$ routes other than the route l and in view of condition (2) no two of these routes have an additional stop in common (otherwise two routes would have two stops in common). Also, each route belongs to the set of routes just mentioned. It follows that there are altogether $n(n-1)+1$ routes.

286. Suppose there are exactly n stops on a route l. The solution of Problem 285 tells us that there are exactly n routes through any stop B not on the route l. We show that there are exactly n stops on any route l' different from l. By condition (3) there are at least three stops on l' and by condition (2) only one of these stops can be a stop on the route l. There are $n-1$ stops on the route l which are not on the route l'. We claim that there is an additional stop which is neither on l' nor on l. Let A_1 be one of the $n-1$ stops on l not on l' and let C_1 be one of the stops on l' not on l (there are at least two such stops). By condition (1) there is a route l'' passing through A_1 and C_1, and by condition (3) there is at least one stop C_1 on l'' which is different from A_1 and B_1. This stop is neither on l' nor on l. The solution of Problem 285 tells us that there are n routes through the stop B_1. Each of these n routes intersects the route l' in a unique point. On the other hand, there is at least one route linking each stop on l' to the stop B_1. Hence the number of stops on l' is equal to the number of routes through B_1, that is, n.

The solution of Problem 285 implies that under the circumstances the number of routes is $n(n-1)+1$. In our case this number is 57. The equality $n(n-1)+1 = n^2 - n + 1 = 57$ implies that $n = 8$.

287. This is possible. We consider 10 lines in the plane no two of which are parallel and no three of which are concurrent and regard the lines as bus routes and the points of intersection of the lines as bus stops. Also, we say that one can travel from one stop to another directly or by transfer according as the two stops are on the same line or on different lines. Even if we remove 1 of our lines, it is still possible to go from any one stop to any other stop with the aid of at most one transfer. If we remove 2 of our lines, then there is a stop, namely the point of intersection of the rejected lines, which cannot be reached from any of the remaining routes and, conversely, there is no way of getting from this stop to any other stop.

288. A sphere can be tangent to a plane on one of two sides. It can be tangent to another sphere either externally or internally. Hence the answer to our problem is $2^4 = 16$.

289. Let A be one of the 3 given points. Each of the m lines through A intersects $2m$ lines. Hence the lines through A intersect the remaining lines in $2m^2$ points. The total number of points of intersection (exclusive of the 3 given points) is $3m^2$.

290. Let $A_1,...,A_m$ be the m coplanar points and $B_1,...,B_{n-m}$ the remaining points. Each plane is determined by three points. Three of the given points may include 0, 1, 2, or 3 of the points $A_1,...,A_m$. It follows that the number of admissible planes is

$$1 + C_m^2 C_{n-m}^1 + C_m^1 C_{n-m}^2 + C_{n-m}^3 .$$

291. There are $n + p$ points of intersection on each line through A, $m + p$ points of intersection on each line through B, and $m + n$ points of intersection on each line through C. The total number of points of intersection is therefore equal to

$$\tfrac{1}{2}[m(n + p) + n(m + p) + p(m + n)] = mn + mp + np.$$

Of these, 3 points can be selected in $C_{mn+mp+np}^3$ ways. However, $mC_{n+p}^3 + nC_{m+p}^3 + pC_{m+n}^3$ of these triples of points are collinear. Hence the number of triangles is

$$C_{mn+mp+np}^3 - mC_{n+p}^3 - nC_{m+p}^3 - pC_{m+n}^3 .$$

292. The first vertex of our triangle can be chosen in n ways. Call this vertex A. The remaining 2 vertices must be chosen out of the $n - 3$ vertices of our n-gon not adjoining A. Also, these 2 points must not be adjoining vertices of our n-gon. Such a choice of the 2 points can be effected in $P(n - 6, 2) = C_{n-4}^2$ ways (see p. 000). Since each of the 3 vertices of the triangle can be regarded as the vertex A, the number of triangles is

$$\frac{n}{3} C_{n-4}^2 = \frac{n(n - 4)(n - 5)}{6} .$$

293. We separate the triangles into two classes. In one class we put triangles whose vertices lie on different lines and in the other we put triangles with 2 vertices on the same line. There are $p^3 C_n^3$ triangles in the first class (there are C_n^3 triples of lines containing the vertices, and for any triple of lines there are p^3 ways of choosing the 3 vertices). There are $p^2(p - 1)\, n(n - 1)/2$ triangles in the second class (we choose the line containing the 2 vertices, then we choose 2 points on this line, then we choose a line containing the third vertex and a point on that line). The total number of triangles is

$$p^3 C_n^3 + \tfrac{1}{2} p^2(p - 1)n(n - 1) = \frac{n(n - 1)p^2(pn + p - 3)}{6} .$$

294. Each interior point of intersection of the diagonals is uniquely determined by the choice of 4 vertices of the n-gon (the 4 end points of the intersecting diagonals). It follows that the number of interior points of intersection of the diagonals is C_n^4. We compute the number of all the points of intersection of the diagonals). The number of diagonals issuing from each vertex is $n - 3$. This gives a total of $n(n - 3)/2$ diagonals. Each diagonal AB intersects all the diagonals connecting vertices different from A and B. Hence the diagonal AB intersects the remaining diagonals in

$$\frac{n(n - 3)}{2} - 2(n - 3) + 1 = \frac{(n - 3)(n - 4)}{2} + 1$$

points. Since the number of diagonals is $n(n - 3)/2$, it follows that the diagonals intersect in

$$\frac{n(n - 3)[(n - 3)(n - 4) + 2]}{8}$$

points. Subtracting the number of internal points of intersection, we obtain for the number of external points of intersection the figure of

$$\frac{n(n - 3)(n - 4)(n - 5)}{12}.$$

295. Each r-gon is determined by choosing an r-sample of the n given points. Cyclic permutation of the points of the sample as well as reversal of the order in the sample yield an r-sample which determines the same r-gon. Hence the number of r-gons is $A_n^r/2r$. It follows that the number of all polygons is $\sum_{r=3}^n A_n^r/2r$. The number of convex polygons is $\sum_{r=3}^n C_n^r$.

296. m parallel lines divide the plane into $m + 1$ strips. Each new line adds a number of regions equal to the number of parts into which it is divided by its predecessors. Hence the number of regions is

$$m + 1 + (m + 1) + \cdots + (m + n) = \frac{n(2m + n + 1)}{2} + m + 1.$$

297. We separate the circles into classes according to the number of the 5 given points they contain. One circle (the given one) contains all of the 5 points, $C_5^2 C_6^1$ circles contain 2 of these points, $C_5^1 C_6^2$ contain 1 of these points, and C_6^3 circles contain none of these points. The total number of circles is

$$1 + C_5^2 C_6^1 + C_5^1 C_6^2 + C_6^3 = 156.$$

298.   Each set of 3 of our lines determines 4 circles which are tangent to them. Hence the number of circles is $4C_{10}^3 = 480$.

299.   We choose s successive vertices $A_1, ..., A_s$ of our n-gon and separate the admissible k-gons into two classes. In the first class we put all those k-gons which have a vertex belonging to the set $\{A_1, ..., A_s\}$. In the second class we put the remaining k-gons. Next we subdivide the k-gons in the first class into subclasses according to which of the points $A_1, ..., A_s$ they share. It is clear that these subclasses are disjoint.

We compute the number of k-gons in the subclass associated with a point A_m. We ignore the vertex A_m and the following (in the clockwise sense) s vertices of our n-gon (none of these s vertices is a vertex of our k-gons). From the remaining $n - s - 1$ vertices of the n-gon we must select $k - 1$ vertices of our k-gon so that any selected vertex is followed by at least s vertices which are not selected. This can be done in C_{n-ks-1}^{k-1} ways (see Problem 250). This is also the number of k-gons in the subclass determined by A_m. Hence the number of k-gons in the first class is sC_{n-ks-1}^{k-1}.

To compute the number of k-gons in the second class, we cut the circle between the vertices A_s and A_{s+1}. For each of our k-gons we must select k vertices so that each is followed by at least s vertices which are not selected (this condition rules out the vertices $A_1, ..., A_s$). This can be done in C_{n-ks}^k ways. Hence the total number of admissible k-gons is

$$sC_{n-ks-1}^{k-1} + C_{n-ks}^k.$$

300.   Each parallelogram is determined by 2 pairs of parallel lines. Hence the answer is $(C_{r+2}^2)^2$.

301.   Let the vertices of the n-gon be $A_1, ..., A_n$. We draw all the diagonals through the vertex A_1, then all the diagonals through $A_2, ...,$ all the diagonals through A_n. The number of new regions introduced by a particular diagonal is equal to the number of segments into which the preceding diagonals divide the diagonal in question, that is, one more than the number of its (internal) points of intersection with these diagonals. Since each point of intersection comes up exactly once, the number of regions at each step is one more than the number of diagonals and points of intersection (the "one more" is due to the fact that we have one region in the absence of any diagonals). Since the number of diagonals is $n(n - 3)/2$ and the number of their internal points of intersection is C_n^4 (see Problem 294), it follows that the required number of regions is

$$1 + \frac{n(n - 3)}{2} + \frac{n(n - 1)(n - 2)(n - 3)}{24} = \frac{n(n - 3)(n^2 - 3n + 14)}{24} + 1.$$

302. Suppose n even, $n = 2k$. Then n can be written as a sum of 2 terms in the following ways:

$$n = 1 + (2k - 1) = 2 + (2k - 2) = \cdots = k + k.$$

A card with the number 1 can be withdrawn in 1 way, a card with the number 2 in 2 ways, and so on. Two cards with the number $k > 1$ can be withdrawn in C_k^2 ways. Hence the sum $n = 2k$ can be obtained in

$$1(2k - 1) + 2(2k - 2) + \cdots + (k - 1)(k + 1) + \frac{k(k - 1)}{2}$$

$$= \sum_{s=1}^{k-1} s(2k - s) + \frac{k(k - 1)}{2} = \frac{2k(k^2 - 1)}{3} = \frac{n(n^2 - 4)}{12}$$

ways. In n is odd, $n = 2k - 1$, then

$$n = 1 + (2k - 2) = 2 + (2k - 3) = \cdots = (k - 1) + k,$$

and the corresponding number of ways is

$$\sum_{s=1}^{k-1} s(2k - s - 1) = \frac{k(k - 1)(2k - 1)}{3} = \frac{n}{12}(n^2 - 1).$$

303. Consider a selection with k of the n like objects. This selection contains $n - k$ different objects selected out of $2n + 1$ different objects. Such a selection can be obtained in C_{2n+1}^{n-k} ways. Summing on k, we obtain the answer

$$C_{2n+1}^n + C_{2n+1}^{n-1} + \cdots + C_{2n+1}^0 = \frac{1}{2} \sum_{k=0}^{2n+1} C_{2n+1}^k = 2^{2n}.$$

304. Let the terms of the progression be a, $a + d$, $a + 2d$. Then $a + 2d \leqslant 2n$. For a given d, this inequality has $2n - 2d$ solutions. Hence the total number of solutions is

$$(2n - 2) + (2n - 4) + \cdots + 2 = n(n - 1).$$

By reversing the increasing progression a, $a + d$, $a + 2d$ we obtain the decreasing progression $a + 2d$, $a + d$, a. Hence the number of progressions is $2n(n - 1)$. The corresponding number for the sequence $1, 2,..., 2n + 1$, is $2n^2$.

305. The proof is by induction on the number s of curves. For $s = 1$, the conclusion is obvious since there are no points of intersection and the number of regions is 1. Suppose the assertion true for s curves. This means that if the s

curves have n_2 points of intersection of multiplicity 2 (the multiplicity of a point of intersection is the number of curves which intersect at the point), n_3 points of intersection of multiplicity 3, and so on, then the number of regions described in the statement of our problem is $1 + n_2 + 2n_3 + \cdots + rn_r + \cdots$. Now consider an additional curve (this brings the total of curves to $s + 1$) which intersects the s curves in k_2 points of multiplicity 2, k_3 points of multiplicity 3,..., k_{r+1} points of multiplicity $r + 1$,... . Then the new curve intersects the s old curves in $k_2 + k_3 + \cdots + k_{r+1} + \cdots$ points. These points divide the new curve into $k_2 + k_3 + \cdots + k_{r+1} + \cdots$ parts. Each of these parts corresponds to a new region. Hence the number of regions is now equal to

(*) $(1 + n_2 + \cdots + rn_{r+1} + \cdots) + (k_2 + k_3 + \cdots + k_{r+1} + \cdots).$

Observe that if the new curve passes through a point of intersection of the original curves of multiplicity r, then the multiplicity of this point is changed to $r + 1$.

Let n_r' denote the number of points of intersection of multiplicity r in the new system of curves. Then $n_r' = n_r - k_{r+1} + k_r$ (the number of points of multiplicity r in the old system must be decreased by the number of points whose old multiplicity r was changed to $r + 1$ and increased by the number of points whose old multiplicity $r - 1$ was changed to r). But then

$$1 + n_2' + 2n_3' + \cdots + rn_{r+1}' + \cdots$$
$$= 1 + (n_2 - k_3 + k_2) + 2(n_3 - k_4 + k_3) + \cdots + r(n_{r+1} - k_{r+2} + k_{r+1}) + \cdots$$
$$= (1 + n_2 + 2n_3 + \cdots + rn_{r+1} + \cdots) + (k_2 + k_3 + \cdots + k_{r+1} + \cdots).$$

This equality states that $1 + n_2' + 2n_3' + \cdots + rn_{r+1}' + \cdots$ is equal to the number of regions associated with the $s + 1$ curves (see relation (*)). This completes the proof.

306.   The lines of the first pencil divide the plane into $2m$ regions. The first line of the second pencil intersects all the m lines of the first pencil and so contributes $m + 1$ new regions. Each of the remaining $n - 1$ lines of the second pencil intersects the preceding lines in $m + 1$ points and so contributes $m + 2$ new regions. Hence the total number of regions is

$$2m + m + 1 + (n - 1)(m + 2) = nm + 2n + 2m - 1.$$

307.   No, since the number of connecting lines would be $77 \cdot 15/2$.

308.   The sum of the coefficients is equal to the value of our expression for $x = y = z = 1$, that is, -1.

309.   The largest number of balls among which there are no 15 balls of one color is 74 (10 white, 10 black, 12 yellow, 14 red, 14 green, and 14 blue). Any selection of 75 or more balls is certain to include 15 balls of one color.

310. We classify the painting schemes in accordance with the number of white faces. There is 1 scheme with no white faces and 1 scheme with 1 white face. There are 2 schemes with 2 white faces: either the 2 white faces have an edge in common or are parallel. There are 2 schemes with 3 white faces: either the 3 white faces have a common vertex or 2 of them are parallel. The remaining cases can be reduced to the cases already considered by changing "white" to "black." In all, we obtain $1 + 1 + 2 + 2 + 2 + 1 + 1 = 10$ geometrically different ways of painting.

311. There is 1 scheme with no white vertices, 1 scheme with 1 white vertex, 3 schemes with 2 white vertices (the 2 white vertices are on the same edge, on the same diagonal of a face, or on the same diagonal of the cube), 3 schemes with 3 white vertices, and 5 schemes with 4 white vertices (see Fig. 39). The remaining

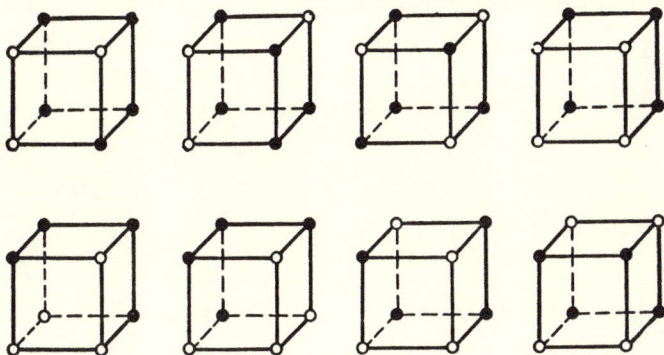

Fig. 39

cases can be reduced to the cases already considered by changing "white" to "black." There are in all $1 + 1 + 3 + 3 + 5 + 3 + 3 + 1 + 1 = 21$ ways of painting.

312. Eleven different cutouts. See Fig. 40.

313. Exactly 4. For proof see Problem 42 in "A Hundred Problems in Elementary Mathematics" by H. Steinhaus published in 1964 by Basic Books, New York.

314. See Problem 44 in "A Hundred Problems in Elementary Mathematics" by H. Steinhaus (see above).*

315. We prove first that after 2^n units of time there remain only 2 particles

* The English translation of the Steinhaus book leaves out this problem and its (very long) solution (Translators).

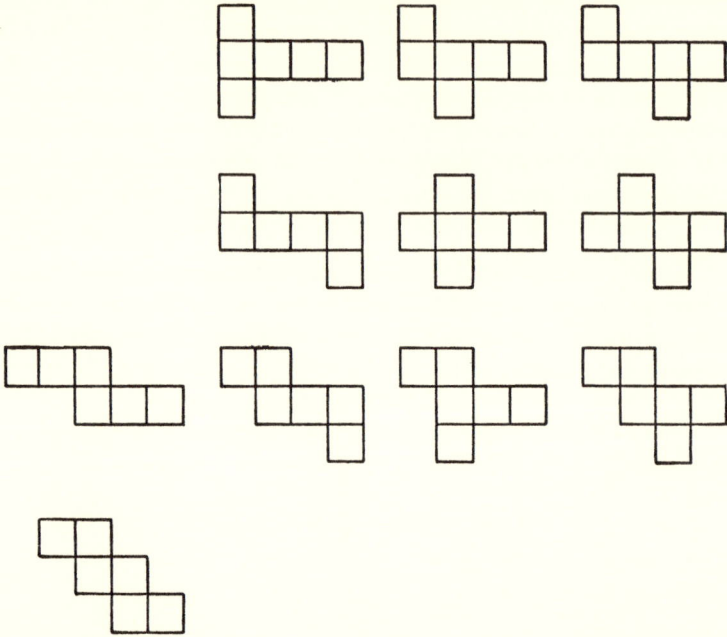

Fig. 40

located at the points with coordinates 2^n and -2^n. For $n = 1$, this is obvious. Suppose it true for $n = k$. This means that after 2^k units of time we have one particle at the point -2^k and one particle at the point 2^k. Note that during the next $2^k - 1$ steps of the process the particles derived from these two "centers" do not interact. By the induction assumption, after 2^k additional steps (that is at time 2^{k+1}) the particle at 2^k will have been replaced by a pair of particles located at 0 and 2^{k+1}. Similarly, the particle at -2^k will have been replaced by a pair of particles located at 0 and -2^{k+1}. Since the two particles at the origin vanish, it follows that after 2^{k+1} units of time we are left with a particle at the point -2^{k+1} and a particle at the point 2^{k+1}. This proves our assertion for $n = k + 1$, and thus for all positive integral n.

In view of the result just proved, after $128 = 2^7$ steps we are left with 2 particles, 1 at the point with coordinate 128 and the other at the point with coordinate -128. After 129 steps we have 4 particles located at the points with coordinates $-129, -127, 127, 129$.

If $n = 2^{k_1} + 2^{k_2} + \cdots + 2^{k_s}$, $k_1 > k_2 > \cdots > k_s$, then we obtain 2^s particles located at the points with coordinates $\pm 2^{k_1} \pm 2^{k_2} \pm \cdots \pm 2^{k_s}$ (all combinations of signs must be included). This assertion is easy to prove by induction on s. We proved it earlier for the case $s = 1$. Suppose that it is true for $s < m$ and that

$n = 2^{k_1} + 2^{k_2} + \cdots + 2^{k_m}$. After the $(n - 2^{k_m})$th step, we obtain 2^{m-1} particles located at the points with coordinates $\pm 2^{k_1} \pm 2^{k_2} \pm \cdots \pm 2^{k_{m-1}}$. The minimal distance between particles is $2^{k_{m-1}+1}$. Hence over the next $2^{k_{m-1}} - 1$ steps, the particles generated by the various "centers" do not interact, and after 2^{k_m} steps each center will have given rise to 2 particles which are removed from it $\pm 2^{k_m}$ units. In other words, we obtain particles at the points with coordinates $\pm 2^{k_1} \pm \cdots \pm 2^{k_m}$. This proves our assertion for $s = m$, and thus for all positive integral s.

316. To decode the word, it suffices to introduce a space before each letter consisting of two symbols. Since the total number of possible spaces is 13 and no space can be introduced after the 11th symbol or after the 12th symbol, it is clear that spaces can be introduced at 11 locations. Also, a space must not be introduce in 2 successive locations. If the word is to contain p two-symbol letters, then we must introduce p spaces. This can be done in C_{12-p}^{p} ways.* It follows that the 12-symbol word can be decoded in

$$C_{12}^{0} + C_{11}^{1} + C_{10}^{2} + C_{5}^{3} + C_{8}^{4} + C_{7}^{5} + C_{6}^{6} = 233$$

ways.

317. There are $8 \cdot 9^{p-1}$ p-digit numbers written without the digit 1. This means that there are

$$8(1 + 9 + 9^2 + 9^3 + 9^4 + 9^5 + 9^6) = 9^7 - 1 = 4{,}782{,}968$$

numbers from 1 to 10,000,000 inclusive written without the digit 1. This is less than $10^7/2$.

318. There are 8 ways of setting down the first 3 symbols of a word. We show that the maximal number of words which agree on the first 3 symbols but have different symbols in at least 3 places is 2. In fact, consider 3 such words written one below the other. In each of the last 4 columns of these 3 words a symbol must appear twice. But then certain 2 of these 3 words must have the same symbol in at least 2 of the last 4 positions. This proves that if a set of words is such that any 2 words are different in at least 3 places, then this set contains at most 16 words. The elements of such a set are written out in Fig. 41.

319. Since p is prime, the rotated pattern coincides with the original pattern if and only if the loop is painted one color. The remaining $n^p - n$ patterns form $(n^p - n)/p$ classes of different patterns with patterns in the same class differing only by a cyclic permutation. In all, there are $(n^p - n)/p + n$ different patterns. Incidentally, our argument proves the so-called little theorem of Fermat, which asserts that for p prime and integral n the number $n^p - n$ is divisible by p.

* To see this, add a space at the end and see Problem 250 (Translators).

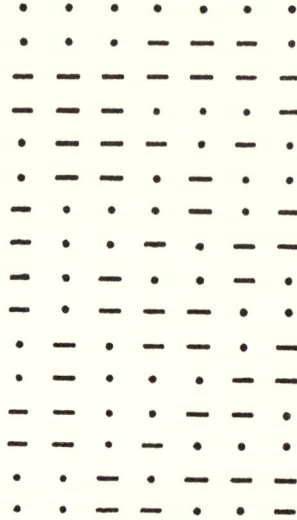

Fig. 41

320. The number 1 must be placed in a corner. Suppose we place 1 in the upper left-hand corner. Then 2 must be placed in the next horizontal or vertical square. Suppose we place 2 in the horizontal square next to 1. Then the numbers 1, 2,..., n must be placed in the first row. After that, $n + 1$ must be placed in the first square of the second row, and so on. Since the position of the number 1 can be chosen in 4 ways and the numbers 1, 2,..., n can occupy either the first row or the first column, the number of ways is 8.

321. Otherwise the population of Moscow would exceed 9,300,000.

322. Every choice of an odd number of objects leaves behind an even number of objects.

323. The number of ways of changing a dollar into 2-cent and 5-cent coins is equal to the number of nonnegative integral solutions of the equation $2x + 5y = 100$. Clearly, y can take on any of the 11 even values from 0 to 20 inclusive. On the other hand, in case of the equation $3x + 5y = 100$, y can take on only the 7 values 2, 5, 8, 11, 14, 17, 20.

324. We must find the number of nonnegative integral solutions of the equation $x + 2y + 5z = 20$, or equivalently, of the inequality $2y + 5z \leqslant 20$. Clearly, z can take on only the values 0, 1, 2, 3, 4. The number of corresponding values of y is 11, 8, 6, 3, and 1. This gives a total of 29 solutions of our original equation.

325. Since $3 = 2 + 1, 4 = 2 + 2, 6 = 5 + 1, 7 = 5 + 2, 8 = 5 + 2 + 1,$ $9 = 5 + 2 + 2,$ our weights suffice for any number of milligrams from 1 to 9 inclusive. The same argument applies to a weight expressed in tens, hundreds, and so on, milligrams.

326. The average value of the last digit is $(0 + 2 + 4)/3 = 2$; of the third digit, $5/2$; of the second digit, $5/2$; of the first digit, 3. The number of our numbers is $5 \cdot 6 \cdot 6 \cdot 3 = 540$. Hence the required sum is

$$540(3000 + 250 + 25 + 2) = 1,769,580.$$

327. The assertion is obvious for $r = 1$. Indeed, if $p \leqslant n$, then the card numbered p is shifted to $2p$, and if $p > n$, then this card is shifted to $2p - 2n - 1$. In either case the new position of our card is the remainder in the division of $2p$ by $2n + 1$. Suppose our assertion true for r, that is, after r shuffles the card numbered p is shifted to position x, where $2^r p = k(2n + 1) + x$. After an additional shufflle, this card will occupy the position y, where $2x = l(2n + 1) + y$, $l = 0$ or 1. But then

$$2^{r+1}p = 2k(2n + 1) + 2x = (2k + l)(2n + 1) + y.$$

where $y < 2^{r+1}p$. Hence y is the remainder in the division of 2^{r+1} by $2n + 1$. This proves our assertion for $r = 1, 2,\ldots$.

328. Follows directly from Problem 327.

329. Follows directly from Problem 327.

330. Indeed, in this case division of $2^x p$ by $2n + 1$ yields the remainder p.

331. Indeed, in this case the card originally in position $2n$ will be topped by $2n - 1$ even-numbered cards.

332. The assertion concerning the card 8 follows from the preceding problem. The remaining assertions can be checked directly.

333. Under each card we write its number after a shuffle:

$$(*)$$

1	2	3	4	5	6	7	8	9	10	11	12	13	14	15	16
9	8	10	7	11	6	12	5	13	4	14	3	15	2	16	1

The permutation (*) tells us what happens to a card after a certain number of shuffles. Thus one shuffle takes 1 into 9, a second shuffle takes it into 13, a third shuffle into 15, a fourth shuffle into 16 and a fifth shuffle takes it back into position 1; this can be briefly summarized in the form of the cycle (1, 9, 13, 15, 16, 1). In addition to this cycle, we have the cycles (2, 8, 5, 11, 14, 2), (3, 10, 4, 7,

12, 3), and the cycle consisting of the number 6. Thus each cycle contains one or five different digits. This implies that after five shuffles each card must return to its original position. Similar reasoning applies to the remaining questions.

334. We are free to paint the squares in the first row in any of the 24 possible ways. After that we are free to paint the unpainted three squares in the first column in any of the possible 6 ways. It is no restriction of generality

white	black	red	blue
black	white	blue	red
red	blue	white	black
blue	red	black	white

to assume that the first row and first column have been painted in the manner shown in the above table. Then the second row can be painted in one of the following 3 ways: black, white, blue, red; black, red, blue, white; black, blue, white, red. Suppose the second row is painted in accordance with the first of these 3 patterns. Then the pattern of the second column is uniquely determined and there remain only 2 ways of painting the remaining 4 squares. The same is true of the other 2 possible ways of painting the second row. In all, there are $4!3! \cdot 2 \cdot 3 = 144$ admissible ways of painting our squares.

335. Consider an arrangement of the 15 boys in 5 rows of 3. Each row of 3 determines 3 unordered pairs (for example, the triple a, b, c determines the 3 pairs ab, ac, bc), and so an arrangement of 15 boys determines 15 pairs none of which must appear in an alternative arrangement. Since there are $C_{15}^2 = 105$ ways of selecting 2 out of 15 boys, the number of admissible arrangements cannot exceed $105/15 = 7$. The following table shows that this maximum can be realized.

klo	ino	jmo	ilm	jln	ijk	kmn
iab	jac	lad	nae	kaf	mag	oah
ncd	mdb	kbc	ocg	mch	lce	icf
mef	keg	ieh	jfb	obe	ofd	jde
jgh	lhf	nfg	khd	idg	nhb	lbg

336. The number $(n^2)!/(n!)^{n+1}$ is an integer because it is the number of ways of dividing n^2 (different) objects into a collection of n unordered batches of n elements each. The number $(mn)!/[(m!)^n n!]$ is an integer because it is the number of ways of dividing mn objects into n unordered batches of m elements each. Similarly, the number $(mn)!/[(n!)^m m!]$ is an integer. As a product of integers, the number $\{(mn)!/[(m!)^{(n+1)/2}(n!)^{(m+1)/2}]\}^2$ must be an integer. Since m and n are odd, the number $(mn)!/[(m!)^{(n+1)/2}(n!)^{(m+1)/2}]$ is a rational number whose square is an integer. This implies that this number must itself be an integer.

337. See p. 57.

338. This number is equal to the coefficient of x^m in the polynomial

$$(x^l + x^{l+1} + \cdots + x^n)^p = x^{lp}(1 - x^{n-l+1})^p (1 - x)^{-p}.$$

Using the binomial expansion, we find that the value of this coefficient is

$$C^m_{m-(l-1)p-1} - C^1_p C^{m+l-n-1}_{m-(l-1)(p-1)-n-1} + C^2_p C^{m+2(l-n-1)}_{m-(l-1)(p-2)-2n-1} + \cdots.$$

339. Let x, y, z be the number of books of the first, second, and third kind, respectively, obtained by the first person. Then $x + y + z = 12$ and $0 \leqslant x \leqslant 7$, $0 \leqslant y \leqslant 8$, $0 \leqslant z \leqslant 9$. The number of admissible integral solutions of our equation is equal to the coefficient of t^{12} in the expansion of

$$(1 + t + \cdots + t^7)(1 + t + \cdots + t^8)(1 + t + \cdots + t^9).$$

This product can be rewritten in the form

$$\frac{(1 - t^8)(1 - t^9)(1 - t^{10})}{(1 - t)^3} = (1 - t^8 - t^9 - t^{10} + t^{17} + \cdots)$$
$$\times (1 + 3t + 6t^2 + 10t^3 + 15t^4 + \cdots + 91t^{12} + \cdots).$$

It is clear that after removing brackets we obtain as the coefficient of t^{12} the number 60. This, then, is the number of ways of dividing the books.

340. Since the number of n-combinations with repetitions of n letters is C^n_{2n-1}, the number of letters in these combinations is nC^n_{2n-1}. Since each letter appears the same number of times, each letter appears C^n_{2n-1} times.

341. The sum of the two different digits on a sign must be 9. Hence the only pairs of different digits on admissible signs are 0, 9; 1,8; 2,7; 3,6; 4,5. Each of these pairs gives rise to 8 admissible signs (for example, the pair 1,8 gives rise to the signs (111; 888), (118; 881), (181; 818), (811; 188), and to 4 more signs with the same numbers in reversed order). Hence there are $5 \cdot 8 = 40$ admissible signs.

342. The required number is $\sum_{p=0}^{m} \sum_{q=0}^{n} P(p, q) - 1$. (We reject the empty arrangement.) Since $\sum_{q=0}^{n} P(p, q) = P(p + 1, q)$, we have

$$\sum_{p=0}^{m} \sum_{q=0}^{n} P(p, q) - 1 = \sum_{p=0}^{m} P(p + 1, n) - 1 = P(m + 1, n + 1) - 2.$$

343. In view of Problem 342, there are $P(k + 1, n + 1) - P(k, n + 1)$ arrangements with exactly k white balls. Hence the required sum is

$$\sum_{k=1}^{m} k[P(k + 1, n + 1) - P(k, n + 1)].$$

This expression can be rewritten as follows:

$$\sum_{k=1}^{m} kP(k + 1, n + 1) - \sum_{k=0}^{m-1} (k + 1)P(k + 1, n + 1)$$

$$= mP(m + 1, n + 1) - \sum_{k=0}^{m-1} P(k + 1, n + 1)$$

$$= mP(m + 1, n + 1) - P(m, n + 2) + 1$$

$$= 1 + \frac{mn + m - 1}{n + 2} P(m + 1, n + 1).$$

The corresponding sum for the black balls is computed in a similar manner.

344. The required number is equal to the sum

$$\sum_{p=0}^{m} \sum_{q=0}^{n} (p + q + 1)P(p, q).$$

Since

$$\sum_{q=0}^{n} (p + q + 1)P(p, q) = (p + 1) \sum_{q=0}^{n} P(p, q) + \sum_{q=1}^{n} qP(p, q)$$

$$= (p + 1) \left[P(p + 1, n) + \sum_{q=1}^{n} P(p + 1, q - 1) \right]$$

$$= (p + 1)[P(p + 1, n) + P(p + 2, n - 1)]$$

$$= (p + 1)P(p + 2, n),$$

this sum is equal to

$$\sum_{p=0}^{m} (p + 1)P(p + 2, n) = \sum_{p=0}^{m} (p + 2)P(p + 2, n) - \sum_{p=0}^{m} P(p + 2, n)$$

$$= (n + 1)P(m + 1, n + 2) - P(m + 2, n + 1) + 1$$

$$= 1 + \frac{mn + m + n}{m + n + 4} P(m + 2, n + 2).$$

345. The required answer is the sum of the answers in Problem 342 and in Problem 344.

346. There are $C_7^2 = 21$ ways of choosing 2 out of 7 people. Since each triple a, b, c determines 3 pairs $((a, b), (a, c), (b, c))$, each of the 21 pairs will be represented exactly once. Since there are 7 people and the total number of guests on the 7 occasions is $3 \cdot 7 = 21$, each person will appear in 3 triples which have exactly this person in common.

The number of ways of selecting 3 triples which have just a certain one of the 7 friends in common is equal to $6!/[3!(2!)^3]$, the number of ways of grouping 6 people into 3 pairs without common elements. After that, the number of ways of selecting 3 triples which have just a certain other one of the 7 friends in common is equal to $4!/[2!(2!)^2] - 1 = 2$ (for example, if the first friend appears in the triples $(1, 2, 3)$, $(1, 4, 5)$, $(1, 6, 7)$, then the second can appear only in the triples $(2, 4, 6)$, $(2, 5, 7)$ or in the triples $(2, 4, 7)$, $(2, 5, 6)$). The triples which have in common just one of the remaining 5 friends are uniquely determined. In view of the possibility of permuting the 7 triples, we obtain the answer

$$\frac{6!}{(2!)^3 3!} \cdot 2 \cdot 7! = 151{,}200.$$

347. Note that $C_5^3 = 10$ and $C_4^3 = 4$. This shows that in order to have a schedule of 7 different triples we must have at least 5 people. Since $C_7^3 = 35$, there are altogether A_{35}^7 ways of making up schedules of 7 different triples. Similarly, since $C_6^3 = 20$, there are A_{20}^7 ways of making up such a schedule if a certain friend is left out. Finally, since $C_5^3 = 10$, there are A_{10}^7 ways of making up such a schedule if certain two friends are left out. Applying the principle of inclusion and exclusion, we see that the number of admissible schedules is $A_{35}^7 - 7A_{20}^7 + 21A_{10}^7$.

348. If we do not reject schedules with a certain person included in all 7 triples, then we have A_{35}^7 schedules. The number of schedules with a certain person included in all 7 triples is $7A_{15}^7$ (there are $C_6^2 = 15$ pairs which do not include this person). Hence the number of admissible schedules is $A_{35}^7 - 7A_{15}^7$.

349.

$$A_n^n + A_n^{n-1} + \cdots + A_n^1 = n! + \frac{n!}{1!} + \frac{n!}{2!} + \cdots + \frac{n!}{(n-1)!}$$

$$= n! \left[2 + \frac{1}{2!} + \cdots + \frac{1}{(n-1)!} \right].$$

On the other hand,

$$en! - 1 = n! \left[2 + \frac{1}{2!} + \cdots + \frac{1}{(n-1)!} \right]$$

$$+ \left[\frac{1}{n+1} + \frac{1}{(n+1)(n+2)} + \frac{1}{(n+1)(n+2)(n+3)} + \cdots \right].$$

For every natural number $n \geqslant 2$,

$$\frac{1}{n+1} + \frac{1}{(n+1)(n+2)} + \frac{1}{(n+1)(n+2)(n+3)} + \cdots$$

$$< \frac{1}{n+1} + \frac{1}{(n+1)^2} + \frac{1}{(n+1)^3} + \cdots = \frac{1}{n}.$$

The assertion of the problem follows.

350. The number of objects in all the samples is

$$nA_n^n + (n-1)A_n^{n-1} + \cdots + A_n^1$$

$$= n! \left[n + \frac{n-1}{1!} + \cdots + \frac{1}{(n-1)!} \right]$$

$$= (n-1)n! \left[\left(1 + 1 + \frac{1}{2!} + \cdots + \frac{1}{(n-1)!} \right) \right.$$

$$\left. + \frac{1}{n-1} \left(1 - \frac{1}{2!} - \frac{2}{3!} - \cdots - \frac{n-2}{(n-1)!} \right) \right].$$

It is easy to see that

$$1 - \frac{1}{2!} - \frac{2}{3!} - \frac{3}{4!} - \cdots - \frac{n-2}{(n-1)!} = \frac{1}{(n-1)!}.$$

Since each object appears the same number of times, each object appears

$$N = (n-1)(n-1)! \left[\left(1 + 1 + \frac{1}{2!} + \cdots + \frac{1}{(n-1)!} \right] + 1 \right.$$

times.

On the other hand,

$$(n-1)(n-1)!\,e = (n-1)(n-1)!$$
$$\times \left[1 + 1 + \frac{1}{2!} + \cdots + \frac{1}{(n-1)!} + \frac{1}{n!} + \frac{1}{(n+1)!} + \cdots\right]$$
$$= (n-1)(n-1)! \left[1 + 1 + \frac{1}{2!} + \cdots + \frac{1}{(n-1)!}\right]$$
$$+ (n-1) \left[\frac{1}{n} + \frac{1}{n(n+1)} + \cdots\right],$$

and therefore

$$N - (n-1)(n-1)!\,e = 1 - (n-1)\left[\frac{1}{n} + \frac{1}{n(n+1)} + \cdots\right]$$
$$= \frac{1}{n}\left[1 - \frac{1}{n+1} - \frac{1}{(n+1)(n+2)} - \cdots\right] < \frac{1}{2}.$$

This proves that $N = E[(n-1)(n-1)!\,e]$, as asserted.

351. See p. 59.

352. One of the 3 people gets n books. These n books can be selected in C_{3n}^n ways. Each of the remaining $2n$ books can be given to either of the remaining 2 people. This means that there are 2^{2n} ways of distributing these books. Since n books can be given to any one of the 3 people, the total number of ways is $3 \cdot 2^{2n} C_{3n}^n$.

353. The number of arrangements in which the letters in a certain k pairs of like letters are not separated is

$$\underbrace{P(2,..., 2,}_{n-k \text{ times}} \underbrace{1,..., 1)}_{k \text{ times}} = 2^{k-n}(2n-k)!$$

These k pairs can be selected in C_n^k ways. Application of the principle of inclusion and exclusion yields the required result.

354. There are $(n+p-k)^r$ distributions in which k people obtain nothing. Now use the principle of inclusion and exclusion to obtain the required result.

355. The number of ways of putting n different objects into r different boxes with no box left empty is $r!\ \Pi_n^r$. This number is the product of $n!$ and the coefficient of x^n in the expansion of $(e^x - 1)^r$ in a power series. It follows that

(*) $$n!\ [1 - \Pi_n^2 + 2!\ \Pi_n^3 - 3!\ \Pi_n^4 + \cdots]$$

is the coefficient of x^n in the power series expansion of the series

$$(e^x - 1) - \tfrac{1}{2}(e^x - 1)^2 + \tfrac{1}{3}(e^x - 1)^3 - \tfrac{1}{4}(e^x - 1)^4 + \cdots.$$

Since

$$x - \tfrac{1}{2}x^2 + \tfrac{1}{3}x^3 - \tfrac{1}{4}x^4 + \cdots = \ln(1 + x).$$

the sum of this series is $ln[1 + (e^x - 1)] = x$. Hence for $n > 1$, the expression (*) must vanish.

356. There are C_{kn}^n ways of withdrawing n objects from kn different objects. Hence the answer is

$$C_{2n}^n C_{3n}^n \cdots C_{mn}^n = \frac{(mn)!}{(n!)^m}.$$

357. We must prove the inequality

$$C_{2n+r}^n C_{2n-r}^n \leqslant (C_{2n}^n)^2.$$

This is equivalent to the inequality

$$\frac{(2n + r)(2n + r - 1) \cdots (2n + 1)}{(n + r)(n + r - 1) \cdots (n + 1)} \leqslant \frac{2n(2n - 1) \cdots (2n - r + 1)}{n(n - 1) \cdots (n - r + 1)}.$$

The latter inequality follows from the fact that for $0 \leqslant k < n$ we have $(2n + k)/(n + k) < (2n - k)/(n - k.)$

358. We compute the angle sum of all the triangles. The sum of the angles with vertex at an interior point is 360°. Since the number of interior points is 500, the corresponding angles add up to $360° \cdot 500$. The sum of the angles whose vertices coincide with the vertices of the 1000-gon is equal to the sum of the interior angles of the 1000-gon, that is, $180° \cdot 998$. Thus the angle sum of our triangles is $180° \cdot 1998$. Hence the required number of triangles 1998.

359. Each player plays 4 out of the total of 5 games. Suppose that in the first game the pair (a, c) plays against the pair (b, d). Then in the next 3 games the partners of a are, respectively, b, d, e. Also, a does not play in the fifth game. Player e plays in all games except the first. In the second and third games, e opposes a. The partner of e in the second game can be c or d, and his partner in the third game can be b or c. If the partner of e in the second game is d, then his partner in the third game must be c (otherwise c would miss two games); but then d must not play in the fourth game, and b and c are partners (in the fourth game). This implies that in the fifth game we have the pairs (c, d) and (e, b). If in the second game we choose c (as a partner of e), then in the third game we must choose b (otherwise e and c would be partners twice), in the fourth game we must choose the pair (c, d), and in the fifth game the pair (b, c) must oppose the pair (d, e). We see that after the choice of pairs in the first game we can continue the match in 2 ways. If we consider order in the last 4 games, then we have $2 \cdot 4! = 48$ possibilities. If we note that there are 15 ways of choosing the players for the first game (15 is the number of ways of choosing 2 pairs and an extra out of 5 players), then we obtain a total of 720 possibilities. If the order of the games is disregarded, then there remain 6 possibilities.

360. The number of such polygonal lines is $(C_{2n}^n)^2$ (see p. 112).

361. We number the horizontal lines and the vertical lines using the numbers $1, 2,..., n$. This numbering assigns a pair of coordinates to each of the n^2 points of intersection of the two families of lines. Now let $(a_1 ,..., a_n)$ and $(b_1 ,..., b_n)$ be two permutations of the numbers $1, 2,..., n$. With these permutations we form the following two sequences of $2n$ points each:

$$(a_1 , b_1), \quad (a_1 , b_2), \quad (a_2 , b_2), \quad (a_2 , b_3),..., (a_n , b_1)$$

and

$$(a_1 , b_1), \quad (a_2 , b_1), \quad (a_2 , b_2), \quad (a_3 , b_2),..., (a_1 , b_n).$$

Each of these sequences of points defines an admissible polygonal line whose vertices are the points of the defining sequence. Since there are $(n!)^2$ pairs of permutations $(a_1 ,..., a_n)$ and $(b_1 ,..., b_n)$ of $1, 2,..., n$, we obtain $2(n!)^2$ admissible curves. However, we can traverse each curve in one of two directions starting at any one of it $2n$ points. This reduces the number of curves to $2(n!)^2/4n = (n!)^2/2n$.

362. We divide the rattles into classes and put into the mth class those rattles for which the smallest number of blue balls between two red ones is m. For $m = 0$, we have 4 types of rattles (the third red ball adjoins the other two or is separated from them by 1, 2, or 3 blue balls). For $m = 1$ two red balls are separated by a blue one. The remaining red ball is separated from its nearest red neighbor by means of 1, 2, or 3 blue balls. Hence for $m = 1$, there are 3 types of rattles. Finally, for $m = 2$ there is just 1 type of rattle. Hence the total number of types of rattles is 8.

363. Suppose person X in the group has m acquaintances $a_1 ,..., a_m$. It is clear from the problem that any two of the people $a_1 ,..., a_m$ are strangers (since they have X as a common acquaintance). Hence each pair (a_i , a_j) has an additional acquaintance other than X who is a stranger to X. Also, the additional acquaintances associated with different pairs must be different (if someone were a common acquaintance of different pairs (a_k , a_l) and (a_i , a_j), then that someone and X would share at least three common acquaintances). Hence the number of people who are strangers to X is at least as large as the number of pairs of people selected from among $a_1 ,..., a_m$, that is, at least equal to C_m^2.

On the other hand, each person who is a stranger to X shares with X exactly two common acquaintances who necessarily belong to the group $a_1,...,a_m$. Also, the pairs of common acquaintances determined by different strangers to X are different (if the same pair (a_i, a_j) were determined by two people, then a_i and a_j would have more than two acquaintances, for they are acquainted with X). It follows that the number of strangers to X is not greater than C_m^2, and thus equal to $C_m^2 = m(m-1)/2$. But then the number n of people in the group is $1 + m + m(m-1)/2$. If we consider the equality $n = 1 + m + m(m-1)/2$ as a quadratic equation in m, then we see that this equation has a single positive root. This shows that the number m of acquaintances of different people has a fixed value.

364. We can easily verify the fact that permuting two neighboring letters A and B has no effect on the product (it suffices to consider arrangements $AABA$, $BABB$, and $AABB$). This means that we may suppose all letters A to be grouped together and all letters B to be grouped together. But in this case the assertion is obvious.

365. There is a rook in every row and column. Hence each of the numbers a, b, c, d, e, f, g, h, and each of the numbers $1, 2, 3, 4, 5, 6, 7, 8$ enters our product exactly once. It follows that the value of this product is $8!abcdefgh$.

366. Suppose 5 members of the organizing committee meet. They do not have a key to at least one lock. Also, each of the remaining members must have a key to any such lock. Since this is true for every group of 5 members, the number of locks is $C_{11}^5 = 462$. Since there are 6 keys to each lock, the number of keys is $462 \cdot 6 = 2772$, and each member of the organizing committee has $2772/11 = 252$ keys.

If the committee consists of n members and the presence of m members is necessary and sufficient for opening the safe, then the number of locks is C_n^{m-1} and the number of keys kept by each member of the committee is $[(n - m + 1)/n] \, C_n^{m-1}$.

367. We find the maximal length of a chain with the property that after k of its links are opened the resulting pieces can be grouped so as to give any weight from 1 to n. To do this we consider the optimal location of the links to be opened. If we open k links, then we can group them so as to get any weight from 1 to k but not $k + 1$. Hence it would be nice to have a piece of chain consisting of $k + 1$ links. Then we can put together any weight from 1 to $2k + 1$. We see that next in line of handy weights are the weights $2(k + 1), 4(k + 1),..., 2^k(k + 1)$. With all of these weights we can obtain any weight from 1 to

$$n = k + [(k + 1) + 2(k + 1) + 4(k + 1) + \cdots + 2^k(k + 1)]$$
$$= k + (k + 1)(2^{k+1} - 1) = 2^{k+1}(k + 1) - 1.$$

We see that if $2^k k \leqslant n < 2^{k+1}(k+1)$, then we can manage by opening k links but not by opening $k - 1$ links. In particular, since $2^3 \cdot 3 \leqslant 60 \leqslant 2^4 4 - 1$, it suffices to open up 3 links so as to obtain pieces of chain weighing 4, 8, 16, and 29 oz.

If we use a balance, then to k opened links we add a piece of chain weighing $2k + 1$ (by putting this weight on one scale and an appropriate number of the open links on the other scale, we can obtain any weight from $k + 1$ to $2k$, and by putting this weight and an appropriate number of open links on the same scale, we can obtain any weight from $2k + 1$ to $3k + 1$). The subsequent pieces of chain should weigh $3(2k + 1)$, $9(2k + 1)$,..., $3^k(2k + 1)$. Then we can obtain any weight from 1 to

$$k + [(2k + 1) + 3(2k + 1) + \cdots + 3^k(2k + 1)] = \tfrac{1}{2}[(2k + 1) \, 3^{k+1} - 1].$$

In particular, for a chain weighing 60 oz. we must open 2 links and obtain pieces of chain weighing 5, 15, and 38 oz.

368. If the remainders in the division of x by 7 are, respectively, 0, 1, 2, 3, 4, 5, 6, then the corresponding remainders in the division of x^2 by 7 are 0, 1, 4, 2, 2, 4, 1. Hence $x^2 + y^2$ is divisible by 49 if and only if x and y are divisible by 7. It follows that the number of ordered admissible pairs is $[E(1000/7)]^2 = 142^2 = 20{,}164$. The corresponding number of unordered pairs is $\bar{C}^2_{142} = 10{,}153$.

369. If the number is $10a + b$, then reversing the order of the digits yields the number $10b + a$. The sum of these two numbers is $11(a + b)$. Since this is a perfect square and $2 \leqslant a + b \leqslant 18$, it follows that $a + b = 11$. This implies the following 8 possibilities: 29, 38, 47, 56, 65, 74, 83, 92.

370. Three digits are arbitrary and the fourth digit can be selected in one of two ways (depending on the remainder in the division of the 3-digit number composed of the selected digits). This means that after the choice of one of the 4 digits the remaining digits can be selected in $6^2 \cdot 2 = 72$ ways. It follows that the sum of the digits in the first column is $72(1 + 2 + 3 + 4 + 5 + 6) = 1512$ and the sum of all the numbers is

$$1512 + 15{,}120 + 151{,}200 + 1{,}512{,}000 = 1{,}679{,}832.$$

371. The last digit is 0, 2, or 4. Each of those 3 choices involves $5 \cdot 6 \cdot 6 = 180$ choices for the remaining digits. Hence the first column in our sum is $(0 + 2 + 4) \cdot 180 = 1080$. The corresponding count for the second column is $(0 + 1 + 2 + 3 + 4 + 5) \, 900 = 13{,}500$, for the third column 135,000, and for the last column $(1 + 2 + 3 + 4 + 5) \, 108{,}000 = 1{,}620{,}000$. Hence the required sum is 1,769,580.

372. The equation $x + y = k$ has $k - 1$ integral solutions with $1 \leqslant x$, $1 \leqslant y$. Hence the inequality $|x| + |y| \leqslant 1000$ has

$$4 \sum_{k=2}^{1000} (k - 1) = 1{,}998{,}000$$

solutions with $x \neq 0$, $y \neq 0$. In addition there are 3996 solutions with just one nonzero entry and 1 solution with $x = 0$, $y = 0$. The total number of solutions is 2,001,997.

373. We call the polygons in the first class *polygons with A_1* and the remaining polygons, *polygons without A_1*. By adding A_1 to the vertices of a polygon without A_1, we obtain a polygon with A_1; this establishes a one-to-one correspondence between the polygons without A_1 and some of the polygons with A_1. This correspondence does not involve triangles with A_1. Hence the class of polygons with A_1 is larger than the class of polygons without A_1.

374. The color of the squares reached by a knight in an even number of moves is the same as the color of the square originally occupied by the knight. We find it convenient to rotate the board through 45° and to represent only squares of the relevant color with each square replaced by its center. Then the squares our knight can reach in 2 moves are represented by the pattern of black dots shown in Fig. 42. The number of such squares is 33. Each of these squares

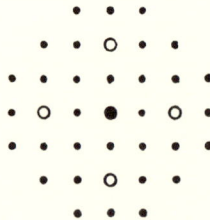

Fig. 42

is the center of a similar pattern which represents the squares reached by the knight in the next 2 moves starting at the square in question. The union of these patterns is shown in Fig. 43. It consists of a square containing $9^2 = 81$ dots and 4 trapezoids each of which consists of $7 + 5 = 12$ dots. In all we have $81 + 4 \cdot 12 = 129$ dots.

After $2n$ moves, we obtain a pattern consisting of a square with side $4n$ which contains $(4n + 1)^2$ dots and 4 trapezoids each of which contains

$$(4n - 1) + (4n - 3) + \cdots + (2n + 1) = 3n^2$$

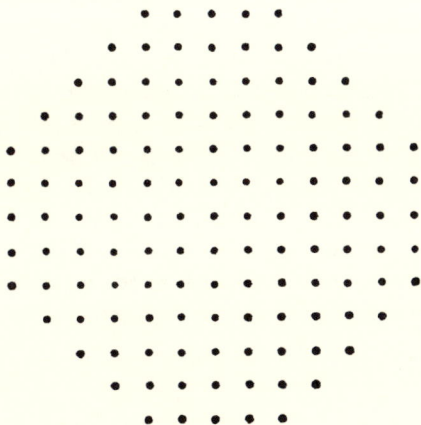

Fig. 43

dots. In all we obtain

$$12n^2 + (4n + 1)^2 = 28n^2 + 8n + 1$$

dots. In other words, in $2n$ moves, $n > 1$, the knight can reach one of $28n^2 + 8n + 1$ squares.

375. The set of triples which have in common an element a is an admissible set of triples and contains $C_{1954}^2 = 1,907,481$ triples. We show that it is impossible to construct a larger admissible set of triples. The proof is by contradiction. Suppose there is an admissible set containing $N > C_{1954}^2$ triples and that (a, b, c) is one of them. Since each of the remaining $N - 1$ triples must have an element in common with the triple (a, b, c), there are at least $(N - 1)/3$ triples which contain, say, the element a; note that $(N - 1)/3 > 635,826$. Among our $N - 1$ triples there are at most 3906 triples which contain in addition to a either b or c. Hence our $N - 1$ triples must contain a triple (a, d, e), with $d \neq b$ and $e \neq c$. Similarly, it must be possible to find triples (a, f, g) and (a, h, j) with f and g different from b, c, d, e, and h and j different from b, c, d, e, f, g.

Each of our N triples must have at least one element in common with each of the 4 triples (a, b, c), (a, d, e), (a, f, g), (a, h, j). If a were not such a common element, then we would have a triple with four different elements. Hence a is common to all of our N triples. Since $N > C_{1954}^2$ we obtain a contradiction.

376. Our number consists of

$$9 + 2 \cdot 90 + 3 \cdot 900 + \cdots + 8 \cdot 90,000,000 + 9$$

digits. We compute the number of zeros in the sequence $1, 2,..., 10^9$. We write each of the numbers from 1 to $10^9 - 1$ as a 9-digit number by preceding the numbers with less than nine digits with an appropriate number of zeros (for example, we write 000000003 instead of 3), and replace 10^9 with 000000000. As a result we obtain $9 \cdot 10^9$ digits and each digit appears with the same frequency. This means that our new sequence contains $9 \cdot 10^8$ zeros. This figure includes $8 \cdot 9$ zeros added to the 1-digit numbers of the original sequence, $7 \cdot 90$ zeros added to the 2-digit numbers of the original sequence, and so on. If we delete these added zeros, then we obtain $9 \cdot 10^8 - 8 \cdot 9 - 7 \cdot 90 - \cdots - 9 \cdot 10^7$ zeros. It is easy to see that this number is equal to the number $2 \cdot 9 + 2 \cdot 90 + \cdots + 8 \cdot 9 \cdot 10^7$. This proves the assertion of the problem.

377. If the sum of the first 2 digits is k, then for $k \leqslant 9$ we have $(k + 1)^2$ admissible numbers, and for $k > 9$ we have $(19 - k)^2$ admissible numbers. Summing over k from $k = 1$ to $k = 18$, we obtain the answer

$$2(1^2 + \cdots + 9^2) + 10^2 = 670.$$

378. Let A_a denote the set of subjects in which student a received grades of "very good." Each of these sets contains at most $2n$ elements. The statement of our problem implies that two such sets are never comparable, that is, neither of the two sets is contained in the other. We separate these sets into classes and put in the kth class all of the sets consisting of k elements. Let r denote the least value of k. We are about to show that if $r < n$, then we can replace our collection of sets with another collection of sets such that:

(a)   No two sets in the new collection are comparable;
(b)   The new collection contains more sets than the old one;
(c)   The new minimal value of k is $r + 1$.

To do this we form out of each set in the rth class of the old collection $2n - r$ new sets by adding to it in all possible ways one of the $2n - r$ elements not in it. We leave the remaining sets of the old collection alone. It is clear that the minimal value of k for the collection of sets obtained in this way is $r + 1$. Also, it is not difficult to see that any two sets in the new collection are not comparable. In fact, if an unchanged set B contained a new set A', then it would certainly contain the set A in the rth class from which A' arose by the addition of an element, and this would contradict the nature of the old collection. Conversely, if a new set A' obtained from an old set A by the addition of an element x coincided with an old set B, then B would contain A, contrary to the nature of the old collection.

It remains to show that the new collection contains more sets than the old collection. We noted above that each set A in the old rth class gave rise to $2n - r$ new sets. Some of these sets could easily coincide (for example, by adding c to (a, b) and a to (b, c), we obtain twice the set (a, b, c)). Nevertheless, a set of $r + 1$ elements can arise from a set of r elements in no more than $r + 1$ ways. Hence, if the old rth class contained m elements and if they gave rise to p new sets, then $m(2n - r) \leqslant p(r + 1)$. Since $r < n$, $2n - r > r + 1$. Hence, $m < p$, that is, the new collection contains more sets than the old collection.

Repetition of this procedure enables us to replace all sets containing fewer than n elements with sets containing n elements in such a way that condition (a) above remains in force and the number of new sets is greater than the number of old sets. In a similar way, it is possible to get rid of sets with more than n elements (by successively replacing them with sets obtained by deletion of an element). In the end we obtain a collection of sets consisting of n elements each and containing more sets than the original collection. Since the number of n-combinations of $2n$ objects is C_{2n}^n , the number of sets in our newest collection is $\leqslant C_{2n}^n$. But then the number of sets in our original collection is $< C_{2n}^n$.

379. We refer to the first m elements as elements of the first kind, and to the last n elements as elements of the second kind. We separate all r-samples of $m + n$ elements into classes and put in the kth class all those r-samples of $m + n$ elements which contain exactly k elements of the first kind. We claim that the kth class contain $C_r^k A_m^k A_n^{r-k}$ samples. To see this, note that there are C_r^k ways of selecting the positions of the elements of the first kind, A_m^k ways of filling the selected positions with elements of the first kind, and A_n^{r-k} ways of filling the remaining $r - k$ positions with elements of the second kind. It follows that the number of r-samples of $m + n$ elements is $\sum_{k=0}^r C_r^k A_m^k A_n^{r-k}$ or, to use the notation in the problem, $\sum_{k=0}^r C_r^k M_k N_{r-k}$. But this is precisely the result of expanding $(M + N)^r$ and changing exponents to indices.

The following is another way of computing the number of samples in the kth class: We select k elements of the first kind and $r - k$ elements of the second kind and permute these r elements in all possible ways. This can be done in

$$P(k, r - k) A_m^k A_n^{r-k} = C_r^k A_m^k A_n^{r-k}$$

ways.

380. The exponent 8 can be obtained from the exponents 2 and 3 in the following ways: $8 = 2 + 2 + 2 + 2 = 2 + 3 + 3$. Hence $x^8 = 1^5 (x^2)^4 = 1^6 (x^2)^1 (x^3)^2$. The coefficients associated with these two representations of x^8 are $P(5, 4) = 126$ and $P(6, 1, 2) = 252$ and their sum is 378 (see the multinomial expansion formula, p. 154).

381. We have

$$(1 + x)^k + \cdots + (1 + x)^n = \frac{(1 + x)^{n+1} - (1 + x)^k}{x} .$$

Hence the coefficient of x^m is $C_{n+1}^{m+1} - C_k^{m+1}$ for $m < k$, and C_{n+1}^{m+1} for $m \geqslant k$.

382. We have $17 = 7 + 5 + 5$. On the other hand, 18 is not decomposible into a sum of multiples of 5 and 7. Hence the coefficient of x^{17} is $C_{20}^1 C_{19}^2 = 3420$ and the coefficient of x^{18} is zero.

383. We have $17 = 2 + 2 + 2 + 2 + 2 + 2 + 2 + 3 = 2 + 2 + 2 + 2 + 3 + 3 + 3 = 2 + 3 + 3 + 3 + 3 + 3$. It follows that the coefficient of x^{17} in the expansion of $(1 + x^2 - x^3)^{1000}$ is

$$-C_{1000}^7 C_{993}^1 - C_{1000}^4 C_{996}^3 - C_{1000}^1 C_{999}^5,$$

and in the expansion of $(1 - x^2 + x^3)^{1000}$ it is

$$-C_{1000}^7 C_{993}^1 + C_{1000}^4 C_{996}^3 - C_{1000}^1 C_{999}^5.$$

384. Let

$$(*) \qquad (1 + x + x^2)^n = a_0 + a_1 x + a_2 x^2 + \cdots + a_{2n} x^{2n}.$$

We show that $a_k = a_{2n-k}$. We put $x = 1/y$ and multiply both sides of the equality by y^{2n}. Then we obtain the equality

$$(**) \qquad (y^2 + y + 1)^n = a_0 y^{2n} + a_1 y^{2n-1} + \cdots + a_{2n}.$$

Comparison of $(*)$ and $(**)$ shows that $a_k = a_{2n-k}$, as asserted.

Next we put $-x$ in place of x and obtain

$$(***) \qquad (1 - x + x^2)^n = a_0 - a_1 x + a_2 x^2 - \cdots + a_{2n} x^{2n}.$$

Multiplying $(*)$ and $(***)$, we obtain

$$(****) \quad (1 + x^2 + x^4)^n = \sum_{k=0}^{4n} (-1)^k (a_0 a_k - a_1 a_{k-1} + \cdots + a_k a_0) x^k.$$

It is clear that the expansion of the left-hand side of this equality contains only even powers of x, so that the coefficient of x^{2n-1} is zero. On the right-hand side of our equality the coefficient of x^{2n-1} is

$$-(a_0 a_{2n-1} - a_1 a_{2n-2} + a_2 a_{2n-3} - \cdots - a_{2n-1} a_0)$$
$$= -(a_0 a_1 - a_1 a_2 + a_2 a_3 - \cdots - a_{2n-1} a_{2n}).$$

This proves (a).

Observe that in view of $(*)$ we can write the expansion $(****)$ in the form

$$(1 + x^2 + x^4)^n = a_0 + a_1 x^2 + a_2 x^4 + \cdots + a_{2n} x^{4n}.$$

Hence the coefficient of x^{2n} in this expansion is a_n. On the other hand, the expansion (****) shows that this coefficient is equal to

$$a_0 a_{2n} - a_1 a_{2n-1} + a_2 a_{2n-2} - \cdots + a_{2n} a_0 = 2a_0^2 - 2a_1^2 + 2a_2^2 - \cdots + (-1)^n a_n^2.$$

This implies the equality (b).

We rewrite (*) in the form

$$(1 - x^3)^n = (1 - x)^n (a_0 + a_1 x + a_2 x^2 + \cdots + a_{2n} x^{2n}).$$

This equality implies the equality

$$1 - C_n^1 x^3 + C_n^2 x^6 - \cdots + (-1)^n C_n^n x^{3n}$$

$$= (1 - C_n^1 x + C_n^2 x^2 - \cdots + (-1)^n C_n^n x^n)(a_0 + a_1 x + a_2 x^2 + \cdots + a_{2n} x^{2n}).$$

If r is not divisible by 3, then the coefficient of x^r on the right-hand side of this equality is zero and on the left-hand side it is

$$a_r - C_n^1 a_{r-1} + C_n^2 a_{r-2} - \cdots + (-1)^r C_n^r a_0.$$

It follows that this expression has the value 0 if r is not divisible by 3 and the value $(-1)^k C_n^k$ if $r = 3k$. This proves (c).

If we put $x = 1$ in (*), then we obtain

$$a_0 + a_1 + a_2 + \cdots + a_{2n} = 3^n.$$

If we put $x = 1$ in (***) we obtain

$$a_0 - a_1 + a_2 - \cdots + a_{2n} = 1.$$

Addition and subtraction of these equalities yields the relations (d).

385. There are C_n^1 terms of the form x_k^3, $2C_n^2$ terms of the form $x_j^2 x_k$, $j \neq k$, and C_n^3 terms of the form $x_i x_j x_k$, $i \neq j$, $i \neq k$, $j \neq k$; in all, $C_n^1 + 2C_n^2 + C_n^3$ terms.

386. We have

$$(1 + x + \cdots + x^{n-1})^2 = \frac{(x^n - 1)^2}{(x - 1)^2} = (x^n - 1)^2 (x - 1)^{-2}$$

$$= (x^{2n} - 2x^n + 1)(1 + 2x + 3x^2 + \cdots + mx^{m-1} + \cdots).$$

Hence the coefficient of x^k is $k + 1$, if $0 \leqslant k \leqslant n - 1$, and $2n - k - 1$, if $n \leqslant k \leqslant 2n - 2$. In either case, the coefficient in question is equal to $n - |n - k - 1|$.

387. Since

$$C_{n+1}^{r+1} = \frac{n+1}{r+1}\, C_n^r, \qquad C_n^r = \frac{n}{r}\, C_{n-1}^{r-1},$$

we can write

$$\frac{\dfrac{n}{r}\left(\dfrac{n+1}{r+1}-1\right)(C_{n-1}^{r-1})^2}{\left(\dfrac{n^2}{r^2}-\dfrac{(n+1)n}{(r+1)r}\right)(C_{n-1}^{r-1})^2} = \frac{\dfrac{n(n-r)}{r(r+1)}}{\dfrac{n(n-r)}{r^2(r+1)}} = r.$$

388. The number of 3-samples of n kinds of elements is n^3. We separate these samples into classes and put in the kth class all 3 samples with exactly k different types of elements. The number of samples in the first class is C_n^1, the number of samples in the second class is $6C_n^2$ (there are n ways of choosing the repeated element, $n-1$ ways of choosing the remaining element, and there are three ways of permuting these elements), and the number of samples in the third class is $A_n^3 = 6C_n^3$. The total number of these samples is $C_n^1 + 6C_n^2 + 6C_n^3$. This proves the first relation. To prove the second relation we use a similar approach except that now we put in one class permutations with repetitions with at least one element of a fixed type. Then we see that

$$(n+1)^3 - n^3 = 1 + 6C_n^1 + 6C_n^2.$$

Adding these two equalities we obtain the second relation.

389. The proof is similar to that in Problem 388 except that we consider 4-samples of elements of n kinds.

390. Consider the equality

$$\left(-\frac{1}{2}+i\,\frac{\sqrt{3}}{2}\right)^n = \left(\cos\frac{2\pi}{3}+i\sin\frac{2\pi}{3}\right)^n = \cos\frac{2n\pi}{3}+i\sin\frac{2n\pi}{3}.$$

By the binomial expansion, we have

$$\frac{(-1)^n}{2^n}\{1 + C_n^1(-i\,\sqrt{3}) + C_n^2(-i\,\sqrt{3})^2 + C_n^3(-i\,\sqrt{3})^3 + \cdots\}$$

$$= \frac{(-1)^n}{2^n}\{1 - 3C_n^2 + 9C_n^4 - \cdots - i\,\sqrt{3}[C_n^1 - 3C_n^3 + \cdots]\}.$$

Equating corresponding real and corresponding imaginary coefficients, we obtain the required relations.

391. We consider the identity

$$(1+x)^n = C_n^0 + C_n^1 x + C_n^2 x^2 + C_n^3 x^3 + \cdots + C_n^n x^n$$

and put for x successively the values $x = 1, \varepsilon, \varepsilon^2$, where

$$\varepsilon = \cos \frac{2\pi}{3} + i \sin \frac{2\pi}{3},$$

so that $\varepsilon^2 + \varepsilon + 1 = 0$. Then we obtain the relations

$$2^n = C_n^0 + C_n^1 + C_n^2 + \cdots + C_n^n,$$

$$(1 + \varepsilon)^n = C_n^0 + C_n^1 \varepsilon + C_n^2 \varepsilon^2 + \cdots + C_n^n \varepsilon^n,$$

$$(1 + \varepsilon^2)^n = C_n^0 + C_n^1 \varepsilon^2 + C_n^2 \varepsilon^4 + \cdots + C_n^n \varepsilon^{2n}.$$

If k is not divisible by 3, then $1 + \varepsilon^k + \varepsilon^{2k} = 0$, and if k is divisible by 3, then $1 + \varepsilon^k + \varepsilon^{2k} = 3$. Therefore,

$$2^n + (1 + \varepsilon)^n + (1 + \varepsilon^2)^n = 3(C_n^0 + C_n^3 + C_n^6 + \cdots).$$

Since

$$1 + \varepsilon = -\varepsilon^2 = -\left(\cos \frac{2\pi n}{3} + i \sin \frac{2\pi n}{3}\right) = \cos \frac{\pi}{3} + i \sin \frac{\pi}{3},$$

$$1 + \varepsilon^2 = -\varepsilon = \cos \frac{\pi}{3} - i \sin \frac{\pi}{3},$$

it follows that

$$2^n + (1 + \varepsilon)^n + (1 + \varepsilon^2)^n = 2^n + 2 \cos \frac{n\pi}{3}.$$

Hence

$$C_n^0 + C_n^3 + C_n^6 + \cdots = \frac{1}{3}\left(2^n + 2 \cos \frac{n\pi}{3}\right).$$

Relations (b) and (c) are proved in a similar manner by considering the sums

$$2^n + \varepsilon(1 + \varepsilon)^n + \varepsilon^2(1 + \varepsilon^2)^n, \qquad 2^n + \varepsilon^2(1 + \varepsilon)^n + \varepsilon(1 + \varepsilon^2)^n.$$

Relation (d) is proved in the same way be considering the expression $(1 + i)^n$.

392. We have

$$(1 + x)^n + (1 - x)^n = 2 \sum_{k=0}^{E(n/2)} C_n^{2k} x^{2k}.$$

The coefficients of this polynomial are positive. Hence its values for $|x| < 1$ are dominated by 2^n, its value for $x = 1$.

393. We have

$$\sum_{x=0}^{n} \frac{n!\,(m-x)!}{m!\,(n-x)!} = \frac{1}{C_m^n} \sum_{x=0}^{n} C_{m-x}^{m-n} = \frac{C_{m+1}^{m-n+1}}{C_m^n} = \frac{m+1}{m-n+1}$$

and

$$\sum_{x=0}^{n} \frac{C_n^x C_n^r}{C_{2n}^{x+r}} = \frac{n!\,C_n^r}{(2n)!} \sum_{x=0}^{n} \frac{(x+r)!\,(2n-x-r)!}{x!\,(n-x)!}$$

$$= \frac{(n!)^2}{(2n)!} \sum_{x=0}^{n} C_{x+r}^r C_{2n-x-r}^{n-r} = \frac{(n!)^2}{(2n)!} C_{2n+1}^{n+1} = \frac{2n+1}{n+1}.$$

394. The sum on the left-hand side reduces to

$$\sum_{k=1}^{n} C_{m+k-1}^k = C_{m+n}^m - 1.$$

This is also the value of the sum of the right-hand side.

395. We have

$$\sum_{x=1}^{n} \frac{C_{n-1}^{x-}}{C_{2n-1}^x} = \frac{(n-1)!}{(2n-1)!} \sum_{x=1}^{n} \frac{x(2n-x-1)!}{(n-x)!}$$

$$= \frac{2n}{(2n-1)C_{2n-2}^{n-1}} \sum_{x=1}^{n} C_{2n-x-1}^{n-1} - \frac{1}{C_{2n-1}^{n-1}} \sum_{x=1}^{n} C_{2n-x}^n$$

$$= \frac{2nC_{2n-1}^{n-1}}{(2n-2)C_{2n-2}^{n-1}} - \frac{C_{2n}^{n-1}}{C_{2n-1}^{n-1}} = \frac{2}{n+1}.$$

396. We have

$$\sum_{x=1}^{n} \frac{C_{n-1}^{x-1}}{C_{n+q}^x} = \frac{(n-1)!}{(n+q)!} \sum_{x=1}^{n} \frac{x(n+q-x)!}{(n-x)!}$$

$$= \frac{(n+q+1)(n-1)!\,q!}{(n+q)!} \sum_{x=1}^{n} C_{n+q-x}^{n-x}$$

$$- \frac{(n-1)!\,(q+1)!}{(n+q)!} \sum_{x=1}^{n} C_{n+q-x+1}^{n-x}$$

$$= \frac{(n+q+1)(n-1)!\,q!}{(n+q)!} C_{n+q}^{n-1} - \frac{(n-1)!\,(q+1)!}{(n+q)!} C_{n+q+1}^{n-1}$$

$$= \frac{n+q+1}{q+1} - \frac{n+q+1}{q+2} = \frac{n+q+1}{(q+1)(q+2)}.$$

397. We have

$$\sum_{x=1}^{n} \frac{C_{n-2}^{x-2}}{C_{n+q}^{x}} = \frac{(n-2)!}{(n+q)!} \sum_{x=1}^{n} \frac{x(x-1)(n+q-x)!}{(n-x)!}.$$

Furthermore, using the identity

$$x(x-1) = (n+q-x+1)(n+q-x+2) \\ + (n+q+1)[n+q-2(n+q-x+1)].$$

we find that our sum is equal to

$$\frac{(n-2)!}{(n+q)!} \left[(q+2)! \sum_{x=1}^{n} C_{n+q-x+2}^{n-x} \right.$$

$$- 2(n+q+1)(q+1)! \sum_{x=1}^{n} C_{n+q-x+1}^{n-x}$$

$$+ (n+q)(n+q+1)q! \sum_{x=1}^{n} C_{n+q-x}^{n-x} \Bigg]$$

$$= \frac{(n-2)! \, q!}{(n+q)!} \left[(q+1)(q+2)C_{n+q+2}^{n-1} \right.$$

$$- 2(n+q+1)(q+1)C_{n+q+1}^{n-1} + (n+q)(n+q+1)C_{n+q}^{n-1}].$$

Substitution of factorials for C_{n+q+2}^{n-1}, C_{n+q+1}^{n-1}, C_{n+q}^{n-1} leads to the required formula.

398. We have $C_{n-1}^{k-1} = (k/n) \, C_n^k$. Since

(*) $$(1+x)^n = 1 + C_n^1 x + \cdots + C_n^k x^k + \cdots + C_n^n x^n,$$

it follows that

(**) $$n(1+x)^{n-1} = C_n^1 + \cdots + kC_n^k x^{k-1} + \cdots + nC_n^n x^{n-1}$$

(the reader familiar with the differential calculus can obtain (**) by differentiating (*)).

Multiplying (*) and (**) we see that

$$n(1+x)^{2n-1} = (1 + C_n^1 x + \cdots + C_n^n x^n)(C_n^1 + \cdots + nC_n^n x^{n-1}).$$

By equating the coefficients of x^{n-1} on both sides of this equality we obtain the required relation.

399. The number of n-combinations with repetitions of elements of n types is C_{2n-1}^n. We separate these combinations into classes and put in the kth class the combinations with exactly k kinds of elements. The number of combinations in the kth class is $C_n^k C_{n-1}^{n-k}$ (there are C_n^k ways of selecting the k types of elements which appear in the combinations in this class, and there are C_{n-1}^{n-k} n-combinations with repetitions containing elements of each of the selected k types). Hence $C_{2n-1}^n = \sum_{k=1}^n C_n^k C_{n-1}^{n-k}$. Replacing C_{2n-1}^n, C_n^k, C_{n-1}^{n-k} with factorials yields the required result.

400. The relation which we are about to prove can be put in the following form:

$$C_{n+r-1}^r - C_n^1 C_{n+r-3}^{r-2} C_n^2 C_{n+r-5}^{r-4} - \cdots = C_n^r.$$

For proof, we consider r-combinations with repetitions of elements of n types and compute in two ways the number of such combinations which consist solely of elements of different types. On the one hand, their number is C_n^r. On the other hand, the number of r-combinations with repetitions of elements of n types which contain at least 2 elements of each of a certain k types is $C_{n+r-2k-1}^{r-2k}$; also, the k types can be selected in C_n^k ways. Application of the principle of inclusion and exclusion yields the desired result.

401. (a) Put $S_n = C_n^1 + 2C_n^2 + 3C_n^3 + \cdots + nC_n^n$. Since $C_n^k = C_n^{n-k}$, we have $S_n = nC_n^0 + (n-1)C_n^1 + \cdots + C_n^{n-1}$. Adding both expressions for S_n, we see that

$$2S_n = n[C_n^0 + C_n^1 + \cdots + C_n^n] = 2^n n,$$

and therefore $S_n = 2^{n-1}$.

      (b) In much the same way we prove that $S_n = (n+1)2^{n-1}$.
      (c) $S_n = (n-2)2^{n-1} + 1$.
      (d) $S_n = (n+1)2^n$.
      (e) $S_n = 0$.
      (f) We have

$$S_n = 4(C_n^1 + 2C_n^2 + \cdots + nC_n^n) - (C_n^1 + C_n^2 + \cdots + C_n^n) = 2^{n-1}n - 2^n + 1.$$

      (g) We have $C_n^k = C_{n-1}^{k-1} + C_{n-1}^k$. Hence

$$S_n = C_{n-1}^0 + C_{n-1}^1 - 2(C_{n-1}^1 + C_{n-1}^2)$$

$$+ 3(C_{n-1}^2 + C_{n-1}^3) - \cdots + (-1)^{n-1} nC_{n-1}^{n-1}$$

$$= C_{n-1}^0 - C_{n-1}^1 + C_{n-1}^2 - \cdots + (-1)^{n-1} C_{n-1}^{n-1}.$$

This sum is 1 for $n = 1$, and 0 for $n > 1$.

(h) This sum is equal to

$$S_n = \frac{1}{n+1}[C_{n+1}^1 + C_{n+1}^2 + \cdots + C_{n+1}^{n+1}] = \frac{2^{n+1} - 1}{n+1}.$$

(i) Since $C_n^k = [(k+2)(k+1)/(n+1)(n+2)]\, C_{n+2}^{k+2}$, this sum is equal to

$$S_n = \frac{1}{(n+1)(n+2)}[C_{n+2}^2 + 2C_{n+2}^3 + \cdots + (n+1)C_{n+2}^{n+2}]$$

$$= \frac{1}{(n+1)(n+2)}\{[C_{n+2}^1 + 2C_{n+2}^2 + \cdots + (n+2)C_{n+2}^{n+2}]$$

$$- (C_{n+2}^1 + \cdots + C_{n+2}^{n+2})\}.$$

Applying the results in (a) and (b), we get

$$S_n = \frac{1}{(n+1)(n+2)}[2^{n+1}(n+2) - 2^{n+2} + 1] = \frac{2^{n+1}n + 1}{(n+1)(n+2)}.$$

(j) Rewrite the sum in the form

$$S_n = \frac{1}{n+1}[C_{n+1}^1 - C_{n+1}^2 + \cdots + (-1)^n C_{n+1}^{n+1}] = \frac{1}{n+1},$$

(the expression in square brackets is 1).

(k) If n is odd, then $S_n = 0$, and if $n = 2k$ is even, then $S_n = (-1)^k C_{2k}^k$. For proof, it suffices to multiply the expansions of $(1+x)^n$ and $(1-x)^n$ and to find the coefficient of x^n in the product.

402. The largest coefficient in the first expansion is the coefficient of $a^3 b^3 c^4$ (or $a^3 b^4 c^3$, $a^4 b^3 c^3$). Its value is $P(3, 3, 4) = 4200$. In the second expansion the largest coefficient is the coefficient $P(4, 4, 3, 3)$ of $a^3 b^3 c^4 d^4$.

403. The binomial expansion yields

$$(*) \qquad (1 - 4x)^{-1/2} = 1 + \sum_{n=1}^{\infty} \frac{(-\frac{1}{2})(-\frac{3}{2}) \cdots (-\frac{1}{2} - n + 1)}{n!} (-4x)^n.$$

Hence the coefficient Y_n of x^n is equal to

$$Y_n = \frac{1 \cdot 3 \cdots (2n-1) \cdot 2^n}{n!} = \frac{(2n)!}{(n!)^2} = C_{2n}^n.$$

For $(1 - 4x)^{1/2}$ we have the expansion

$$(1 - 4x)^{1/2} = 1 + \sum_{n=1}^{\infty} \frac{\frac{1}{2}(-\frac{1}{2})(-\frac{3}{2}) \cdots (-\frac{1}{2} - n + 1)}{n!} (-4x)^n$$

$$= 1 + \sum_{n=1}^{\infty} \frac{Y_n}{1 - 2n} x^n.$$

Since

$$\frac{Y_n}{1 - 2n} = -\frac{C_{2n}^n}{2n - 1} = -\frac{2n!}{(n!)^2(2n - 1)}$$

$$= -\frac{2}{n} \frac{(2n - 2)!}{[(n - 1)!]^2} = -\frac{2}{n} Y_{n-1},$$

we have

(**)
$$(1 - 4x)^{1/2} = 1 - 2 \sum_{n=1}^{\infty} \frac{Y_{n-1}}{n} x^n,$$

where we put $Y_0 = 1$.

404. (a) If we multiply the expansions (*) and (**), then we find that

$$1 = \left(1 + \sum_{n=1}^{\infty} Y_n x^n\right)\left(1 - 2 \sum_{n=1}^{\infty} \frac{Y_{n-1}}{n} x^n\right)$$

$$= 1 + \sum_{n=1}^{\infty} \left[Y_n - 2\left(Y_{n-1} + \frac{1}{2} Y_{n-2}Y_1 + \cdots + \frac{1}{n} Y_{n-1}\right)\right] x^n.$$

This implies the required result.

(b) If we square the expansion (*), then we find that

$$(1 - 4x)^{-1} = \left(1 + \sum_{n=1}^{\infty} Y_n x^n\right)^2$$

$$= 1 + (Y_0 Y_1 + Y_1 Y_0)x + (Y_0 Y_2 + Y_1 Y_1 + Y_2 Y_0)x^2$$
$$+ \cdots + (Y_0 Y_n + Y_1 Y_{n-1} + \cdots + Y_n Y_0)x^n + \cdots.$$

Since

$$(1 - 4x)^{-1} = 1 + 4x + 4^2 x^2 + \cdots + 4^n x^n + \cdots,$$

we readily obtain the required equality.

(c) Square the expansion (**).

405. Let E denote even numbers and O odd numbers. Consider the first 4 elements of each row beginning with the third row. In the third row these elements fit the pattern OEOE; in the fourth row, the pattern OOEO; in the fifth row, the pattern OEEE; in the sixth row, the pattern OOOE; and in the seventh row, the pattern OEOE. After that the cycle of patterns is repeated (this is so because the first 4 elements of each row are determined by the first 4 elements of the preceding row). Hence each row contains at least one even number.

406. We show that every row of the triangle is an arithmetical progression and that the sum of the elements equidistant from the end elements is divisible by 1958. The proof is by induction on the number of the row. For the first row the assertion is obvious. Suppose it holds for the nth row. Take three neighboring elements a, $a + d$, $a + 2d$ of the nth row. These three elements determine two definite neighboring elements of the $(n + 1)$th row, and their values are $2a + d$ and $2a + 3d$. This shows that the $(n + 1)$th row is an arithmetical progression with $2d$ the difference of neighboring terms. Let a, b be the first two elements of the nth row and c, d its last two elements. Then the sum of the first and last elements of the $(n + 1)$th row is $(a + b) + (c + d) = 2(a + d)$ and, by the induction assumption, is divisible by 1958. It follows that the sum of the first and last elements of any row is divisible by 1958. In particular this is true of the sum of the two elements which make up the last but one row, that is, of the last element in the table.

407. The proof is by induction on $n + m$. Assume (a) for all k and s for which $k + s < n + m$. Then we have

$$u_{n+m} = u_{n+m-1} + u_{n+m-2}$$
$$= u_{n-1}u_{m-1} + u_n u_m + u_{n-1}u_{m-2} + u_n u_{m-1}$$
$$= u_{n-1}(u_{m-1} + u_{m-2}) + u_n(u_m + u_{m-1}) = u_{n-1}u_m + u_n u_{m+1}. \qquad (*)$$

Since (*) is readily verified for the case when $n + m = 1$, it follows that (*) holds for $n + m = 1, 2, 3, \ldots$.

(b) The proof is by induction on k. For $k = 1$, our assertion is trivially true. Suppose that u_{km} is divisible by u_m. In view of (*) we have

$$u_{(k+1)m} = u_{km+m} = u_{km-1}u_m + u_{km}u_{m+1},$$

so that $u_{(k+1)m}$ is divisible by u_m. Hence u_{nm} is divisible by u_m for $n = 1, 2, 3, \ldots$.

(c) Suppose u_n and u_{n+1} are divisible by $k \neq 1$. Then the same is true of $u_{n-1} = u_{n+1} - u_n$. Continuing, we arrive at the false statement that $u_1 = 1$ is divisible by $k \neq 1$.

408. Let (a, b) denote the greatest common divisor of a and b. The equality

$u_{m+n} = u_{n-1}u_m + u_nu_{m+1}$ shows that (u_{m+n}, u_n) is a divisor of $u_{n-1}u_m$ and, since u_n and u_{n-1} are relatively prime, a divisor of u_m. Conversely, (u_m, u_n) divides u_{m+n}. It follows that $(u_m, u_n) = (u_{m+n}, u_n)$. But then with $n = km + q$, we have $(u_m, u_n) = (u_m, u_q)$. Application of the Euclidean algorithm shows that $(u_m, u_n) = u_{(m,n)}$. In particular, $(u_{1000}, u_{770}) = u_{10} = 55$.

409. By putting enough zeros in front of the early Fibonacci numbers, we may say that each Fibonacci number has at least 4 digits. Now we associate with each of the 10^8 pairs of neighboring Fibonacci numbers $u_0, u_1; u_1, u_2; \ldots;$ u_{10^8-1}, u_{10^8}, 8 digits, namely, the last 4 digits of the first number in the pair followed by the last 4 digits of the second number in the pair. Since there are only 10^8 different ordered sets of 8 digits (namely, 00000000,..., 99999999) and not all such sets can arise from our sequence of 10^8 pairs of neighboring Fibonacci numbers, it follows that our sequence of pairs must contain two pairs (u_m, u_{m+1}) and (u_n, u_{n+1}), $n > m$, such that u_m and u_n have the same last 4 digits, and u_{m+1} and u_{n+1} also have the same last 4 digits. But then the numbers $u_n - u_m$ and $u_{n+1} - u_{m+1}$ end in 4 zeros. Since

$$u_{n-1} - u_{m-1} = (u_{n+1} - u_{m+1}) - (u_n - u_m),$$

$u_{n-1} - u_{m-1}$ also ends in 4 zeros. By successive reduction of the indices we find that $u_{n-m} - u_{m-m} = u_{n-m}$ ends in 4 zeros.

410. Let $u_n, u_{n+1}, u_{n+2}, \ldots, u_{n+7}$ be the selected numbers. If we express them in terms of u_n and u_{n+1}, then we have

$$u_{n+2} = u_n + u_{n+1}, \qquad u_{n+3} = u_n + 2u_{n+1}, \qquad u_{n+4} = 2u_n + 3u_{n+1},$$
$$u_{n+5} = 3u_n + 5u_{n+1}, \qquad u_{n+6} = 5u_n + 8u_{n+1}, \qquad u_{n+7} = 8u_n + 13u_{n+1}.$$

It follows that the sum of these numbers is $21u_n + 33u_{n+1}$. Now $u_{n+8} = 13u_n + 21u_{n+1}$, $u_{n+9} = 21u_n + 34u_{n+1}$. The inequality

$$u_{n+8} < 21u_n + 33u_{n+1} < u_{n+9}$$

shows that $21u_n + 33u_{n+1}$ is not a Fibonacci number.

411. Assertion (a) is proved by induction. It is obvious for $n = 1$. Suppose it holds for n. Then

$$u_2 + u_4 + \cdots + u_{2n} = u_{2n+1} - 1.$$

We add u_{2n+2} to both sides of this equality. Since $u_{2n+2} + u_{2n+1} = u_{2n+3}$, we obtain $u_2 + u_4 + \cdots + u_{2n+2} = u_{2n+3} - 1$. This comples the proof. Similar reasoning proves (b).

Assertion (c) is also proved by induction.

To prove (d) note that

$$u_{n+1}^2 - u_n u_{n+2} = u_{n+1}^2 - u_n^2 - u_n u_{n+1} = u_{n+1}(u_{n+1} - u_n) - u_n^2 = u_{n-1}u_{n+1} - u_n^2 .$$

Hence $u_{n+1}^2 - u_n u_{n+2} = (-1)^n [u_1^2 - u_0 u_2] = (-1)^n$.

We prove (e) and (f) simultaneously. Both assertions are obvious for $n = 1$. We suppose that they are true for $n = k$. With the aid of (d) we obtain the relations

$$u_1 u_2 + u_2 u_3 + \cdots + u_{2k}u_{2k+1} + u_{2k+1}u_{2k+2}$$

$$= u_{2k+1}^2 - 1 + u_{2k+1}u_{2k+2} = u_{2k+1}u_{2k+3} - 1 = u_{2k+2}^2$$

and

$$u_1 u_2 + \cdots + u_{2k+1}u_{2k+2} + u_{2k+2}u_{2k+3}$$

$$= u_{2k+2}^2 + u_{2k+2}u_{2k+3} = u_{2k+2}u_{2k+4} = u_{2k+3}^2 - 1.$$

This shows that (e) and (f) hold for $n = k + 1$ and thus for all n.

To prove (g) note that by (a) and (b), $u_1 + u_2 + \cdots + u_{n+1} = u_{n+3} - 1$. Now, (g) holds for $n = 1$. If we make the usual induction assumption, then we have

$$(n + 1) u_1 + n u_2 + \cdots + 2u_n + u_{n+1}$$
$$= u_{n+4} - (n + 3) + u_{n+3} - 1 = u_{n+5} - (n + 4).$$

This completes the proof.

Assertion (h) follows readily from the equality

$$\frac{u_{3n+2} - 1}{2} + u_{3n+3} = \frac{u_{3n+5} - 1}{2}.$$

To prove (i) we put $m = n$ in the formula $u_{n+m} = u_{n-1}u_m + u_n u_{m+1}$. Then we find that $u_{2n} = u_{n-1}u_n + u_n u_{n+1} = u_{n+1}^2 - u_{n-1}^2$. Similarly, $u_{2n+1} = u_n^2 + u_{n+1}^2$. If we put in the same formula $m = 2n$, then we see that

$$u_{3n} = u_{n-1}u_{2n} + u_n u_{2n+1}$$

$$= u_{n-1}(u_{n+1}^2 - u_{n-1}^2) + u_n(u_n^2 + u_{n+1}^2) = u_{n+1}^3 + u_n^3 - u_{n-1}^3 .$$

412. Let $u_n \leqslant N < u_{n+1}$. Then $0 \leqslant N - u_n < u_{n-1}$. Hence there is an $s < n - 1$ such that $u_s \leqslant N - u_n < u_{s+1}$. But then $0 \leqslant N - u_n - u_s < u_{s-1}$, and $s - 1 < n - 2$. After a sequence of such steps, we find that

$$N = u_n + u_s + u_p + \cdots + u_r ,$$

where the neighboring indices $n, s, p, \ldots, r$ differ by at least 2.

413. The number of ways is equal to the coefficient of x^s in the expansion of

$$(1 + x + \cdots + x^p)(1 + x + \cdots + x^q)(1 + x + \cdots + x^r)$$

$$= (1 - x^{p+1})(1 - x^{q+1})(1 - x^{r+1})(1 - x)^{-3}$$

$$= (1 - x^{p+1} - x^{q+1} - x^{r+1} - \cdots)(1 + 3x + 6x^2 + \cdots + C_{n+2}^2 x^n + \cdots).$$

Since $p < q + r$, we have $p < s$, $q < s$, $r < s$, and the coefficient in question is given by

$$C_{s+2}^2 - C_{s-p+1}^2 - C_{s-q+1} - C_{s-r+1}$$

$$= \frac{(s+2)(s+1)}{2} - \frac{(s-p+1)(s-p)}{2}$$

$$- \frac{(s-q+1)(s-q)}{2} - \frac{(s-r+1)(s-r)}{2}.$$

If we bear in mind that $p + q + r = 2s$, then this expression reduces ultimately to $s^2 + s + 1 - \frac{1}{2}(p^2 + q^2 + r^2)$.

414. If $q + r < p$, then $q < s$, $r < s$, and $p \geqslant s$. Therefore the required coefficient is $C_{s+r}^2 - C_{s-q+1}^2 - C_{s-r+1}^2$. This implies the required answer.

415. All the objects can be permuted in $(pq + r)!$ ways. Selection of r (out of p) people who obtain $q + 1$ objects each can be accomplished in C_p ways. Each of the permutations defines a distribution of the $p + q + r$ objects into shares of q and $q + 1$ objects, respectively. Since the result is not affected by the order of the elements in each share, we must divide $C_p^r(pq + r)!$ by

$$(q!)^{p-r}[(q+1)!]^r = (q!)^p (q+1)^r.$$

416. Since

$$\sum_{i_0=1}^{i_1} 1 = i_1 = C_{i_1}^1, \qquad \sum_{i_1=1}^{i_2} \sum_{i_0=1}^{i_1} 1 = \sum_{i_1=1}^{i_2} C_{i_1}^1 = C_{i_2+1}^2.$$

Furthermore,

$$\sum_{i_2=1}^{i_3} \sum_{i_1=1}^{i_2} \sum_{i_0=1}^{i_1} 1 = \sum_{i_2=1}^{i_3} C_{i_2+1}^2 = C_{i_3+2}^3.$$

Hence the value of our sum is C_{n+m}^{n+1}.

417. We separate the permutations of m white and n black balls into classes. The class $(k_1, \ldots, k_m)$ consists of those permutations in which there are k_1 single white balls, k_2 pairs of white balls, k_3 triples of white balls,..., k_m m-tuples of white balls. Clearly, $k_1 + 2k_2 + \cdots + mk_m = m$. We compute the number of permutations in the class $(k_1, \ldots, k_m)$. Since there are n black balls, there are $n + 1$ places for locating groups of white balls. Of these, k_1 are taken up by the single white balls, k_2 by the pairs of white balls,..., k_m by the m-tuples of white balls. The number of unoccupied places is $n - k_1 - \cdots - k_m + 1$. Hence the number of ways of distributing the white balls, that is, the number of permutations in the class $(k_1, \ldots, k_m)$ is $P(k_1, \ldots, k_m, n - k_1 - \cdots - k_m + 1)$. Since the number of permutations of m white balls and n black balls in C_{n+m}^m, the assertion of the problem follows.

418. (a) The roots of the characteristic equation $r^2 - 7r + 12 = 0$ are $r_1 = 3$, $r_2 = 4$. Hence the general solution is of the form $a_n = C_1 3^n + C_2 4^n$. (b) The general solution is $a_n = C_1 2^n + C_2(-5)^n$. (c) $a_n = C_1(2 + 3i)^n + C_2(2 - 3i)^n$. (d) $a_n = C_1(3i)^n + C_2(-3i)^n$. (e) $r_1 = r_2 = -2$. Hence $a_n = (-2)^n (C_1 + C_2 n)$. (f) The characteristic equation is $r^3 - 9r^2 + 26r - 24 = 0$. Its roots are $r_1 = 2$, $r_2 = 3$, $r_3 = 4$. Hence $a_n = C_1 2^n + C_2 3^n + C_3 4^n$. (g) $r_1 = r_2 = r_3 = -1$. But then $a_n = (-1)^n (C_1 + C_2 n + C_3 n^2)$. (h) The characteristic equation is $r^4 + 4 = 0$. Its roots are $r_{1,2} = 1 \pm i$, $r_{3,4} = -1 \pm i$. Hence

$$a_n = 2^{1/2}[C_1(1 + i)^n + C_2(1 - i)^n + C_3(-1 + i)^n + C_4(-1 - i)^n].$$

419. (a) The roots of the characteristic equation $r^2 - 5r + 6 = 0$ are $r_1 = 2$, $r_2 = 3$, and therefore $a_n = C_1 2^n + C_2 3^n$. Putting $n = 1$ and $n = 2$, we obtain the following system of equations for C_1 and C_2:

$$2C_1 + 3C_2 = 1, \qquad 4C_1 + 9C_2 = -7.$$

The solution of this system of equations is $C_1 = 5$, $C_2 = -3$, so that $a_n = 5 \cdot 2^n - 3^{n+1}$.

(b) We have $a_n = 2^n(C_1 + C_2 n)$. Putting successively $n = 1, 2$, we obtain the system of equations $C_1 + C_2 = 1$, $C_1 + 2C_2 = 1$ with solution $C_1 = 1$, $C_2 = 0$. Hence $a_n = 2^n$.

(c) $\quad a_n = \dfrac{1}{2^{n+2}} [(-1 + i\sqrt{3})^n + (-1 - i\sqrt{3})^n].$

(d) $\quad a_n = 2^n + 3^n - 4^n.$

420. The characteristic equation $r^2 - 2r \cos \alpha + 1 = 0$ has roots $r_{1,2} = \cos \alpha \pm i \sin \alpha$. Hence $a_n = C_1(\cos \alpha + i \sin \alpha)^n + C_2(\cos \alpha - i \sin \alpha)^n$. Putting $n = 1, 2$, we obtain the system of equations

$$(C_1 + C_2) \cos \alpha + (C_1 - C_2) i \sin \alpha = \cos \alpha,$$
$$(C_1 + C_2) \cos 2\alpha + (C_1 - C_2) i \sin 2\alpha = \cos 2\alpha.$$

Hence

$$C_1 = C_2 = \tfrac{1}{2}, \qquad na_n = \tfrac{1}{2}[(\cos\alpha + i\sin\alpha)^n + (\cos\alpha - i\sin\alpha)^n].$$

By Moivre's theorem, $a_n = \cos n\alpha$.

421. For proof, we need only note that the characteristic equation

$$r^k - C_k^1 r^{k-1} + C_k^2 r^{k-2} + \cdots + (-1)^k = 0$$

can be written as $(r-1)^k = 0$, and therefore has the root 1 with multiplicity k. This implies that one solution of the recurrence relation is $a_n = n^k$ (see p. 138).

422. $a_n = \dfrac{n}{12} \cdot 2^n + C_1(-4)^n + C_2 2^n$

423. We have

(*) $(1+x)^p = 1 + C_p^1 x + C_p^2 x^2 + \cdots + C_p^m x^m + \cdots + C_p^p x^p,$

(**) $(1+x)^{-k-1} = 1 - C_{k+1}^1 x + C_{k+2}^2 x^2 - \cdots + (-1)^s C_{k+s}^s x^s + \cdots,$

(***) $(1+x)^{p-k-1} = 1 - C_{p-k-1}^1 x + \cdots + (-1)^n C_{p-k-1}^n x^n + \cdots.$

We multiply the expansions (*) and (**) and find that the coefficient of x^n in this product is

$$\sum_s (-1)^{n-s} C_{k+n-s}^{n-s} C_p^s = \sum_s (-1)^s C_{k+s}^s C_p^{n-s}.$$

On the other hand, (***) shows that this coefficient is equal to $(-1)^n C_{p-k-1}^n$. The assertion of the problem follows. Problems 424–438 are solved similarly.

439. Proof by induction on n.

Index

295